CALLOSAL AGENESIS

A Natural Split Brain?

ADVANCES IN BEHAVIORAL BIOLOGY

Recent Volumes in This Series

Volume 31 TOBACCO SMOKING AND NICOTINE: A Neurobiological Approach
Edited by William R. Martin, Glen R. Van Loon, Edgar T. Iwamoto, and Layten Davis

Volume 32 THE BASAL GANGLIA II: Structure and Function—Current Concepts
Edited by Malcolm B. Carpenter and A. Jayaraman

Volume 33 LECITHIN: Technological, Biological, and Therapeutic Aspects
Edited by Israel Hanin and G. Brian Ansell

Volume 34 ALTERATIONS IN THE NEURONAL CYTOSKELETIN IN ALZHEIMER'S DISEASE
Edited by George P. Perry

Volume 35 MECHANISMS OF CEREBRAL HYPOXIA AND STROKE
Edited by George Somjen

Volume 36 NOVEL APPROACHES TO THE TREATMENT OF ALZHEIMER'S DISEASE
Edited by Edwin M. Meyer, James W. Simpkins, and Jyunji Yamamoto

Volume 37 KINDLING 4
Edited by Juhn A. Wada

Volume 38A BASIC, CLINICAL AND THERAPEUTIC ASPECTS OF ALZHEIMER'S AND PARKINSON'S DISEASES
Volume 1
Edited by Toshiharu Nagatsu, Abraham Fisher, and Mitsuo Yoshida

Volume 38B BASIC, CLINICAL AND THERAPEUTIC ASPECTS OF ALZHEIMER'S AND PARKINSON'S DISEASES
Volume 2
Edited by Toshiharu Nagatsu, Abraham Fisher, and Mitsuo Yoshida

Volume 39 THE BASAL GANGLIA III
Edited by Giorgio Bernardi, Malcolm B. Carpenter, Gaetano Di Chiara, Micaela Morelli, and Paolo Stanzione

Volume 40 TREATMENT OF DEMENTIAS: A New Generation of Progress
Edited by Edwin M. Meyer, James W. Simpkins, Jyunji Yamamoto, and Fulton T. Crews

Volume 41 THE BASAL GANGILA IV: New Ideas and Data on Structure and Function
Edited by Gérard Percheron, John S. McKenzie, and Jean Féger

Volume 42 CALLOSAL AGENESIS: A Natural Split Brain?
Edited by Maryse Lassonde and Malcolm A. Jeeves

A Continuation Order Plan is available for this series. A continuation order will bring delivery of each new volume immediately upon publication. Volumes are billed only upon actual shipment. For further information please contact the publisher.

CALLOSAL AGENESIS

A Natural Split Brain?

Edited by

Maryse Lassonde

Université de Montréal
Montréal, Québec, Canada

and

Malcolm A. Jeeves

University of St. Andrews
St. Andrews, Scotland, United Kingdom

PLENUM PRESS • NEW YORK AND LONDON

Library of Congress Cataloging-in-Publication Data

Callosal agenesis : a natural split brain? / edited by Maryse Lassonde and Malcolm A. Jeeves.
p. cm. -- (Advances in behavioral biology ; v. 42)
"Proceedings of an International Brain Research Organization satellite symposium, held August 10-13, 1991, in Quebec City, Quebec, Canada"T.p. verso.
Includes bibliographical references and index.
ISBN-13:978-1-4612-7592-3 e-ISBN-13:978-1-4613-0487-6
DOI: 10.1007/978-1-4613-0487-6

1. Corpus callosum--Abnormalities--Congresses. 2. Split brain--Congresses. I. Lassonde, Maryse. II. Jeeves, Malcolm A., [DATE]
III. International Brain Research Organization. IV. Series.
[DNLM: 1. Corpus Callosum--abormalities--congresses. 2. Corpus Callosum--physiopathology--congresses. W3 AD215 v.42 1993 / WL 307 C163 1993]
RC395.C35 1993
612.8'25--dc20
DNLM/DLC
for Library of Congress 93-46518
CIP

Proceedings of an International Brain Research Organization Satellite Symposium, held August 10–13, 1991, in Québec City, Québec, Canada

ISBN-13:978-1-4612-7592-3

Softcover reprint of the hardcover 1st edition 1994

A Division of Plenum Publishing Corporation
233 Spring Street, New York, N.Y. 10013

PREFACE

This book is devoted to the description of agenesis of the corpus callosum, a congenital malformation of midline structures in the brain that may be regarded as a natural model of the "split-brain." First reported by Reil in 1812, this anomaly has since been described by several investigators. Interest in this malformation was revived in the 1970s by studies of interhemispheric transfer in Bogen and Vogel's commissurotomized patients and the subsequent findings by Sperry that acallosal patients were devoid of the typical disconnection deficits found in patients with surgical transection of the corpus callosum. Since this seminal work, the bulk of neuropsychological research on callosal agenesis has focused on the particulars of interhemispheric transfer and integration. An ever-growing literature has emerged on the subject, attempting to specify the extent and limits of neural plasticity in a nervous system that has evolved in the absence of the most important interhemispheric pathway.

Whilst callosal agenesis proves to be an excellent model of cerebral plasticity, it has to be pointed out that this anomaly is often associated with other malformations and neurological diseases that may result in different degrees of mental retardation or other cognitive and sensorimotor deficits. In this context, neurological research on callosal agenesis has concentrated on the description of various syndromes associated with this pathology as well as on the attempt to specify its neurobehavioral manifestations. However, it is important to identify the pattern of functioning specific to callosal agenesis in the absence of gross concomitant pathology of the central nervous system in order to assess the real consequences of this condition.

As for the distribution of this malformation, callosal agenesis is far from being an isolated phenomenon. It has been described throughout the world but its frequency may vary from one country to another. With the advent of new neuroradiological techniques, including prenatal sonographic examination, an increasing number of cases is being reported. Yet, there are no recent accounts of the clinical and neurobehavioral aspects of this malformation. The purpose of this book is to provide a complete up-to-date analysis of this pathology, including description of the medical (diagnosis, etiology and clinical manifestations), anatomical (based on animal models) and behavioral (sensorimotor and cognitive analysis) components of callosal agenesis.

Part I deals with the medical description of the syndrome. It includes a presentation of the various neurological syndromes that are frequently associated with callosal agenesis as well as a description of the clinical, pathological and genetic features of each syndrome. This section is followed by a complete account of the neuroradiological characteristics of this pathology. Thus, prenatal and postnatal neuroradiagnostic procedures that permit early detection of this cerebral malformation are reported along with the description of computed tomography and magnetic resonance imaging techniques. Part I also includes the analysis of sleep patterns found in infant and adult acallosal patients.

Part II presents animal models of brain organization and development in callosal agenesis together with their implications for human studies. Part III follows with a detailed assessment of basic sensory and sensorimotor skills as well as cognitive functioning of acallosal patients. From this section, it appears that earlier work may have overemphasized the apparent lack of a disconnection syndrome in callosal agenesis. Disconnection signs similar to those found in split-brain patients are now being reported and specific cognitive and sensorimotor deficits and, to some extent, psychiatric features are also being identified as characteristic of callosal agenesis. A final review chapter draws out the theoretical implications of the empirical work reported, focusing especially on the evidence for a diversity of functions for the forebrain commissures in normal cognition and behavior.

In editing this book, our intention was to address a multidisciplinary readership. Various professionals dealing with callosal agenesis patients, be it for diagnosis, clinical manifestations and rehabilitation, will benefit from the presentation of different aspects of the syndromes described in this book. Experimental and cognitive neuropsychologists are also provided with valuable information regarding cerebral reorganization following early brain damage.

As with other works of this type, a number of people and organizations were involved in its realization. The book was based upon an International Brain Research Organization (IBRO) satellite symposium held in Quebec City in August 1991. The symposium was made possible by funding from the Medical Research Council of Canada, Le Fonds de la Recherche en Santé du Québec, Merk Frosst Canada, the Département de Psychologie, the Faculté des Arts et Sciences and the Vice-Rectorat à l'Enseignement et à la Recherche de l'Université de Montréal. We wish to thank Drs Frederick Andermann, M.D., Pradeep G. Bhide, Ph.D., Richard Desbiens, M.D., Augustin M. O'Gorman, M.D. and Hannelore C. Sauerwein, Ph.D., for their helpful comments on various chapters. Our thanks also go to Ms. Diane Poulin and Ms. Debbie Kavanagh, E.P.M., for their secretarial and technical help, Ms. Lauren S. Ptito, B.A. for her revision of the texts, and finally to Ms. Alessandra Schiavetto, M.Sc., and especially Dr. Hannelore C. Sauerwein, Ph.D., for the many hours they spent in helping editing and formatting this book.

Maryse Lassonde and
Malcolm A. Jeeves

CONTENTS

PART I: NEUROCLINICAL FINDINGS

A. Clinical Description and Related Disorders

B. Neuroradiological Characteristics

CALLOSAL AGENESIS: REVIEW OF CLINICAL, PATHOLOGICAL, AND CYTOGENETIC FEATURES

Krystyna E. Wisniewski and Joseph S. Jeret

Institute for Basic Research in Developmental Disabilities, 1050 Forest Hill Road, Staten Island New York, 10314-6399, U.S.A.

INTRODUCTION

Agenesis of the corpus callosum (ACC) was first described at autopsy in 1812 (Reil, 1812). The antemortem diagnosis by pneumoencephalography has been made since 1934 (Davidoff and Dyke, 1934). Subsequently, agenesis of the corpus callosum has been identified by computerized tomography (Larsen et al., 1982), magnetic resonance imaging (Davidson et al., 1985; Han et al., 1985), ultrasonography (Hernanz-Schulman et al., 1985), and even prenatal fetal ultrasonography (Sandri et al., 1988; Hilpert and Kurtz, 1990).

ACC has been the subject of many case literature reports. Our review article (encompassing 705 cases) was published several years ago. More than 50 different disorders have been reported to be associated with ACC. When clinically symptomatic, neuropathological changes may be demonstrable in the cortical mantle. The reader is referred to our paper for further study and an exhaustive bibliography (Jeret et al., 1987). We will stress some of the more recent reports on this topic and focus on the clinicopathological heterogeneity in both the presentation and etiology of callosal agenesis.

CLINICAL MANIFESTATIONS

Agenesis of the corpus callosum is not an entity unto itself. ACC may be clinically asymptomatic and occur as an isolated malformation discovered at autopsy or by neuroimaging. More often, it is associated with other developmental abnormalities, both within the central nervous system and other organ systems. Subtle, nonspecific neuropsychological, cognitive, or electrophysiological abnormalities may be demonstrated, even in cases where no gross abnormalities are evident and where neuroimaging and autopsy disclose no other findings besides callosal agenesis.

Symptomatic acallosal patients may manifest mental retardation, seizures, motor or cerebellar impairment (e.g. hypotonia, spasticity), or ocular abnormalities. These findings are usually nonspecific and occur in association with numerous other defects in

Callosal Agenesis, Edited by M. Lassonde and M.A. Jeeves, Plenum Press, New York, 1994

the central nervous system (CNS), peripheral nervous system, or other organ systems as specific syndromes (Table 1).

Five disease entities are nearly constantly associated with agenesis of the corpus callosum (Table 1). These include Aicardi (Aicardi et al., 1965, 1969), Andermann (Andermann et al., 1972, 1975, 1981; Larbrisseau et al., 1984), Shapiro (Shapiro et al., 1969), Acrocallosal (Schinzel 1979, 1982; Schinzel and Schmid 1980; Jeret et al., 1987; Moeschler et al., 1989; Casamassima et al., 1989; Temtamy and Meguid 1989; Hendriks et al., 1990; Turolla et al., 1990) and Menkes disease (Menkes et al., 1962; Danks et al., 1972a, 1972b) syndromes. These various entities are described in detail in other chapters of this volume.

In addition callosal agenesis is sometimes seen in association with other entities, often at a frequency higher than can be explained solely by chance (Jeret et al., 1987; DeLeon et al., 1987, 1989; Lewis et al., 1988; Cohen and Kreiborg, 1990; Swayze et al., 1990; see table 1).

Table 1. Entities associated with callosal agenesis.

I. Usually have ACC:

1. Aicardi Syndrome
2. Andermann Syndrome [P]
3. Shapiro Syndrome
4. Acrocallosal Syndrome
5. Menkes Disease [P]

II. Sometimes have ACC:

A. Environmental pathogens:

1. Fetal Alcohol Syndrome
2. Maternal riboflavin, folate, or niacin deficiency
3. Intrauterine infections (ToRCH)
4. Post-radiation encephalopathy

B. Inborn errors of metabolism:

1. Krabbe Disease [P]
2. Leigh Disease [P]
3. Hurler Syndrome [P]
4. Mucolipidosis type III [P]
5. Mucolipidosis type IV [P]

C. Other:

1. Dandy-Walker Syndrome
2. Arnold-Chiari II Malformation
3. Holoprosencephaly
4. Hydrocephalus
5. Oral-Facial-Digital Syndrome Type I
6. Fukuyama Congenital Cerebromuscular Dystrophy [P]
7. Apert Syndrome
8. Schizophrenia
9. West Syndrome
10. Zellweger syndrome
11. Lowe syndrome [P]
12. Median Cleft Face Syndrome
13. Hirschsprung Disease
14. Oculo-Cerebro-Cutaneous Syndrome
15. Greig Syndrome
16. Familial frontonasal dermoid cyst
17. Noonan Syndrome
18. Neurofibromatosis
19. Tuberous sclerosis

[P] denotes progressive

In our review of 705 cases, the most common clinical findings were mental retardation (73-85%), seizures (42%), and hyper- or hyporeflexia (approximately 50%). Ocular abnormalities (e.g. chorioretinal lacunae, strabismus, colobomata) were frequently present and were invariable in patients with Aicardi Syndrome. Midline orofacial abnormalities--particularly hypertelorism, low-set ears, and high or cleft palate--were also common. Hydrocephalus is present in 23%, and costovertebral abnormalities are seen in 24% of cases (Jeret et al., 1987).

Pathologically, these patients demonstrated gyral abnormalities (32%), most often heterotopias and/or polymicrogyria (23%). The frequencies of these different abnormalities may have been demonstrated to be much higher if morphometric studies or thin serial brain sections had been performed in all cases. In Aicardi Syndrome, the incidence of gyral abnormalities was 75%. Microgyria (6%), lissencephaly (3%), and pachygyria (2%) have also been described. Intracranial cysts or tumors were present in 29%, and were particularly common in cases of partial ACC (Jeret et al., 1987).

BIOLOGICAL BASIS

The corpus callosum is formed early in embryogenesis. During the sixth week, callosal fibers begin crossing between the anterior and hippocampal commissures (Rakic and Yakovlev, 1968; Kendall, 1983; Jeret et al., 1987). Early callosal fibers cross anteroinferior to the foramen of Monro, then superiorly and caudally until a small replica of the crescenteric corpus callosum is formed by the seventeenth week of gestation (Loeser and Alvord, 1968; Jeret et al., 1987). The corpus callosum continues thickening and is completed in the occipital and temporal regions only after birth (Kendall, 1983; Jeret et al., 1987). In cases of total agenesis, longitudinal callosal fibers (Probst bundles) usually run from the frontal to occipital lobes. It seems that some of the fibers that usually cross deviate longitudinally (Ettlinger, 1975; Menkes 1981; Jeret et al., 1987).

Damage during early embryogenesis can cause complete callosal agenesis. Later insults (e.g. tumors or cysts that impair CSF circulation and toxic or metabolic processes) may cause total or partial (posterior) agenesis and hydrocephalus (Jeret et al., 1987; Schwartz and Ghatak, 1990). In rare cases where only a posterior remnant is present, damage to the anterior cerebral artery or its branches should be suspected (Jeret et al., 1987). The sole presence of a splenium has also been described in holoprosencephaly (Barkovich, 1990).

Since the palate fuses during weeks 10-12 of fetal life, it follows that cleft palate and other midline defects may be associated with callosal agenesis. Not surprisingly, 3 patients with Aicardi syndrome have also been described to have facial clefts (McPherson and Jones, 1990).

Myelination of the corpus callosum occurs primarily after birth. Dysmyelination syndromes - e.g. ketotic and nonketotic hyperglycinemia - result in diminished bulk and hypoplasia of corpus callosum (Shuman et al., 1978; Wariyar et al., 1990).

ETIOLOGIC AND CYTOGENETIC CONSIDERATIONS

The etiology of ACC is heterogeneous and multifactorial. Callosal agenesis may be seen in Fetal Alcohol Syndrome, presumably due to the teratogenetic effects of alcohol. There may be a hereditary or familial predisposition, as in Andermann Syndrome (Andermann et al., 1972, 1975; Labrisseau et al., 1984), and chromosomal linkage has recently been done for the X-linked dominant Aicardi Syndrome. Other genetic patterns will undoubtedly be discovered as our sophistication in molecular diagnosis and high-resolution chromosome banding increases.

In one of our previous studies, 35% of patients had a family history of developmental disability (Serur et al., 1988). More recently, chromosomal studies have revealed 102 cases of ACC with chromosomal abnormalities, most commonly Trisomy 13-15, Trisomy 18, Trisomy 8, Trisomy 11q, and Trisomy 17-18 (Jeret and Wisniewski, 1991). These are listed in table 2.

When callosal agenesis occurs in association with abnormal midline structures (e.g. tumors, cysts), a common etiological factor is suggested for both lesions (Schwartz and Ghatak, 1990).

Table 2. Chromosomal abnormalities associated with agenesis of the corpus callosum.

Abnormality	Number of cases	Details
Trisomy B (Chr 3-5)	1	Trisomy 4p
Trisomy C (Chr 6-12)	32	Trisomy 8 (n=21)
		Trisomy C, unspecified (n=5)
		Trisomy 11q (n=6)
Trisomy D (Chr 13-16)	22	Trisomy 13-15 (n=21)
Trisomy E (Chr 17-18)	25	Trisomy 18 (n=17)
		Trisomy E, unspecified (n=8)
Trisomy F (Chr 19-20)	1	Trisomy F, unspecified (n=1)
Trisomy 21	1	Trisomy 21 (Down Syndrome)
Other	20	Klinefelter (XXY) Syndrome (n=3)
		Triploidy (n=2)
		Monosomy 21 (n=2)
		Turner (XO) Syndrome (n=1)
		Ring chromosome (n=2)
		Translocations (n=7)
		Wolf-Hirschhorn (4p-) (n=2)
		Other deletions (n=1)

Selected References: Warkany et al. (1966); Taylor (1968); Parrish et al. (1979); Jellinger et al. (1981); Inagaki et al. (1987); Serur et al. (1988); Jeret and Wisniewski (1991)

EPIDEMIOLOGY

The incidence of agenesis of the corpus callosum depends on both the population studied and the diagnostic technique employed. Among all children, the incidence is .0005% to 0.7% (Grogono, 1968; Larsen and Osborn, 1982). Among the developmentally disabled population it is 2.2% to 2.4% (Melchior et al., 1940; Freytag and Lindberg, 1967; Andermann, 1981; Jeret et al., 1986). A review of the literature on the incidence of agenesis of the corpus callosum was the topic of a previous report (Jeret et al., 1986).

Overall, 53.7% of all patients with agenesis of the corpus callosum are male. When excluding those cases with Aicardi syndrome, which is characterized by sex-linked dominant inheritance with male lethality, the percentage rises to 59.3% (Jeret et al., 1987).

CONCLUSION

Agenesis of the corpus callosum is usually diagnosed during work-up of patients with developmental disabilities or other neurological or neuropsychological problems. Since chromosomal abnormalities are commonly present, high resolution chromosomal analysis is recommended in all cases (Jeret et al., 1989). For women at high risk for fetal malformations, the diagnosis of agenesis of the corpus callosum can easily be made by prenatal fetal ultrasonography. This alerts the obstetrician to the high risk of associated malformations (Sandri et al., 1988; Hilpert and Kurtz 1990; Vergani and Strobelt, this volume).

In addition, any developmentally disabled individual should undergo neuroimaging (preferably MRI), neurophysiological (i.e. EEG, EP, EMG, NCS), and neuropsychological assessments; these will often be abnormal in acallosal patients. Patients with developmental disabilities and dysmorphic features should be evaluated for possible syndrome classification. Chromosomal and molecular genetic studies of the patient and family members are advisable.

REFERENCES

Aicardi, J., Chevrie, J.J., and Rousselie, F., 1969, Le syndrome spasmes en flexion, agénésie calleuse, anomalies chorio-rétiniennes, *Arch. Fr. Paediatr.* 26:1103-1120.

Aicardi, J., Lefebvre, J., and Lerique-Koechlin, A., 1965, A new syndrome: spasm in flexion, callosal agenesis, ocular abnormalities, *Electroencephalogr. Clin. Neurophsiol.* 19:609-610.

Andermann, E., 1981, Agenesis of the corpus callosum. *in:* "Handbook of Clinical Neurology," Vol. 42, P.J. Vinken and G.W. Bruyn, eds., Elsevier, North Holland, Amsterdam.

Andermann, F., Andermann, E., Joubert, M., Karpati, G,. Carpenter, S., and Melanson, D., 1972, Familial agenesis of the corpus callosum with anterior horn cell disease: a syndrome of mental retardation, areflexia, and paraparesis, *Trans. Am. Neurol. Assoc.* 97:242-244.

Andermann, E., Andermann, F., Joubert, M., Melanson, D., Karpati, G., and Carpenter, S., 1975, Three familial midline malformation syndromes of the central nervous system: agenesis of the corpus callosum and anterior horn-cell disease; agenesis of the cerebellar vermis; and atrophy of the cerebellar vermis, *Birth Defects* 11(2):269-293.

Barkovich, A.J., 1990, Apparent atypical callosal dysgenesis: analysis of MR findings in six cases and their relationship to holoprosencephaly, *AJNR* 11:333-339.

Brinciotti, M., Matricardi, M., and Trasatti, G., 1990, Unilateral photoconvulsive response in agenesis of the corpus callosum, *Clin. Elecroencephalogr.* 21:101-103.

Casamassima, A.C., Beneck, D., Gewitz, M.H., et al., 1989, Acrocallosal syndrome: additional manifestations, *Am. J. Med. Genet.* 32:311-317.

Cohen, M.M. and Kreiborg, S., 1990, The central nervous system in the Apert syndrome, *Am. J. Med. Genet.* 35:36-45.

Danks, D.M., Campbell, P.E., Stevens, B.J., Mayne, V., and Cartwright, E., 1972a, Menkes' kinky-hair syndrome: An inherited defect in copper absorption with widespread effects, *Pediatrics* 50:188.

Danks, D.M., Stevens, B.J., Campbell, P.E., et al., 1972b, Menkes' kinky-hair syndrome, *Lancet* 1:1100-1103.

Davidoff, L.M. and Dyke, C.G., 1934, Agenesis of the corpus callosum its diagnosis by pneumoencephalography, *AJNR* 32:1-10.

Davidson, H.D., Abraham, R., and Steiner, R.E., 1985, Agenesis of the corpus callosum: magnetic resonance imaging, *Radiology* 155:371-373.

DeLeon, G.A., DeLeon, G., Grover, W.D., Zaeri, N., and Alburger, P.D., 1987, Agenesis of the corpus callosum and limbic malformation in Apert syndrome, *Arch. Neurol.* 44:979-982.

DeLeon, G.A., DeLeon, G., Grover, W.D., Zaeri, N., and Alburger, P.D., 1989, Agenesis of the corpus callosum in Apert syndrome? *Arch. Neurol.* 46:479 (Letter).

Ettlinger, G., 1975, Agenesis of the corpus callosum, *in:* "Handbook of Clinical Neurology," Vol. 30, P.J. Vinken and G.W. Bruyn, eds., Elsevier, North Holland, Amsterdam.

Freytag, E. and Lindenberg, R., 1967, Neuropathologic findings in patients of a hospital for the mentally deficient. A survey of 359 cases, *Johns Hopkins Med. J.* 121:379-392.

Grogono, J.L., 1968, Children with agenesis of the corpus callosum, *Dev. Med. Child Neurol.* 10:613-616.

Han, J.S., Benson, J.E., Kaufman, B., et al., 1985, MR imaging of pediatric cerebral abnormalities, *J. Comput. Assist. Tomogr.* 9:103-114.

Hendriks, H.J.E., Brunner, H.G., Haagen, T.A.M., and Hamel, B.C.J., 1990, Acrocallosal syndrome, *Am. J. Med. Genet.* 35:443-446.

Hernanz-Schulman, M., Dohan, F.C. Jr, Cayea, P., Wallman, J., and Teele, R.L., 1985, Sonographic appearance of callosal agenesis: correlation with radiographic and pathological findings, *AJNR* 6:361-368.

Hilpert, P.L. and Kurtz, A.B., 1990, Prenatal diagnosis of agenesis of the corpus callosum using endovaginal ultrasound, *J. Ultrasound. Med.* 9:363-365.

Inagaki, M., Ando, Y., Mito, T., et al., 1987, Comparison of brain imaging and neuropathology in cases of Trisomy 18 and 13, *Neuroradiology* 29:474-479.

Jellinger, K., Gross, H., Kaltenback, E., and Grisold, W., 1981, Holoprosencephaly and agenesis of the corpus callosum: frequency of associated malformations, *Acta Neuropathol.* 55:1-10.

Jeret, J.S. and Serur, D., 1991, Fetal alcohol syndrome in adolescents and adults, *JAMA* 266:1077 (Letter).

Jeret, J.S. and Wisniewski, K., 1991, Chromosomal abnormalities in acalossal patients, Poster presented at the *IBRO Satellite Symposium on Callosal Agenesis,* Quebec City, Que., Canada.

Jeret, J.S., Serur, D., and Wisniewski, K., 1989, Agenesis of the corpus callosum and limbic malformations in Apert's Syndrome, *Arch. Neurol.* 46:10 (Letter).
Jeret, J.S., Serur, D., Wisniewski, K., and Fisch, C., 1986, Frequency of agenesis of the corpus callosum in the developmentally disabled population as determined by computerized tomography, *Pediatr. Neurosci.* 12:101-103.
Jeret, J.S., Serur, D., Wisniewski, K.E., and Lubin, R.A., 1987, Clinicopathological findings associated with agenesis of the corpus callosum, *Brain Dev.* 9:255-264.
Kendall, B.E., 1983, Dysgenesis of the corpus callosum, *Neuroradiology* 25:239-256.
Labrisseau, A., Vanasse, M., Brochu, P., and Jasmin, G., 1984, The Andermann syndrome: agenesis of the corpus callosum associated with mental retardation and progressive sensorimotor neuronopathy, *Can. J. Neurol. Sci.* 11:257-261.
Larsen, P.D. and Osborn, A.G., 1982, Computerized tomographic evaluation of corpus callosum agenesis and associated malformations, *J. Comput. Tomogr.* 6:225-230.
Lewis, S.W., Reveley, M.A., David, A.S., and Ron, M.A., 1988, Agenesis of the corpus callosum and schizophrenia: a case report, *Psychol. Med.* 18:341-347.
Loeser, J.D. and Alvord, E.C., 1968, Agenesis of the corpus callosum, *Brain* 91:553-570.
McCusick, V.P., 1983, "Mendelian Inheritance in Man," Johns Hopkins University Press, Baltimore.
McPherson, E. and Jones, S.M., 1990, Cleft lip and palate in Aicardi syndrome, *Am. J. Med. Genet.* 37:318-319.
Menkes, J.H., 1981, Agenesis of the corpus callosum, partial, *in:* "Handbook of Clinical Neurology," Vol. 42, P.J. Vinken and G.W. Bruyn, eds., Elsevier, North Holland, Amsterdam.
Menkes, J.H., Alter, M., Steigleder, G.K., Weakley, D.R., and Sung, J.H., 1962, A sex-linked recessive disorder with retardation of growth, peculiar hair and focal cerebral and cerebellar degeneration, *Pediatrics* 29:764.
Moeschler, J.B., Pober, B.R., Holmes, L.B., and Graham, J.M., 1989, Acrocallosal syndrome: new findings, *Am. J. Med. Genet.* 32:306-310.
Parrish, M., Roessmann, U., and Levinsohn, M.W., 1979, Agenesis of the corpus callosum: a study of the frequency of associated malformations, *Ann. Neurol.* 6:349-354.
Rakic, P. and Yakovlev, P.I., 1968, Development of the corpus callosum and cavum septi in man, *J. Comp. Neurol.* 132:45-72
Reil, J.C., 1812, Mangel des mittleren und freyen Theils des Balkens im Menschengehirn, *Arch. Physiol.* 11:314-344.
Ropers, H.H., Zuffardi, O., Bianchi, E., and Tiepolo, L., 1982, Agenesis of the corpus callosum, ocular, and skeletal anomalies (X-linked dominant Aicardi's syndrome) in a girl with balanced X/3 translocation, *Hum. Genet.* 61:364-368.
Sandri, F., Pilu, G., Cerisoli, M., Bovicelli, L., Alvisi, C., and Salvioli, G.P., 1988, Sonographic diagnosis of agenesis of the corpus callosum in the fetus and newborn infant, *Am. J. Perinatol.* 5:226-231.
Schinzel, A., 1979, Postaxial polydactyly, hallux duplication, absence of the corpus callosum, macrencephaly and severe mental retardation: a new syndrome? *Helv. Paediatr. Acta* 34:141-146.
Schinzel, A., 1982, Four patients including two sisters with the acrocallosal syndrome (agenesis of the corpus callosum in combination with preaxial hexadactyly), *Hum. Genet.* 62:382.
Schinzel, A. and Schmid, W., 1980, Hallux duplication, postaxial polydactyly, absence of corpus callosum, severe mental retardation and additional anomalies in two unrelated patients. A new syndrome, *Am. J. Med. Genet.* 6:241-249.
Schwartz, A.M. and Ghatak, N.R., 1990, Interhemispheric cysts in association with agenesis of the corpus callosum, *Clin. Neuropathol.* 9:177-190.
Serur, D., Jeret, J.S., and Wisniewski, K., 1988, Agenesis of the corpus callosum: Clinical, neuroradiological and cytogenetic studies, *Neuropediatrics* 19:87-91.
Shapiro, W.R., Williams, G.H., and Plum, F., 1969, Spontaneous recurrent hypothermia accompanying agenesis of the corpus callosum, *Brain* 92:423-436.
Shuman, R.M., Leech, R.W., and Scott, C.R., 1978, The neuropathology of the nonketotic and ketotic hyperglycinemias: Three cases, *Neurology* 28:139-146.
Swayze, V.W., Andreasen, N.C., Ehrhardt, J.C., Yuh W.T., Alliger, R.J., and Cohen, G.A., 1990, Developmental abnormalities of the corpus callosum in schizophrenia, *Arch. Neurol.* 47:805-808.
Taylor, A.I., 1968, Autosomal Trisomy syndromes: a detailed study of 27 cases of Edwards' syndrome and 27 cases of Patau's syndrome, *J. Med. Genet.* 5:227-252.
Temtamy, S.A. and Meguid, N.A., 1989, Hypogenitalism in the acrocallosal syndrome, *Am. J. Med. Genet.* 32:301-305.
Turolla, L., Clementi, M., and Tenconi, R., 1990, How wide is the clinical spectrum of the acrocallosal syndrome? Report of a mild case, *J. Med. Genet.* 27:516-518.
Wariyar, U.K., Welch, R.J., Milligan, D.W.A., and Perry, R.H., 1990, Sonographic and pathological features of callosal hypoplasia in non-ketotic hyperglycinaemia, *Arch. Dis. Child.* 65:670-671.
Warkany, J., Passarge, E., and Smith, L.B., 1966, Congenital malformations in autosomal Trisomy syndromes, *Am. J. Dis. Child* 112:502-517.

THE AICARDI SYNDROME

Jean Aicardi and Jean-Jacques Chevrie

Département de Pédiatrie
Hôpital Necker-Enfants Malades
149 rue De Sèvres
757143 Paris, Cedex 15
France

INTRODUCTION

Aicardi syndrome (AS), originally described briefly in 1965 (Aicardi et al., 1965) and more fully in 1969 (Aicardi et al., 1969), is one of the well-defined syndromes that include agenesis of the corpus callosum as one of their major features (Chevrie and Aicardi, 1986). The syndrome consists of a triad, viz. callosal agenesis, chorioretinal 'lacunae', and infantile spasms. Agenesis of the corpus callosum may be partial or complete (Chevrie and Aicardi, 1986; Aicardi et al., 1987). 'Chorioretinal lacunae' appear as white or pinkish areas of retinal discoloration, usually rounded in shape and peripapillary in location. They are generally but not always bilateral. They are variably pigmented especially at their periphery and their appearance is quite different from that of chorioretinitis even though an infectious origin for the syndrome has been suggested (Willis and Rosman, 1980). Infantile spasms are frequently asymmetric and are associated on the EEG not with typical hypsarrhythmia but, characteristically, with a suppression-burst pattern with complete independence of paroxysmal bursts over both hemispheres, the so-called 'split-brain' EEG.

Additional features regularly include mental retardation, almost always of a severe grade; the occurrence in female sex only (or at least in individuals with two X chromosomes) and, very often, other ocular abnormalities such as coloboma of the optic disc and microphthalmia ; abnormalities of ribs and vertebrae (missing or bifid ribs, hemivertebrae, cuneiform or butterfly vertebrae or fused vertebrae) ; subependymal heterotopias ; and absence of familial recurrence (Chevrie and Aicardi, 1986). Occasional features include facial dysmorphy, abnormal dermatoglyphics or cardiac malformations. Microcephaly is not an initial feature but not infrequently appears secondarily after a few months.

The main features of AS, based on a study of 184 cases have been described in detail recently (Chevrie and Aicardi, 1986) and the frequency of the main features in this review are shown in Table 1. Since then, further cases have been reported (Baieri et al., 1988; Yamagata et al., 1990), seen by us or brought to our attention through personal communications. At this point, our total experience with the syndrome amounts to a total of 273 cases (Table 2). Only cases in which clinical information was

Callosal Agenesis, Edited by M. Lassonde and
M.A. Jeeves, Plenum Press, New York, 1994

Table 1. Frequency of main features in 184 cases of Aicardi syndrome.*

Seizures	156	
Infantile spasms only	86	
Infantile spasms and other types	65	
Other types only	5	
Hypsarrhythmia	26	
Split brain, suppression-burst tracing	111	
Undefined type	19	
Ophthalmological abnormalities		
Chorioretinal lacunae	184	(unilateral in 14)
Coloboma of optic disk	81	(unilateral in 50)
Microphthalmia	40	(unilateral in 29)
Other	26	
Neuroradiological findings		
Total agenesis of corpus callosum	137	
Partial	15	
Not indicated	32	
Periventricular heterotopias	81	
Other	37	
Peripheral malformations		
Abnormal ribs and vertebral	71	
Other	39	

* Data from Chevrie and Aicardi (1986)

Table 2. Recorded cases of Aicardi syndrome.

Published cases*	167
Personal cases	46
Personal communications	60
Total	273

* Cases belonging to series of agenesis of the corpus callosum reported only in tabular form are not included

available have been included. Cases reported only in tabular form in series of patients with callosal agenesis (Jeret et al., 1987; Jeret et al., 1986) have not been included.

New cases have amply confirmed earlier reports, in particular the exclusive involvement of persons with two X chromosomes. The only case in a phenotypic male remains that of a patient with an XXY Klinefelter syndrome reported by Hopkins et al. (1979).

This paper concentrates on some clinical, radiological and pathological data that have been recently brought about, in particular by new imaging techniques and on some novel genetic findings.

CLINICAL AND IMAGING DATA

The precise characteristics of the seizures in AS have been described in detail in 7 patients (Bour et al., 1986). The infantile spasms in the syndrome are almost always asymmetrical. They occur in clusters as in spasms of other causes but are usually preceded by a focal clonic or tonic seizure limited to the side in which the spasms will predominate. The EEG manifestation of the focal seizure is a localized discharge of repetitive spikes lasting 10-30 seconds. The focal EEG discharge is followed by a series of slow complexes often with a fast rhythm of low amplitude superimposed on the initial part of the slow wave complex. The complexes are often of a higher amplitude on the side of the focal discharge. They occur 6-20 seconds apart and no paroxysmal activity supervenes between individual complexes. The initial partial seizure and the subsequent series of spasms seem to represent a single seizure that usually starts on awakening or upon changing stage of sleep, especially from REM to slow sleep. However, similar spasms may occur without a preceding focal discharge. Interestingly, this sequence of events was first reported by Horita et al. (1980) in a girl with agenesis of the corpus callosum whose eye fundi were said to be normal. Gobbi et al. (1987) have studied in great detail this type of ictal discharge that they term 'periodic spasms' and showed that although nonspecific it is suggestive of a brain malformation and is consistently associated with a poor prognosis.

The 'chorioretinal lacunae' although essential for the diagnosis, are not the only ophthalmological abnormality. Coloboma of the optic disc was present in one or both eyes in many reported cases (Chevrie and Aicardi, 1986). In addition, a coloboma was present in addition to lacunae in some reported cases and in several personal communications without being mentioned in the description. In all patients personally seen since 1985 an optic disc coloboma was present in one or both eyes and we believe this feature is of diagnostic significance for the diagnosis of atypical cases, especially when lacunae are small and atypical. When large, a coloboma can be visualized by MRI (Figure 1). Ring-like pigment deposits surrounding the colobomatous papilla are

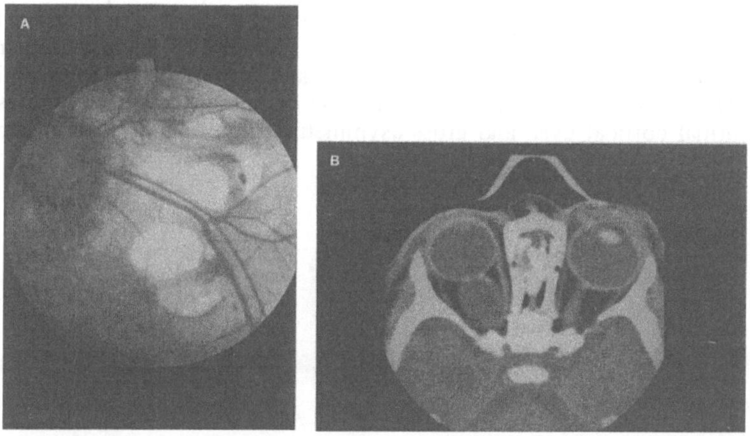

Figure 1. Typical fundoscopic appearance of left fundus in a patient with AS. **A.** Note coloboma of the disc with peripheral emergence of vessels and typical chorioretinal lacunae. **B.** CT scan of the orbit in a patient showing gross dilatation of the right optic nerve due to colobomatous Fundoscopic examination showed coloboma of right eye.

common and contribute to the fundus appearance in many patients. The eye pathology of the syndrome does not feature any degenerative or inflammatory lesion (Chevrie and Aicardi, 1986). There is thinning of the choroid and sclera in the areas corresponding to the lacunae and the pigment epithelium is depigmented or hypopigmented (McMahon et al., 1984) with degeneration of the rods and cones (Ohtsuki et al., 1981). Electroretinogram has been normal or not grossly abnormal in all studied cases although various abnormalities of the visual evoked potentials have been found (Chevrie and Aicardi, 1986). More extensive abnormalities including microcystic degeneration of the outer neuroectodermal layer and the inner plexic layer and folding of the retina have been also mentioned. In one patient (Del Pero et al., 1986), focal areas of attenuation with depigmentation or complete absence of the retinal pigment epithelium were found. In some places, hyperplasia of the retinal pigment epithelium resulted in multilamellar reduplications of the basement membrane. The photoreceptor layer was interrupted by places and folded in the transitional zone between lacunae and normal retina. The external limiting membrane curved around the borders of the folded photoreceptors and merged with the pigment epithelium. This suggests an early developmental disorder of the eye. However, the diagnosis of AS in this patient is not absolutely certain, as there was associated holoprosencephaly and microcephaly. None of our patients appeared to be blind although unilateral blindness is relatively common in patients with unilateral microphthalmia. According to a recent questionnaire survey of 34 patients diagnosed as cases of AS (Raynor and Raynor, 1990), 11 were legally blind 'in at least one eye'. However, no detailed assessment of visual acuity or mapping of the visual field is available.

The overall degree of impairment is usually quite severe although variable. In the same survey, all 34 patients were still having epileptic seizures and 76% of them had never been free of medication. Half of them had been hospitalized at least once for respiratory problems. Twenty-nine per cent were unable to feed themselves and 3 / 4 were fed on baby food or mashed food. Only 8 girls could walk and only 3 could run, while 68 % were wheelchairbound. Only 1 patient was able to speak words but not sentences. We know of only 2 patients who are capable of using 2-word sentences and one such case has been reported (Neidich et al., 1990).

Neuroimaging, especially MRI has allowed for a greater precision for the study of the corpus callosum in partial agenesis and for study of associated CNS abnormalities. It has become increasingly clear that callosal agenesis is virtually never an isolated malformation and that associated features are almost as important -perhaps even more important- than absence of the corpus callosum (Aicardi et al., 1987; Baieri et al., 1988; Hall-Craggs et al., 1990; Hamano et al., 1985). Commonly encountered anomalies include: cysts or tumors of the choroid plexus, cystic formations in the vicinity of the third ventricle or posterior fossa, periventricular or subcortical heterotopias of gray matter, abnormal cortical gyri, and gross asymmetry between the cerebral hemispheres (Figure 2).

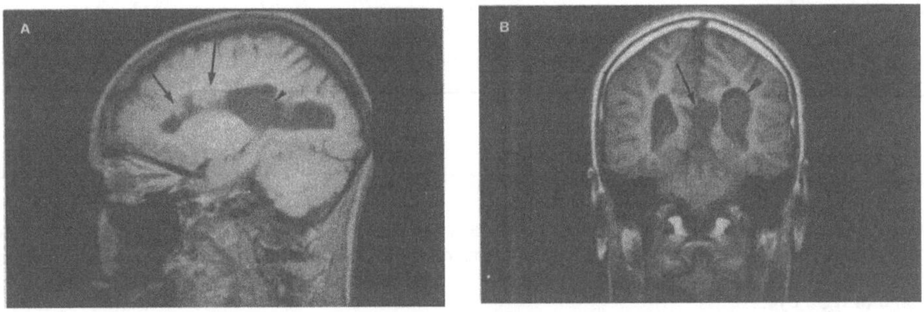

Figure 2. MRI (T1 weighted) of a typical case of AS. **A.** Sagittal cut shows total agenesis of corpus callosum, cyst of the choroid plexus (arrowhead) and periventricular heteretopias of gray matter (arrows). **B.** Coronal cut showing cyst in the region of the third ventricle (arrow), distinct from cyst of left choroid plexus (arrowhead).

Cysts of the choroid plexus were present in 6 of our 10 new patients. They can vary in size from a few millimeters to several centimeters (Chevrie and Aicardi, 1986). They may be multiple (Hall-Craggs et al., 1990; Roland et al., 1989) and unilateral or bilateral. In some cases, tumors, rather than cysts, were found in the choroid plexus. Histologically, those were typical papillomas without signs of malignity (Ferrer et al., 1986).

Cysts in the region of the third ventricle are extremely common and possibly constant, as seen in patients who had MR imaging. They were found in all our recent cases. These cysts are often multiple, usually of less than 2-3 centimeters in diameter. Their signal is different from that of CSF, with a slightly greater intensity on T2-weighted sequences. A frequent location is the pineal region but the cysts are usually not median and may also be found lateral to the third ventricle. The limited pathological data available favor an ependymal origin which has been demonstrated in Ferrer's patient (Ferrer et al., 1986). It is likely that the 'ependymal diverticulum' communicating with a lateral ventricle in Brihaye's patient (Brihaye et al., 1956) was indeed a cyst ruptured into the ventricle as a thin cyst wall is easily torn. Other cystic formations, sometimes of large size, have been observed in the dura mater in 2 cases (Chevrie and Aicardi, 1986) and an arachnoidal cyst was present in the parietal lobe in another patient (De Jong et al., 1976) as well as in one of our own cases. 'Cysts' of the posterior fossa have been reported in several patients (Chevrie and Aicardi, 1986; Hall-Craggs et al., 1990) but their exact nature is not clear. Partial agenesis of the vermis has been reported in two cases (Chevrie and Aicardi, 1986) but the published images are more in favor of arachnoid cysts or a megacitern.

The occurrence of cysts, especially in the choroid plexus and around the third ventricle, is of interest for antenatal diagnosis of AS (Roland et al., 1989). In association with callosal agenesis such cysts strongly suggest the diagnosis of AS in a female fetus.

Gray matter heterotopias also seem to be present in all cases studied by high resolution CT or MRI. Periventricular nodules are easily demonstrated but subcortical heterotopias require high quality MR to be visualized. MR also allows for visualization of areas of abnormal gyration in some cases. In our experience such areas often present in a 'pachygyria-like' manner with a thick cortical ribbon separated from the underlying white matter by a relatively straight border. To our knowledge, a correlation between MRI findings and neuropathological findings has not yet been established but the frequency of micropolygyria in Aicardi syndrome (see below) suggests that such areas are probably microgyric. Focal microgyria has been shown to be associated with 'pachygyria-like' appearance in some cases (Chevrie and Aicardi, 1986; De Jong et al., 1976) and a similar appearance has also been thought to be present in cases of diffuse polymicrogyria (Aicardi, 1991).

Gross asymmetry between the hemispheres is very common and contributes to the peculiar appearance of CT or MR scans as compared with other types of callosal agenesis. Hemiplegia, or unilateral pyramidal tract signs, is often noted on the side opposite the larger hemisphere.

The superiority of MRI over CT for studying the corpus callosum is well established. Most cases of AS so far studied had complete agenesis but partial agenesis involving the posterior (Chevrie and Aicardi, 1986) or both the anterior and posterior (Aicardi et al., 1987) parts of the corpus callosum have been observed. Recently, we observed a patient with typical clinical, electroencephalographic and ophthalmologic findings who had, on MRI, a complete, although hypoplastic, corpus callosum (Figure 3). This patient

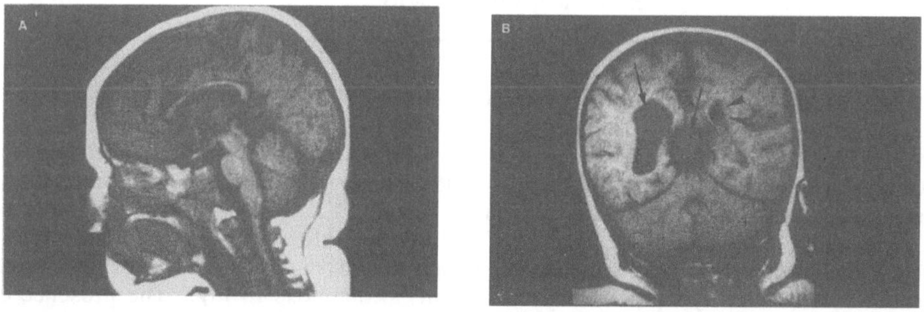

Figure 3. Case of AS without agenesis of the corpus callosum in a girl with infantile spasms, chorioretinal lacunae and vertebral anomalies (MRI T1-weighted sequence). **A.** Sagittal cut shows hypoplastic corpus callosum. **B.** Coronal cut shows cysts of the choroid plexus and the pineal region (arrows) and periventricular heteretopias (arrowheads).

also had heterotopic periventricular nodules, a bilateral cyst of the choroid plexuses and two cystic formations in the vicinity of the third ventricle. Such cases with simple hypoplasia of the corpus callosum would be impossible to diagnose by CT scan. It may even be difficult to separate completely a hypoplastic from a 'normal' corpus callosum as the shape and thickness of this structure vary widely (Barkovich and Norman, 1988).

The nosological situation of more atypical cases such as that of Sato et al (Sato et al., 1987) with associated holoprosencephaly remains uncertain but such cases sem exceedingly rare.

The association of AS with embryonic tumors has been reported in 2 cases (Tagawa et al., 1989; Tanaka et al., 1985) and may suggest a teratogenic action not limited to the CNS and skeleton even though it is rare.

As a consequence of these novel findings, it may be justified to reconsider the diagnostic criteria of Aicardi syndrome, especially with regard to the results of new imaging techniques. Although the classic triad of callosal agenesis, chorioretinal lacunae and infantile spasms remains the cornerstone of diagnosis as long as a chromosomal or other biological marker is lacking, other features such as periventricular or choroid plexus cysts and heterotopias appear to be of increasing significance. They are apparently no less frequent than the classic characteristics in recently studied cases that received adequate ophthalmological and MRI examination and may well be of even greater significance. We propose that new criteria for the syndrome should be established, taking into account the newly described features (Table 3). The diagnosis could be made even in the absence of one component of the classic triad, if two or more of the 'new' major criteria (including heterotopias on MRI) are present.

These new criteria could shed a new light on incomplete forms of AS. Cases resembling the syndrome but in which one element of the triad was missing have been

Table 3. Proposed new diagnostic criteria for the diagnosis of Aicardi syndrome.

Classic triad:	
	Infantile spasms
	Chorioretinal lacunae
	Agenesis of corpus callosum (total or partial)
New major features*:	
	Periventricular and subcortical heterotopia
	Cysts (ependymal ?) around the 3d ventricle and / or choroid plexuses
	Coloboma of optic disc
Supporting features**:	
	Vertebral and costal abnormalities
	Microphthalmia and / or other eye abnormalities
	Split brain' EEG (dissociated suppression-burst tracing)
	Gross hemispheric asymmetry
	Abnormalities of gyration***

* Probably present in all patients recently studied and who received MRI
** Frequent but certainly lacking in some cases
*** Probably constant but may not be possible to evidence

previously reviewed (Chevrie and Aicardi, 1986). These included cases without callosal agenesis. These cases were recognized on CT scanning and no MRI was available so that absence of only a limited portion of the corpus callosum or hypoplasia of this structure could well have been missed. As indicated above, the variability of the corpus callosum is wide and hypoplasia may be difficult to diagnose firmly. The presence of heterotopias, gyral abnormalities and cystic formations is of critical importance in such circumstances as they are more characteristic from a pathological point of view than callosal agenesis itself, so that these new criteria are probably less arbitrary, even though not specific, than mere callosal agenesis. Similarly, in incomplete forms without chorioretinal lacunae, the presence of typical imaging abnormalities and that of colobomas could suggest the possibility of AS, even though such cases are particularly difficult to accept. Cases without infantile spasms pose relatively little difficulty. Many series mention patients without spasms. Such cases are probably infrequent and may escape recognition when other types of seizures are prominent. The long-term evolution of the EEG pattern has not been systematically studied. In our experience, the split-brain, burst-suppression pattern tends to disappear with increasing age. It is not usually replaced by typical hypsarrhythmia but rather by multifocal EEG paroxysms. The clinical spasms may persist until adolescence. The development of a Lennox-Gastaut syndrome is only exceptionally, if ever, observed (Aicardi, 1986).

PATHOLOGICAL DATA

Most available pathological cases have been reported in a rather succinct manner and detailed microscopical examination of the brain was performed only in the case of Ferrer (Ferrer et al., 1986) and in one personal case examined by Billette et al. (1991). Agenesis of the corpus callosum that may be total or partial was never isolated but was always one component of a more complex brain malformation. Other structures may be lacking, e.g. the fornix (Chevrie and Aicardi, 1986; Billette et al., 1991), the first cranial nerves and mamillary bodies. A Probst bundle was present as in other verified cases. The anterior commissure was present in our personal case (Billette et al., 1991) but lacking in that of Ferrer et al. (1986). Abnormalities of gyration were present in all studied cases. Although they may appear in a few cases (Brihaye et al., 1956) as areas of pachygyria or lissencephaly, microgyria is the dominant finding and was present in all cases in which microscopic examination was available. The areas of abnormal gyration are irregularly distributed and usually asymmetrical. Visible heterotopias, either subependymal or in the white matter, were also a constant finding. Microgyria was of the unlayered type, with a thin undulating neuronal ribbon and fusion of facing molecular layers (Billette et al., 1991; Figure 4). Large pyramidal neurons often have an abnormal

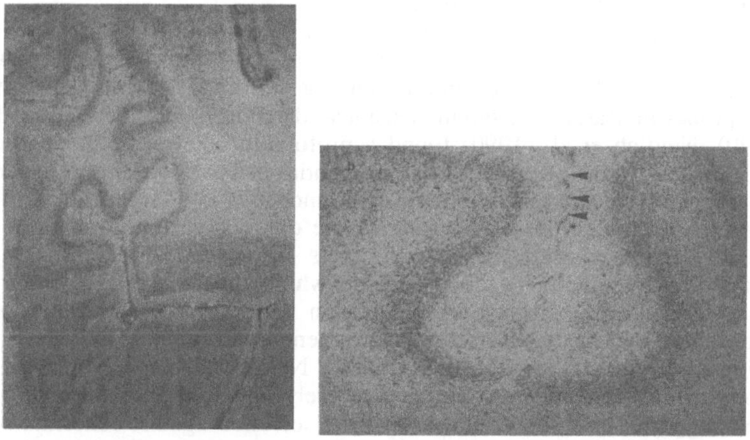

Figure 4. Cerebral cortex in a case of AS. **A.** Unlayered microgyria. **B.** Higher magnification showing the complete absence of lamination of cortical ribbon and fusion of the molecular layers between two microgyric convolutions (arrowhead).

orientation and were distributed randomly through the whole cortical thickness in our case (Billette et al., 1991) while they were placed at the appropriate, deep region in that of Ferrer et al. (1986). It seems, therefore, that the inside-out pattern of neuronal migration is not present in all cases (Billette et al., 1991). Cajal-Retzius and granular cells were found in the molecular layer in association with granular cells. Heterotopias include scattered subcortical neurons blurring the inner cortical margins, nodular heterotopias 2-10 mm in diameter and heterotopic neurons in the pia matter. The association of cortical dysplasia, heterotopias and dysgenesis of the corpus callosum is a rather uncommon pathological finding, even though it has also been observed outside AS (Billette et al., 1991). This strongly suggests an early action of the etiological factor before the end of the migration period. This is consistent with the embryology of the corpus callosum which is completed by 14 weeks gestation.

GENETIC AND CYTOGENETIC DATA

Overall, the basic data on the genetics of AS have been confirmed by recently published reports (Donnenfeld et al., 1989; Yamagata et al., 1990; Neidich et al., 1990). Only female patients are affected with the single exception of a phenotypic male with two X-chromosomes (Hopkins et al., 1979). This apparent exception is, in fact, additional evidence in favor of a sex-linked abnormality, lethal for individuals with a single abnormal X chromosome. The male child with an XY caryotype reported by Curatolo et al. (1980) is too atypic to be considered a genuine case of the syndrome (Aicardi, 1980; Hunter, 1980). Thus, a dominant abnormality (point mutation or chromosomal accident) involving the X chromosome and lethal early in gestation for male conceptuses remains the most likely explanation for the data. It is also consistent with the almost constant sporadic occurrence of the syndrome. However, one familial case in two sisters is on record (Molina et al., 1989). As the clinical and radiological presentation was characteristic, this finding is difficult to explain. No translocation was found in these patients so that the authors suggested the possibility of a germinal mutation or a premutation, as observed occasionally in other conditions. Another finding difficult to reconcile with an X chromosome dominant mutation is the occurrence of AS in only one of a monozygotic twin pair, which the authors consider evidence for a post-zygotic mutation in early embryonic development (Costa et al., 1990). This would not explain the occurrence in sisters. Another explanation could be inactivation of the defective X chromosome in the normal twin.

The strong suspicion of an abnormality of an X chromosome in patients with AS has received some support from recent work, although no deletion has ever been found in many patients studied by high-resolution banding and in at least 8 cases investigated with DNA probes in search of a submicroscopical deletion (Nielsen et al., 1991; Neidich et al., 1990). Neidich et al. (1990) found a profoundly skewed X-inactivation in the lympocytes of 3 of 7 girls with AS. The preferential inactivation of one X chromosome indirectly suggests an abnormality of this chromosome as inactivation is normally a random process (Lyon, 1966) so that one or the other X chromosome is active in an approximately equal number of somatic cells in normal females. Non-random inactivation reflects selection against cells in which the defective X-chromosome is active; thus tissues may consist only of cells in which the defective chromosome is inactive, a pattern that has been observed in patients with other X-linked diseases such as incontinentia pigmenti (Migeon et al., 1989). Neidich also found that patients with non-random inactivation tended to be more severely impaired than those with a random pattern and suggested that this could result from selection against abnormal cells in the developing neural tissue leading to aberrant brain development. Other investigators have found normal inactivation pattern in a few cases of AS (Costa et al., 1990; Wieacker et al., 1985).

A more precise chromosomal location for the syndrome at Xp22.3 was proposed by Ropers et al in 1982 (Ropers et al., 1982). They described a patient with a diagnosis of AS and with a balanced X / 3 translocation, the breakpoint being at band Xp22. They postulated that disruption or deletion of a gene or longer DNA segment at this point could result from the break and indicate the genetic location of the syndrome. However, the diagnosis of AS in this patient cannot be accepted. This girl had agenesis of the corpus callosum, costal malformations and microphthalmia but did not seem to have chorioretinal lacunae. A few other published cases of ocular and cerebral malformation syndromes with some of the features of AS have demonstrated chromosomal abnormalities in the Xp22.3 region. Donnenfeld et al. (1990) reported on a girl with microphthalmia, chorioretinal lesions, seizures and delayed development but a normal corpus callosum and no heterotopias at MR examination, who had an unbalanced X / 3 translocation, also with a breakpoint at Xp22.3. The translocation was inherited from the mother in whom it existed in a balanced form without phenotypic abnormalities. This girl had retinal lesions resembling, but not identical to, chorioretinal lacunae.

Other cases of microphthalmia with associated CNS abnormalities have been found in 4 children with an Xp22.3 abnormality. One of these patients also had agenesis of the corpus callosum (Friedman et al., 1988). Three children (one 46XX boy and 2 girls) had corneal opacities and orbital cysts. These 4 patients also had focal dermal hypoplasia, a sex-linked disorder with lethality for hemigyzous males mapping at Xp22.3 (Curnes et al., 1986). This has led Donnenfeld (Donnenfeld et al., 1990) to suggest that the Xp22.3 locus is likely to be involved in the genesis of both focal dermal hypoplasia and AS, that microphthalmia is a consequence of DNA abnormalities at the same or at a neighboring locus and that their patient may have had a contiguous gene syndrome involving both the Aicardi and the focal dermal hypoplasia loci. Recently, a case closely resembling AS in association with Golz syndrome has been reported (Aughton 1991, written personal communication) adding further support to the hypothesis of contiguous genes, for both syndromes. Even though the evidence pointing to Xp22.3 is interesting and may indicate that DNA sequences in this region are disrupted or deleted with syndromes related to AS, it is clearly not sufficient to allow any firm conclusion.

CONCLUSION

AS is a relatively common cause of infantile spasms, being recognized in about 4% of patients with infantile spasms in several large series. The diagnosis of AS has important implications regarding both prognosis and genetic counseling: individual prognosis for affected infants is exceedingly poor, but the risk of recurrence of the syndrome in the case of a further pregnancy is extremely small. These implications apply to those cases in which the diagnosis is based on the presence of the classic criteria. Whether they will apply to atypical forms defined by the 'new' criteria proposed in this chapter appears likely but remains to be proven.

It seems clear, however, that agenesis of the corpus callosum may not be as essential as initially thought and that the presence of limited hypogenesis or simple hypoplasia does not exclude the diagnosis of AS if the other elements of the MRI constellation of findings are present. Clearly, the issue will be resolved only with the discovery of a marker, most likely a DNA abnormality, a deletion being more probable than a point mutation. The Xp22.3 region is a prime candidate for exploration with DNA probes even though the evidence is far from convincing and the possibility of other accidents such as a somatic mutation cannot be ignored.

REFERENCES

Aicardi J., 1980, The Aicardi syndrome in a male infant, *J. Pediatr.* 97:1040-1041.

Aicardi J., 1986, "Epilepsy in Children," Raven Press, New York.

Aicardi J., 1991, The agyria-pachygyria complex: a spectrum of cortical malformations, *Brain Dev.* 13:1-8.

Aicardi J., Chevrie, J.J., and Baraton, J., 1987, Agenesis of the corpus callosum, *in:* "Handbook of Clinical Neurology," Vol. 6, P.J. Vinken, G.W. Bruyn, and H. Klawaus, eds., North Holland, Amsterdam.

Aicardi, J., Chevrie, J.J., and Rousselie, F., 1969, Le syndrome agénésie calleuse, spasmes en flexion, lacunes choriorétiniennes, *Arch. Fr. Pediatr.* 26:1103-1120.

Aicardi, J., Lefebvre, J., and Lerique -Koechlin, A., 1965, A new syndrome: spasms in flexion, callosal agenesis, ocular abnormalities, *Electroencephalogr. Clin. Neurophysiol.* 19:609-610.

Al-Gazali, L.I., Mueller, R.F., Caine, A., et al., 1988, An XX male and two (X ; Y) females with linear skin defects and congenital microphthalmia: a new syndrome at Xp22.3, *J. Med. Genet.* 25:638-639.

Baieri, P., Marki, A., Thelen, M., and Laub, M.C., 1988, MR imaging in Aicardi syndrome, *AJNR* 9:805-806.

Barkovich, A. and Norman, D., 1988, Anomalies of the corpus callosum: correlation with further anomalies of the brain, *AJNR* 9:493-501.

Billette, T., Robain, O., and Chiron, C., 1991, submitted, Agenesis of the corpus callosum associated with cortical dysplasia, *Acta Neuropathol.*

Bour, F., Chiron, C., Dulac, O., and Plouin, P., 1986, Caractères électrocliniques des crises dans le syndrome d'Aicardi, *Rev. Encephalogr. Neurophysiol. Clin.* 16:341-353

Brihaye, L., Gillet, P., Parmentier, R. and Peetrons, A., 1956, Agénésie de la commissure calleuse, associée à un kyste épendymaire, *Arch. Suisses Neurol. Neurochir. Psychiatr.* 77:415-431.

Chevrie, J.J. and Aicardi, J., 1986, The Aicardi syndrome. *in:* "Recent Advances in Epilepsy," Vol. 3, T.A. Pedley and B.S. Meldrum, eds, Churchill Livingstone, Edinburgh.

Costa, T., Greer, W., Duckworth, Rysiecki, M., Musarella, M., and Ray, P., 1990, Monozygotic twins discordant for Aicardi syndrome, Abstracts of the 41st Annual Meeting of the American Society of Human Genetics, *Am. J. Hum. Genetics, 47 (Suppl. 3)*:A-53 (0202) 14.5.

Curatolo, P., Libutti, G., and Dalla Piccola, B., 1980, Aicardi syndrome in a male infant, *J. Pediatr.* 96:286-287.

Curnes, J.T., Laster, D.W., Koubek, T.D., Moody, D.M., Ball, M.R., and Witkofski, R.L., 1986, MRI of corpus callosal syndromes, *AJNR* 7:617-622.

De Jong, J.G.Y., Delleman, J.W., Houben, M., et al., 1976, Agenesis of the corpus callosum: infantile spasms, ocular anomalies (Aicardi's syndrome). Clinical and pathological findings, *Neurology* 26:1152-1158.

Del Pero, R.A., Mets, M.B., Tripathi, R.C., and Torczynski, E., 1986, Anomalies of retinal architecture in Aicardi syndrome, *Arch. Ophthalmol.* 104:1659-1664.

Donnenfeld, A.E., Graham, J.M., Packer, R.J., Aquino, R,. Berg, S.Z., and Emanuel, B.S., 1990, Microphthalmia and chorioretinal lesions in a girl with an Xp22.2-pter deletion and partial 3p trisomy: clinical observations relevant to Aicardi syndrome gene localization, *Am. J. Med. Genet.* 37:182-186.

Donnenfeld, A.E., Packer, R.J., Zackai, E.H., Chee, C.M., Sellinger, B., and Emmanuel, B.S., 1989, Clinical, cytogenetic and pedigree findings in 18 cases of Aicardi syndrome, *Am. J. Med. Genet.* 32:461-467.

Ferrer, I., Cusi, M.V., Liarte, A., and Campistol, J., 1986, A Golgi study of the poly-microgyric cortex in Aicardi syndrome, *Brain Dev.* 8:518-525.

Friedman, P.A., Rao, K.W., Jeplin, S.W., and Aylsworth, A.S., 1988, Provisional deletion mapping of the focal dermal hypoplasia (FDH) gene to Xp 22.31, *Am. J. Human Genet.* 43:A50.

Gobbi, G., Bruno, L., Pini, A. and Rossi, P.G., and Tassinari, C.A., 1987, Periodic spasms: an unclassified type of epileptic seizure in childhood, *Dev. Med. Child Neurol.* 29:766-775.

Hall-Craggs, M.A., Harbord, M.G., Finn, J.P., Brett, E., and Kendall, B.E., 1990, Aicardi syndrome: MR assessment of brain structure and myelination, *AJNR* 11:532-536.

Hamano, S.I., Yagishita, S., Kawakami, M., et al., 1985, Aicardi syndrome: postmortem findings, *Pediatr. Neurol.* 5:259-261.

Hopkins, I.J., Humphrey, I., Keith, C.G., Susman, M., Webb, G.C., and Turner, E.K., 1979, The Aicardi syndrome in a 47,XXY male, *Aust. Paediatr. J.* 15:278-280.

Horita, H., Kumagai, K., Maekawa, K. and Endo, S., 1980, Overnight polygraphic study of agenesis of the corpus callosum with seizures resembling infantile spasms, *Brain Dev.* 2:379-386.

Hunter, A.G.W., 1980, Aicardi syndrome in a male infant, *J. Pediatr.* 97:1041.

Igidbashian, V., Mahboubi, S., and Zimmerman, R.A., 1987, CT and MR findings in Aicardi syndrome, *J. Comp. Ass. Tomogr.* 11:357-358.

Insley, J., Goodfellow, P.G., and Hulten, M., 1988, An XX male and two (X ; Y) females with linear skin defects and congenital microphthalmia: a new syndrome at Xp22.3., *J. Med. Genet.* 25:638-639.

Jeret, J.S., Serur, D., Wisniewski, K., and Fisch, C., 1986, Frequency of agenesis of the corpus callosum in the developmentally disabled population as determined by computerized tomography, *Pediatr. Neurosci.* 12:101-103.

Jeret, J.S., Serur, D., Wisniewski, K.E., and Lubin, R.A., 1987, Clinicopathological findings associated with agenesis of the corpus callosum, *Brain Dev.* 9:255-264.

Lyon, M.F., 1966, X chromosome inactivation in mammals, *Adv. Teratol,* 1:25-54.

McMahon, R.G., Bell, R.A., Moore, R.W., and Ludwin, S.K., 1984, Aicardi syndrome. A clinicopathologic study, *Arch. Ophthalmol.* 102:250-253.

Migeon, B.R., Axelman, J., De Beur, S.J., Valle, D., Mitchell, G.A., and Rosenbaum, K.N., 1989, Selection against lethal alleles in females heterozygous for incontinentia pigmenti, *Am. J. Hum. Genet.* 44:100-106.

Molina, J.A., Mateos, F., Merino, M., Epifanio, J.L., and Gorrono, M., 1989, Aicardi syndrome in two sisters, *J. Pediatr.* 115:282-283.

Moreno, J.A., Lartigau, M.T., Etcheverria, L., et al., 1986, Sindrome de Aicardi. Presentacion de cuatro casos nuevos y nuevas anomalias asociadas, *Ann. Esp. Pediatr.* 24:53-58.

Neidich, J.A., Nussbaum, R.L., Packer, R.J., Emanuel, B.S., and Puck, J.M., 1990, Heterogeneity of clinical severity and molecular lesions in Aicardi syndrome, *J. Pediatr.* 116:911-917.

Nielsen, K.B., Anvret, M., Flodmark, B., et al., 1991 Aicardi syndrome. Early neuroradiological manifestations and results of DNA studies in one patient, *Am. J. Hum. Genet.* 38:65-68.
Ohtsuki, H., Haebara, H., Takahashi, K. et al., 1981, Aicardi's syndrome: report of an autopsy case, *Neuropediatrics* 12:279-286.
Raynor, M. and Raynor, D., 1990, The Aicardi syndrome survey, *Aicardi Syndrome Newsletter.*
Roland, E.H., Flodmark, O. and Hill, A., 1989, Neurosonographic features of Aicardi syndrome, *J. Child. Neurol.* 4:307-310.
Ropers, H.H., Zuffardi, O., Biancai, E., and Tiepolo, L., 1982, Agenesis of corpus callosum, ocular and skeletal anomalies (X-linked dominant Aicardi's syndrome) in a girl with balanced X / 3 translocation, *Hum. Genet.* 61:364-368.
Sato, N., Matsuishi, T., Utsunomiya, H., et al., 1987, Aicardi syndrome with holoprosencephaly and cleft lip and palate, *Pediatr. Neurol.* 3:114-116.
Tagawa, T., Mimaki, T., Ono, J., et al., 1989, Aicardi syndrome associated with embryonal carcinoma, *Pediatr. Neurol.* 5:45-47.
Tanaka, T., Thkahura, H., Takashima, S., Kodama, T., and Hasegawa, H., 1985, A rare case of Aicardi syndrome with severe brain malformation and hepatoblastoma, *Brain Dev.* 7:507-512.
Wieacker, P., Zimmer, J., and Ropers, H.H., 1985, X-inactivation patterns in two syndromes with probable X-linked dominant, male lethal inheritance, *Clin. Genet.* 28:238-242.
Willis, J. and Rosman, N.P., 1980, The Aicardi syndrome versus congenital infection: diagnostic considerations, *J. Pediatr.* 96:235-239.
Yamagata, T., Momoi, M., Miyamoto, S., Kobayashi, S., and Kamoshita, S., 1990, Multi-institutional survey of the Aicardi syndrome in Japan, *Brain Dev.* 12:760-765.

THE ANDERMANN SYNDROME: AGENESIS OF THE CORPUS CALLOSUM AND SENSORIMOTOR NEUROPATHY

Frederick Andermann[1,2] and
Eva Andermann[1,3]

[1]Departments of Neurology and
Neurosurgery and
[2]Department of Pediatrics
Montreal Neurological Institute and Hospital
3801 University St.
Montreal, Que. H3A 2B4, Canada
[3]Centre for Human Genetics
McGill University, Montreal, Que., H3A 1B1
Canada

INTRODUCTION

The Andermann syndrome is an autosomal recessive disorder which consists of agenesis of the corpus callosum, mild to moderate mental retardation and sensorimotor neuropathy. The eponym was first used to refer to this disorder by Larbrisseau and his colleagues (1984) when they described the clinical features of the patients studied at Ste-Justine Hospital for Children in Montreal.

FIRST CASES

We examined the first two cases about 25 years ago at the Montreal Children's Hospital. These were two brothers who had an obvious neuromuscular disorder and mild to moderate retardation as well as unusual facial features. Accurate distinction between myopathic and neurogenic peripheral disorders could not be made at that time; during their original investigation the nature of the neuromuscular process was not clarified, neither was the retardation explained. Several years later, the children were further investigated and muscle biopsy indicated that the two boys had a peripheral neuropathy. In order to explain the retardation, a pneumoencephalogram was carried out in one of the patients, which showed complete agenesis of the corpus callosum. The pneumoencephalogram of the brother demonstrated identical findings with, in addition, multiple hamartomata protruding into the lumen of the lateral ventricles. The following year, two brothers with a very similar clinical picture were investigated. Their nerve muscle biopsies and pneumoencephalograms showed identical findings. Soon afterwards, a girl with this clinical picture was also seen. A diagnosis of agenesis of the corpus callosum and sensorimotor neuropathy was suggested, and this diagnosis was confirmed by subsequent studies to the surprise of the orthopedic residents caring for her.

Children with a similar neuromuscular disorder but who did have a corpus callosum and otherwise greatly resembled the three families just described were also brought to our attention. The unusual clinical picture differed from the motor neuron diseases and peripheral neuropathies of childhood which were recognized at that time (Andermann et al., 1972, 1975; Andermann, 1981). Because of the retardation, it was difficult to assess sensory function in these patients. Since the wasting was very pronounced, the initial impression was that of a motor neuron disorder (Andermann et al., 1972, 1975). As more patients were examined, it became clear that they had a peripheral neuropathy (Andermann, 1981; Nagy, 1982; Andermann and Andermann, 1990).

Study of the pedigrees of the first three families revealed that the parents or grandparents all originated from the same region, in the Saguenay-Lake St-Jean area of the Province of Québec (see Andermann et al., this volume). The medical staff of the Hospital at Baie St-Paul had been aware of this illness, and they let us examine three affected siblings then in their early 20s, institutionalized in this large center for chronic psychiatric and neurological disorders. Several other sporadic patients with this disorder were also investigated there as well as four affected siblings living nearby. This was the start of extensive genealogical studies carried out by us (Andermann et al., this volume).

The Saguenay-Lake St-Jean region has a population of 300,000 descended from a small number of founding families who settled there in the mid-19th century. Already in the 1950's, pneumoencephalograms performed on children with mental retardation had revealed a series of patients with agenesis of the corpus callosum in this region. There were affected sibships in this series studied by Gerard Leblanc, the pioneer neurosurgeon in this area, who in addition to agenesis of the corpus callosum had a neuromuscular disorder. The peripheral syndrome was initially ascribed to the callosal agenesis. The usual diagnosis made in those early days was either cerebral palsy or muscular dystrophy and, as was then customary, institutional placement was often arranged. We examined larger numbers of patients in the institutions of this area and continued the genealogical studies. Eventually, almost 250 affected children and young adults had been identified (Nagy, 1982; Carpenter et al, 1987; see also Andermann et al., this volume). It then became possible to delineate the clinical spectrum of the disorder.

CLINICAL DESCRIPTION

The children were hypotonic and areflexic from birth. They had a square facies, and tended to resemble one another (Figure 1). Unusual facial features included hypoplastic maxilla, large angle of the mandible, high arched palate, low hairline and narrow forehead; long distance from the lower lip to the chin and a degree of hypertelorism, though never as marked as sometimes found in association with sporadic agenesis (Figures 2 and 3). Eventually they improved from a motor point of view, and were sometimes able to walk with crutches and/or braces around the ages of 4 to 6 years. In the early teens they then lost the ability to walk, developed scoliosis and were confined to a wheelchair (Figure 4). Thus, it was clear early on that the peripheral manifestations were progressive (Figure 5). There was pronounced wasting of the small muscles of the hands with digitalization of the thumbs (Figure 6), but they were able to use their hands for many years after they were unable to stand, until marked contractures developed. The feet were wasted, limp and later contracted. There was purple discoloration of the distal lower extremities suggesting some sympathetic involvement (Figure 7). Scoliosis gradually deteriorated (Figure 5).

Some degree of mental retardation was present in all patients, but there was considerable variation in its severity. Some children appeared to be of borderline normal intelligence, whereas others were severely retarded. However, most were mildly or moderately retarded, in the educable range. In the first two decades, there was little evidence for mental deterioration. We were struck early on by a rather unusual hallucinatory psychosis which developed late in the second or early in the third decade in some of the patients. The first patient who displayed this symptom described castles floating in the air, and there was a paranoid flavor and content to these hallucinatory

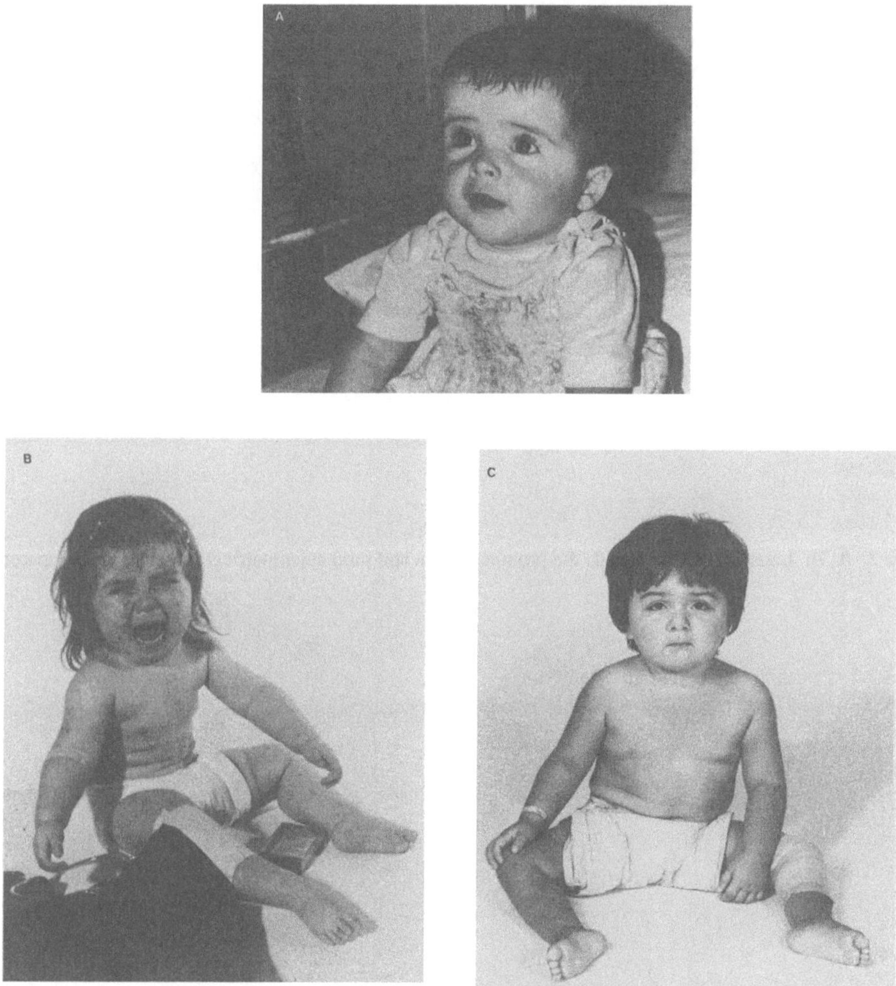

Figure 1. A, B, C. In infancy the patients are hypotonic and areflexic. They have square faces and tend to resemble each other.

phenomena. No auditory or olfactory hallucinations were mentioned. After the onset of these hallucinatory events, the patients became quiet and regressed and this raised the possibility of progressive CNS deterioration. Some more obvious mental deterioration and incontinence of urine and stools appeared in the third decade. Epileptic seizures were infrequent. Other central manifestations consisted of a tremor made worse by intention, likely due to involvement of the dentatorubral system; and, as far as could be assessed, extensor plantar responses. Pyramidal function could not be evaluated further in the presence of flaccidity and absent deep tendon reflexes.

Review of large groups of patients confirmed that some of the affected siblings had a corpus callosum. There were also several patients in the area who had agenesis of the corpus callosum, a large head, mild retardation and an unusual frontal hair whorl. They did not have the peripheral disorder and were not part of the affected families, with the exception of one such patient who had a cousin with the complete syndrome. To this day it is not clear whether agenesis of the corpus callosum of a nonsyndromic type is also more frequent in the Lake St-Jean population or not.

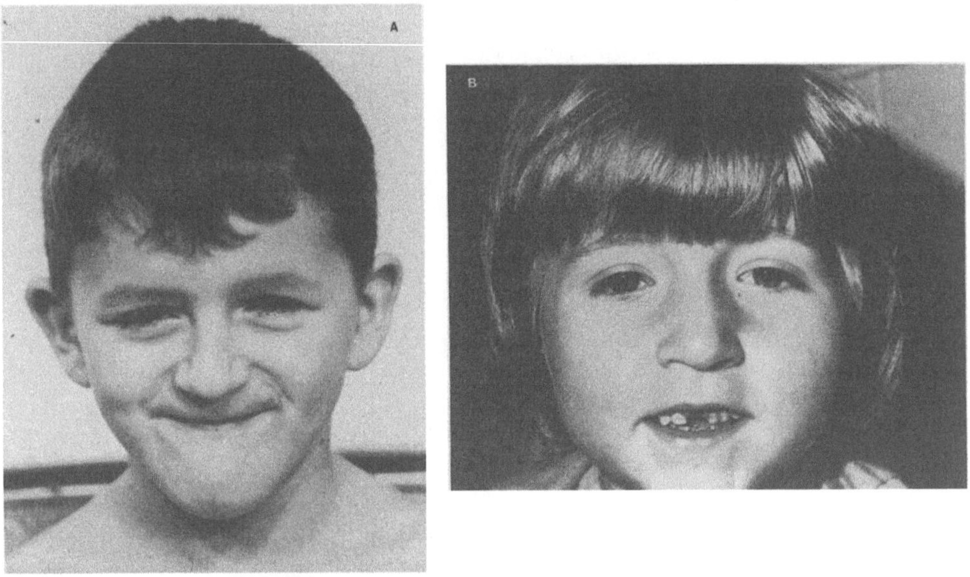

Figure 2. A, B. Later in the first decade the prominent chin and mild asymmetrical ptosis become apparent.

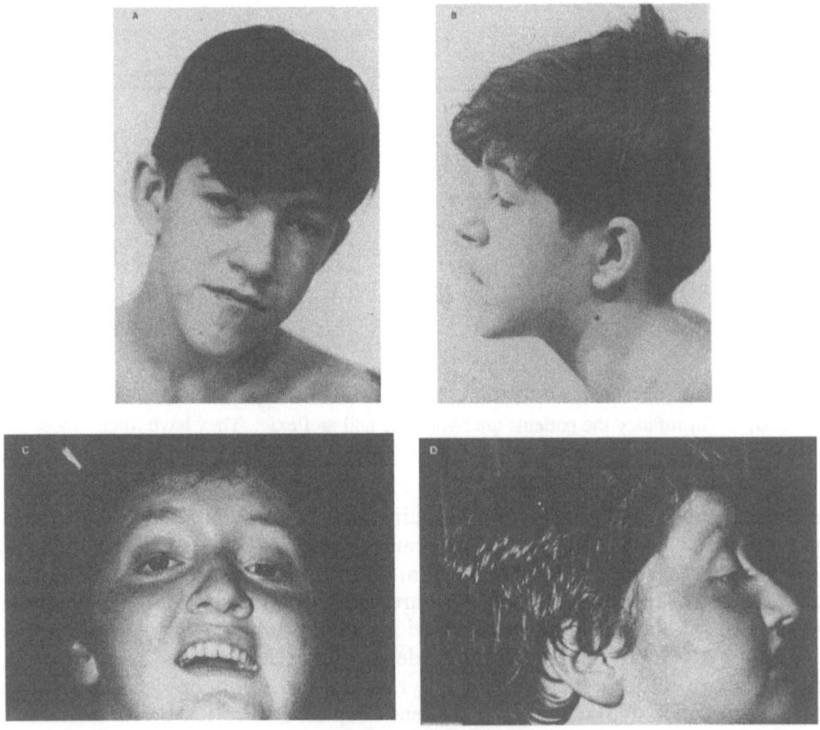

Figure 3. A. Face and profile of J.B., the first patient described, in the second decade of his life. **B.** Another patient, D.J., during her third decade.

A more accurate assessment of the incidence of agenesis, hypoplasia, or presence of the corpus callosum in patients with the syndrome became possible when a CT scanner was introduced in the region. Mathieu and his colleagues were able to determine that the prevalence of complete callosal agenesis in patients with the syndrome was 57.8%. Partial agenesis was found in 9.4%, and the callosum was present in the remaining one-third of patients (Mathieu et al., 1990). CT scanning, however, does not

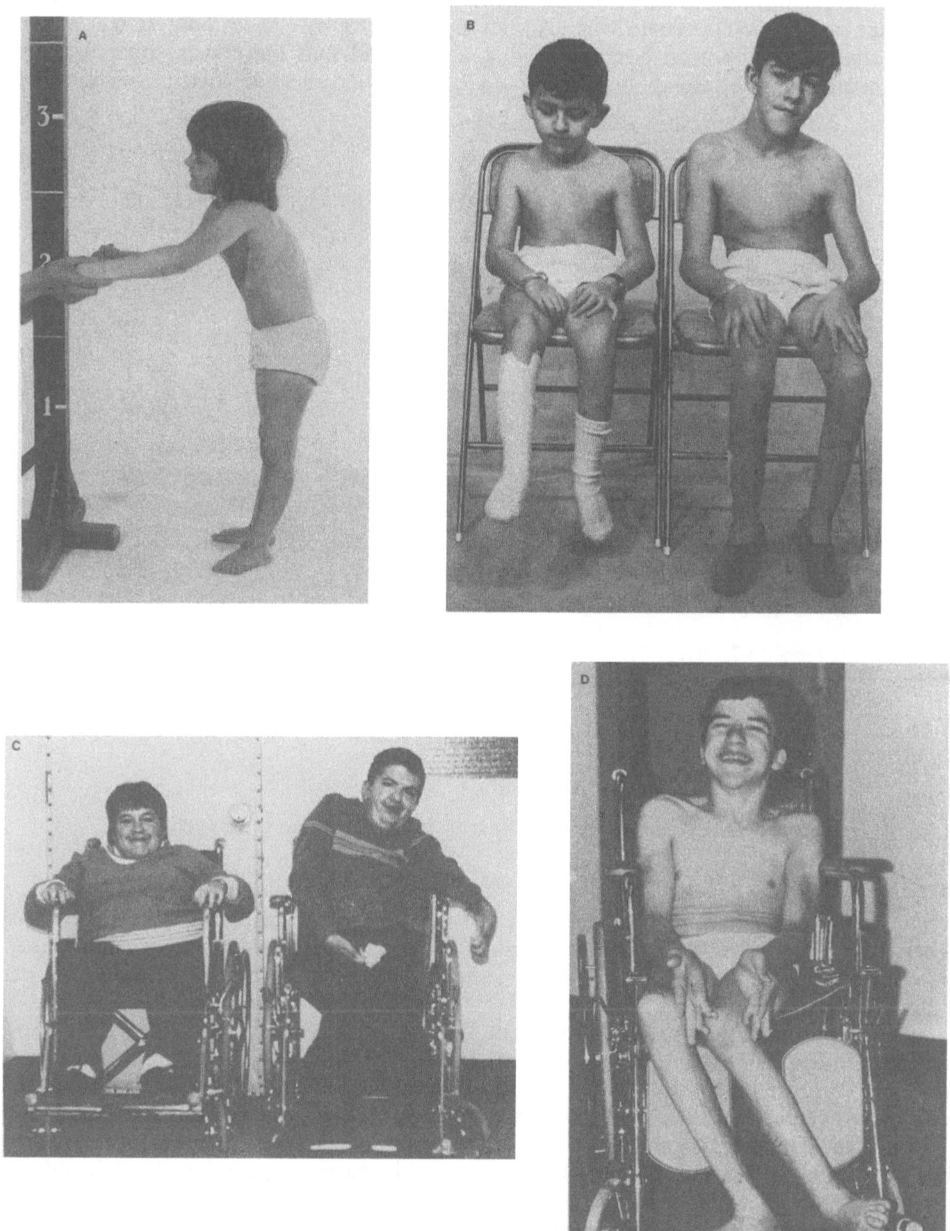

Figure 4. A. At age 4, C.G. is unable to stand alone. **B.** The first patients described: J.B. and F.B. **C.** Older affected siblings confined to a wheelchair. **D.** An older patient showing pronounced distal wasting.

enable one to assess the thickness and texture of this structure. Callosal and extracallosal, usually posterior hemispheral white matter changes can now be visualized by magnetic resonance imaging (Andermann and Melanson, personal observations).

The postmortem examination of patients with the Andermann syndrome is the object of a separate chapter (Carpenter, this volume). Briefly, the study of muscle and nerve biopsies revealed a loss of large myelinated axons; the appearance of the nerve suggested that these axons had been ballooned and then collapsed. The pathological changes were unlike those seen in other genetically determined neuropathies (Carpenter et al., 1987). There was also neurogenic atrophy. Electrodiagnostic studies showed that sensory action potentials were absent and that there was some delay in motor conduction. No metabolic abnormality was found despite extensive screening.

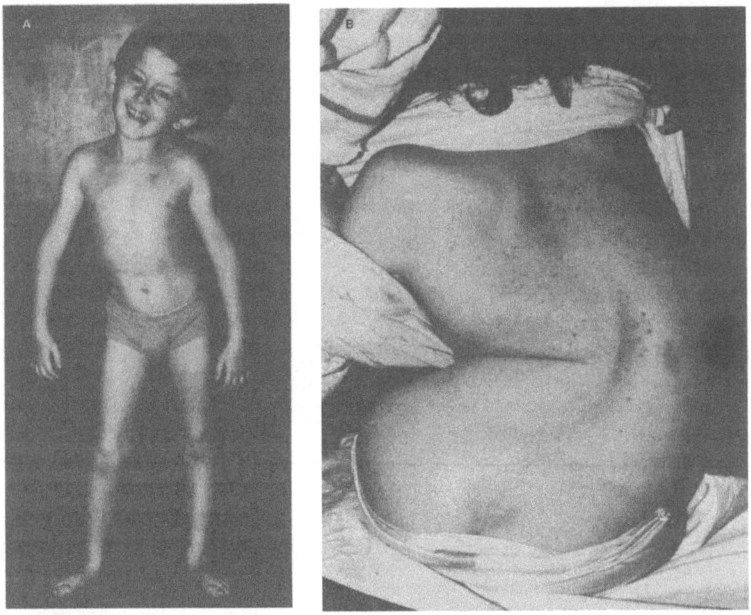

Figure 5. The scoliosis progresses from the first (**A**) to the second decade (**B**).

White matter changes in the CNS similar to those seen in peripheral nerve were also observed. Loss of myelin and ballooning of axons suggesting a neurofilamentous disorder in the callosum and centrum semiovale were described (Carpenter, this volume).

This syndrome, though quite different from giant axonal neuropathy, followed a similar path to clarification. It was obvious early on that the peripheral abnormalities were of a progressive nature, but it took several years before it became clear that progressive central nervous system involvement was also part of the clinical picture.

Since the founding population is of French origin, we wondered about occurrence of this syndrome in France. An early report of this patient aggregate by Giroud and Langevin in the French literature (Andermann and Andermann, 1990) was not echoed by other accounts. The first report outside Quebec describes a typical case in an Italian boy (Battistella et al., 1986). Renault and Dulac then described Algerian siblings, offspring of a consanguineous couple with a quite typical picture of the disorder (Renault et al., 1991). A third family with two affected siblings, American with widely divergent ethnic background, has been studied by Sakubai Naidu at Johns Hopkins University (personal communication) Thus, as expected, the syndrome is not unique in the Quebec population; the larger number of cases occurring there represents a founder effect related to the small number of original settlers (Andermann, 1981; Nagy, 1982; Andermann and Andermann, 1990).

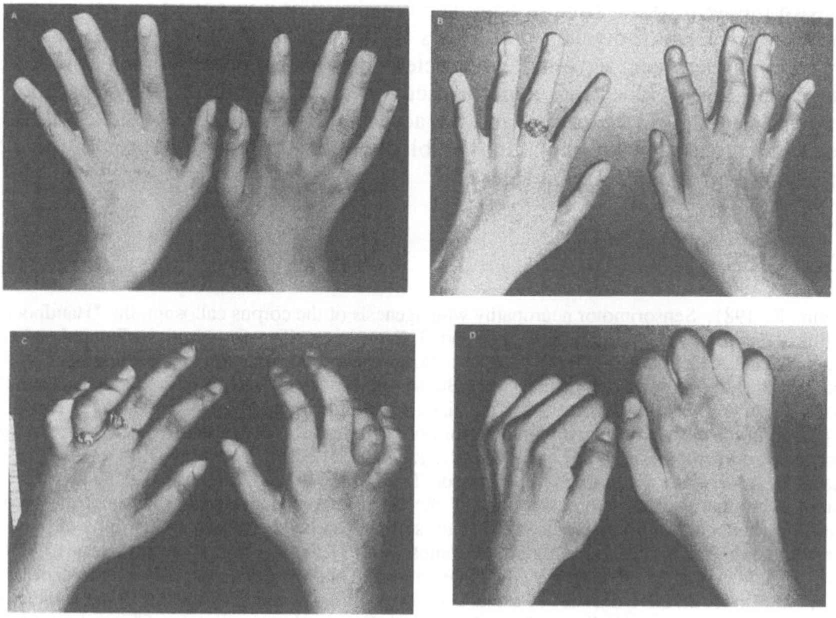

Figure 6. A. Low set digitalized thumbs. **B, C, D.** Contractures of the fingers progressively develop.

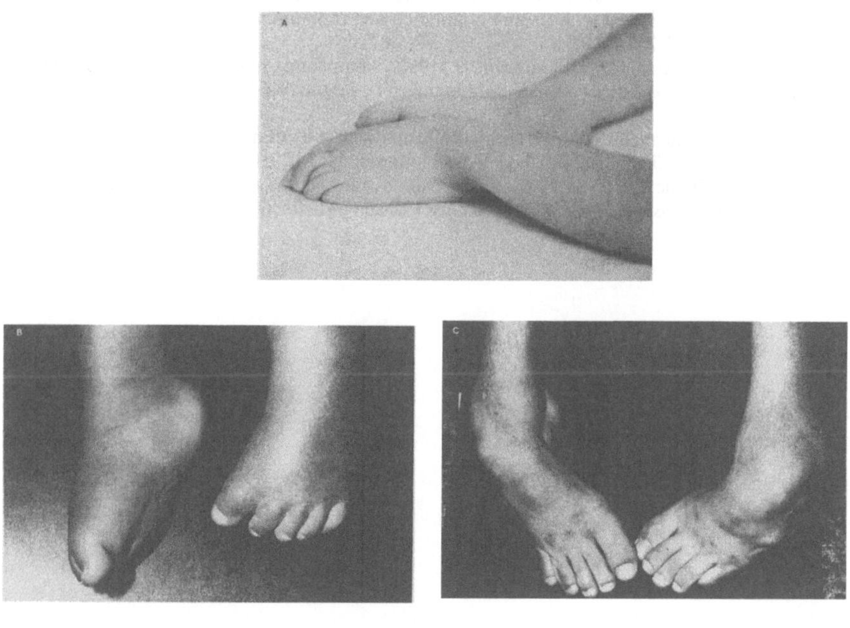

Figure 7. A, B, C. The feet are originally hypotonic; they then develop some signs of sympathetic dysfunction and edema and later are severly contractured.

CONCLUSION

The pathological findings are the first step toward elucidation of the biochemical defect. The molecular biological basis of the disorder is currently being studied. In the meantime, because of awareness of the syndrome and early diagnosis, families with multiple affected children are no longer encountered. Attempts at prenatal diagnosis, mainly with the help of ultrasound, have been made. The presence of a corpus callosum in a number of patients, however, constitutes an important limitation to the utilization of ultrasound in mothers who have had an affected child and in other women at risk.

The recognition, delineation, and clarification of this syndrome have paralleled the development of pediatric neurology and the accessibility of neurological care throughout the area. Identification of the putative biochemical defect and prevention remain challenges for neuroscience in the future.

REFERENCES

Andermann., E., 1981, Sensorimotor neuropathy with agenesis of the corpus callosum, in: "Handbook of Clinical Neurology," Vol. 42, P.J. Vinken and G. Bruyn, eds., Elsevier, North Holland, Amsterdam.

Andermann, E. and Andermann, F., 1990, Corpus callosum agenesis with sensorimotor neuropathy, familial, *in:* "Birth Defects Encycloclopedia," M.L. Buyse, ed., Blackwell Scientific Publications, Cambridge.

Andermann, F., Andermann, E., Joubert, M., Karpati, G., Carpenter, S., and Melanson D., 1972, Familial agenesis of the corpus callosum with anterior horn cell disease. A syndrome of mental retardation, areflexia and paraparesis, *Trans. Am. Neurol. Assoc.* 97:242-244.

Andermann, E., Andermann, F., Joubert, M., Melanson, D., Karpati, G., and Carpenter, S., 1975, Three familial midline malformation syndromes of the central nervous system: Agenesis of the corpus callosum and anterior horn cell disease; agenesis of the cerebellar vermis; and atrophy of the cerebellar vermis, *in:* "Malformation Syndromes, Birth Defects: Original Article Series," Vol. XI, Nr.. 2, D. Bergsma, ed., Excerpta Medica International Congress Series Nr. 340, Amsterdam.

Battistella, P.S., Drigo, P., Laverda, A.M., Casara, G.L., De Martin, G., and Condini, A., 1986, La sindrome di Andermann: primo caso Italiano, *Proc. XIIth Natl. Congr. Neuropediatrics,* Chieti, Italy (Abstract).

Carpenter, S., Karpati G., Andermann, F., and Andermann, E., 1987, Peripheral neuropathy with agenesis of the corpus callosum: A unique clinicopathological entity, *Ann. Neurol.* (Abstract).

Larbrisseau, A., Vanasse M., Brochu, and P., Jasmin, G., 1984, The Andermann syndrome: Agenesis of the corpus callosum associated with mental retardation and progressive sensorimotor neuropathy, *J. Can. Sci. Neurol.* 11:257-261.

Mathieu, J. Bédard, F., Prévost, C., and Langevin, P., 1990, Neuropathie sensitivo-motrice héréditaire avec ou sans agénésie du corps calleux: Étude radiologique et clinique de 64 cas, *J. Can. Sci. Neurol.* 17:103-108.

Nagy, J.R., 1982, Familial agenesis of the corpus callosum with sensorimotor neuropathy: genetic and epidemiological studies, Unpublished Master's thesis, McGill University, Montreal.

Renault, F., Brian, R., Dulac, O., et al., 1991, Andermann Syndrome in Two Algerian Siblings, *IBRO Satellite Symposium on Callosal Agenesis,* Quebec City, Canada (Abstract).

THE PATHOLOGY OF THE ANDERMANN SYNDROME

Stirling Carpenter

Departments of Neurology-Neurosurgery and
Pathology, McGill University
The Montreal Neurological Institute and Hospital
3801 University St.
Montreal, Que. H3A 2B4, Canada

INTRODUCTION

The Andermann syndrome is a familial disease that was initially described in the Saguenay-Lake St-Jean area in Quebec. A few cases have now been reported outside of Quebec (see Andermann and Andermann, this volume). Agenesis of the corpus callosum is usually, but not invariably, part of the syndrome. We have had the opportunity of studying the brains of two patients suffering from Andermann's Syndrome, as well as six sural nerve biopsies.

AUTOPSY STUDIES

The first patient died at the age of 26 years. After 13 years, he had needed a wheelchair. His mentation was said to be normal. The second patient died at the age of 37. A wheelchair was necessary after 28 years. He had moderate mental deficiency. In his 30's, he suffered from psychotic decompensation and frightening visual hallucinations.

The fixed brains of these patients weighed 1400 grams and 1524 grams. The brain of the first patient showed complete agenesis of the corpus callosum with prominent Probst's bundles. The anterior commissure appeared normal in size.

A corpus callosum was present in the brain of the second patient, although posteriorly it was thin. Anteriorly at the level of the corpus striatum it was 5mm thick beneath the cingulate gyrus. Posteriorly, at the level of the pulvinar, it was 1.3mm thick. The anterior commissure could not be seen.

Microscopically, the posterior part of the corpus callosum of patient no. 2 showed increased packing density of oligodendroglial nuclei and a paucity of large axons compared to controls on a Holmes stain. Astrocytes were increased in number in mid portion of the corpus callosum and greatly increased in posterior portions. Pale, pink, oval to tubular structures around 16 μm in diameter were found in the splenium and occasionally as far forward as the middle of the corpus callosum (Figure 1). There were also round to oval sharply demarcated vacuoles up to 55 μm in diameter in the white matter. They were largely lacking in the more damaged portions of the corpus callosum.

Callosal Agenesis, Edited by M. Lassonde and
M.A. Jeeves, Plenum Press, New York, 1994

They sometimes contained pale, granular or filamentous, slightly pink material. The most numerous vacuoles were found in the occipital white matter. Similar vacuoles were found in the white matter of the brain of patient no. 1. The fibers of Probst's bundle appeared normal except for occasional vacuoles. No areas of neuronal loss were seen in either brain, although not all cortical areas have been sampled.

The findings in the peripheral nervous system were identical in the two cases and different from those reported in any other disease of the peripheral nervous system. In paraffin sections of nerve roots, large dilated axons were found which measured up to 120 μm in diameter (Figure 2). They usually stained pink with hematoxin and eosin but were occasionally clear. Their periphery was usually lined by Schwann cell nuclei. The

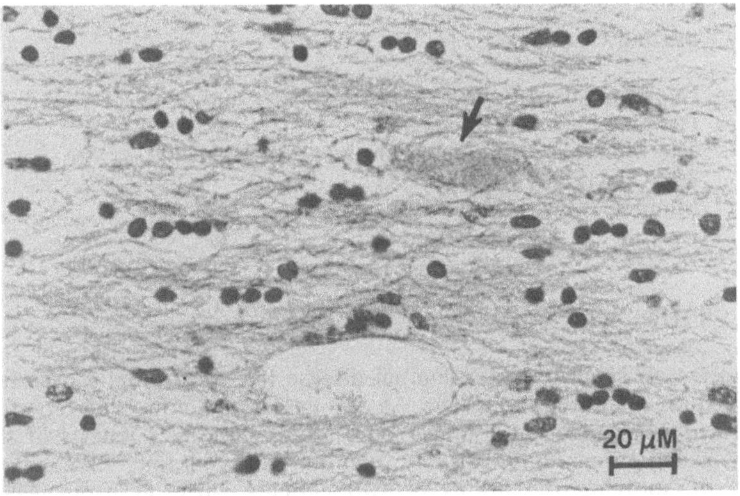

Figure 1. This photomicrograph from the midportion of the corpus callosum of case 1 shows an eosinophilic sausage-shaped axonal swelling (arrow) and an elongated vacuole, presumably the result of marked focal swelling of an axon. Hematoxylin-eosin.

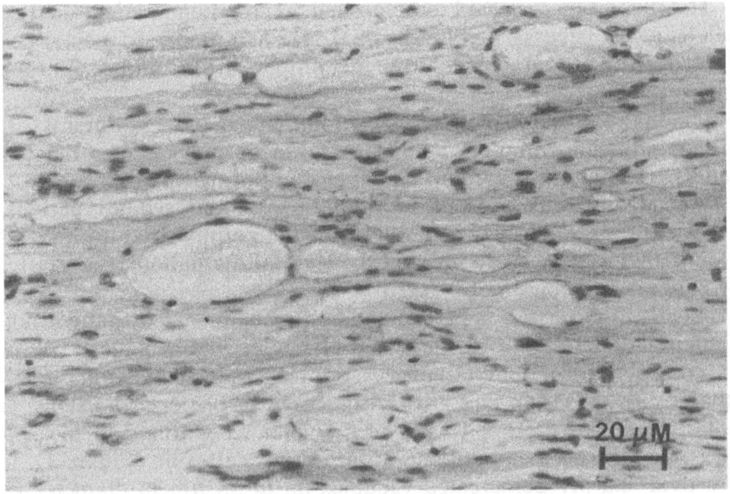

Figure 2. This section from case 1 shows swelling of axons or of periaxonal space in proximal lumbar roots. Hematoxylin-eosin.

contents were usually pale grey with Bodian's stain. Other axons with a lesser degree of enlargement were more frequently seen. They showed an intermediate intensity of Bodian staining. The largest axonal dilatations were not surrounded by myelin. In some areas there are numerous, very tiny axons, sometimes forming neuroma-like tangles. In the roots a few millimeters from the cord, there were many round to oval nodules of cells which measured up to 100 μm in diameter. Moderate axonal loss was present in the roots. Dilated axons were also noted in the oculomotor nerve in case no. 1, along with numerous onion bulbs.

Epoxy resin sections of nerve roots showed moderate reduction in the number of myelinated fibers and the presence of onion bulbs. Thin myelin sheaths surrounded enlarged pale axons. Clustered small myelinated fibers suggestive of axonal regeneration were present.

Teased fibers from motor roots showed the most severe abnormality at roughly 1cm from the spinal cord. Balls of Schwann cells adhered to teased fibers. Myelinated fibers passed through some of these, while in others they formed tangles. Enlarged axons with thin sheaths were observed (Figure 3). Apparently remyelinated internodes were present, and some sheaths showed focal granular breakdown of myelin.

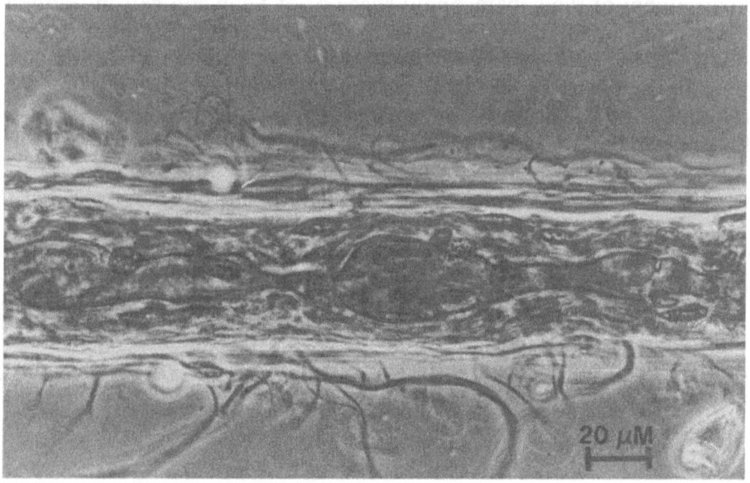

Figure 3. This teased fiber preparation from a proximal lumbar root of case 1 shows a thin, irregularly distended myelin sheath. Osium tetroxide, phase optics.

BIOPSY STUDIES

Sural nerve biopsies were available from 6 patients with ages 2, 2, 3, 6, 7 and 12 years, all of whom showed absence of the corpus callosum on imaging studies, and all of whom were hypotonic with reduced or absent tendon reflexes.

All these biopsies showed a virtually total lack of large myelinated fibers. Small myelinated fibers were present in normal numbers in the younger patients, but were somewhat reduced in the older patients. A few ovoids of Wallerian degeneration were seen in the nerves of the three youngest patients and in one older patient. In five biopsies there were a few large fibers with a star shaped or crumpled outline suggestive of axonal collapse. Isolated axons with diameters beyond the upper range of normal, but with an abnormally thin sheath were seen in two biopsies. Electron microscopy of one such axon disclosed decreased packing density of neurofilaments. All biopsies showed some focal clusters of small myelinated fibers indicative of axonal regeneration. There were very rare onion bulbs.

CONCLUSION

The similar striking abnormalities in the nerve roots in the 2 autopsy cases indicate that both are examples of Andermann's disease, even though the second patient had a corpus callosum. Presence of the corpus callosum in Andermann syndrome has been previously reported and may occur in about 33% of cases (Larbrisseau et al., 1984; Mathieu et al., 1990). The finding of ongoing damage in central axons helps to explain the visual hallucinations and psychosis in the second patient.

On the basis of the abnormalities seen on the nerve roots in the two autopsy cases, we believe that the basic abnormality is a swelling of axons accompanied by relative decrease in packing density of neurofilaments. Normally axonal diameter is proportionate to the number of neurofilaments. This axonal swelling leads to stretching and thinning of the myelin sheath with demyelination and remyelination. It also leads to axonal necrosis. There is proliferation of Schwann cells, which probably occupy the space left by the largest axonal swellings, resulting in balls of Schwann cells. The larger axons are particularly damaged so that they are depleted in sural nerve biopsies.

It is striking that the same axons which show such abnormality in nerve roots appear normal in their course inside the spinal cord. Cell bodies of anterior horn cells likewise appeared normal. This shows that axonal abnormality is particularly brought out in the environment of the subarachnoid space or of Schwann cell investment. On the other hand, scattered swollen axons could be identified in the white matter of both cases. In addition, the second case showed signs that progressive axonal destruction had taken place in the posterior part of the corpus callosum.

The combination of a malformation with progressive neurological disease is highly unusual. In Andermann's Syndrome, it appears likely that those fibers that begin to cross from hemisphere to hemisphere to form the corpus callosum are exposed to some factor in the CSF environment which damages them, resulting in agenesis of the corpus callosum.

Watery swelling of axons has been experimentally produced by intraneural injection of spider and scorpion venoms (Love and Cruz-Höfling, 1986). These toxins are thought to cause excessive influx of sodium. It seems possible that some abnormality of ion channels could lead to the axonal swellings seen in Andermann's Syndrome, although the abnormality is most extensively seen in axons in close proximity to the subarachnoid space, as in the nerve roots.

Acknowledgments

I thank Dr. Jean Mathieu, Dr. Lucien Privé and Dr. André Cholette for the opportunity of studying the brains of patients one and two.

REFERENCES

Larbrisseau, A., Vanasse, M., Brochu, P., and Jasmin, G., 1984, The Andermann Syndrome: Agenesis of the corpus callosum associated with mental retardation and progressive sensorimotor neuronopathy, *Can. J. Neurol. Sci.* 11:257-261.

Love, S. and Cruz-Höfling, M.A., 1986, Acute swelling of nodes of Ranvier caused by venoms which slow inactivation of sodium channels, *Acta Neuropathol.* 70:1-9.

Mathieu, J., Bédard, F., Prévost, C., and Langevin, P., 1990, Neuropathie sensitivo-motrice héréditaire avec ou sans agénésie du corps calleux: Etude radiologique et clinique de 64 cas, *Can. J. Neurol. Sci.* 17:103-108.

GENETIC STUDIES OF THE ANDERMANN SYNDROME

Eva Andermann,[1,2,3] Frederick Andermann,[3]
Richard Nagy,[1] Denis Bergeron,[3]
Jean Mathieu,[3] and Pierre Langevin[3]

[1]Center for Human Genetics
McGill University
1205 Docteur Penfield
Montreal, Que. H3A 1B1, Canada
[2]Neurogenetics Unit and
[3]Departments of Neurology and
Neurosurgery, McGill University
The Montreal Neurological Institute and Hospital
3801 University St.
Montreal, Que. H3A 2B4, Canada

INTRODUCTION

Early observations on a familial syndrome of neuromuscular disease and mental retardation eventually led to recognition of agenesis of the corpus callosum with sensorimotor neuropathy (see Andermann and Andermann, this volume). The disease occurs in patients whose families originate from the north shore of the St. Lawrence River between Quebec City and the Saguenay, and from the Lake St. Jean region, thus from Charlevoix county and the Saguenay area of the province. The geographical location of this region is illustrated in Figure 1.

The earliest French settlements were along the St. Lawrence River banks, moving gradually away from Quebec as the population increased. Eventually, in the last century, the colonists from this region were allowed to settle in the Lake St. Jean area. It has recently been shown that a number of autosomal recessive disorders occur quite frequently in this population. Dr. Claude Laberge described tyrosinemia; Dr. Charles Scriver studied vitamin D dependent rickets; and Dr. Jean Pierre Bouchard identified an unusual form of spastic ataxia. Among the autosomal dominant disorders, myotonic dystrophy has a high incidence in this region with an estimated 600 patients in the Saguenay and 60 patients in Charlevoix county. Disorders which also have a high incidence in this region, but which are not specific to the area, are hemochromatosis, cystic fibrosis, and multiple endocrine neoplasia or Sipple syndrome. Patients with Tay-Sachs disease and with dominantly inherited oculopharyngeal muscular dystrophy also occur in this population group (Table 1).

Callosal Agenesis, Edited by M. Lassonde and
M.A. Jeeves, Plenum Press, New York, 1994

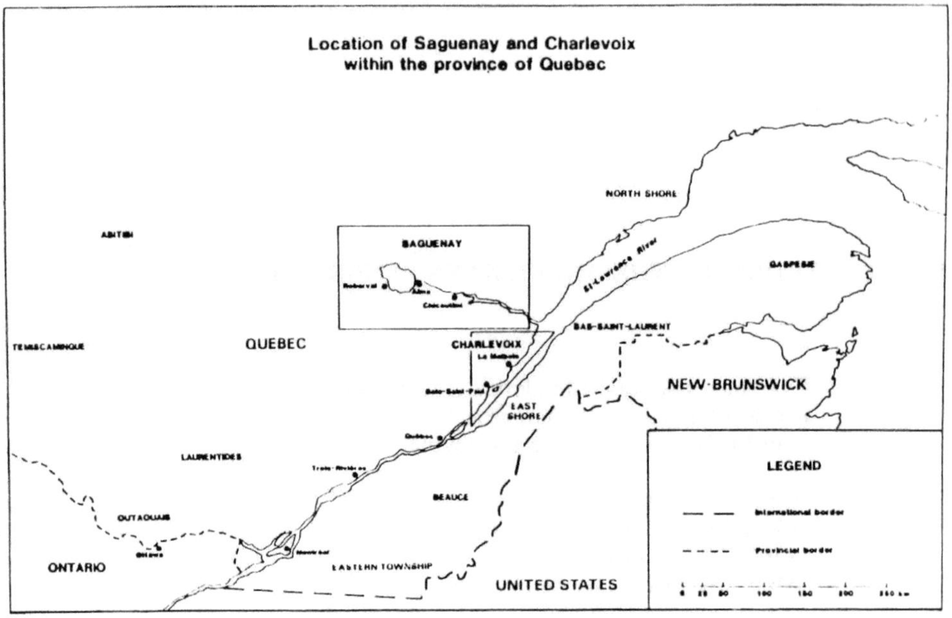

Figure 1. Map of Eastern Quebec.

Table 1. Main Mendelian disorders in the Saguenay and Charlevoix regions of Quebec.

A) DISORDERS SPECIFIC TO SAGUENAY-CHARLEVOIX

1- Recessive
   - a) Tyrosinemia
     (Carriers: 1I15 to1/20)
   - b) Vitamin D dependency rickets
     (Carriers: similar to tyrosinemia)
   - c) Spastic ataxia
     (Carriers: 1I14to1/27)
   - d) Agenesis of the corpus callosum
     (Carriers :1I20 to1/28)

2- Dominant
   Myotonic dystrophy (Steinert disease)
   (Prevalence: 1I480)
   Cases in Sagenay: 600
   Cases in Charlevoix: 63 (estimation)

B) NON SPECIFIC DISORDERS

1- Haemachromatosis
   (Carriers: 1I40)

2- Cystic fibrosis
   (Carriers: 1I23)

3- Multiple endocrine neoplasia, Type II-A (Sipple 's syndrome, hereafter called MEN)
   Cases in Sagenay and Charlevoix: under study

Note: Also present in this populations: cases of Tay-Sachs, multiple intestinal atresia, late onset ptosis and dysphagia.

GENETIC ANALYSIS

The pedigree of the first family with children affected by agenesis of the corpus callosum and neuropathy is illustrated in Figure 2. In the last three or four generations no consanguinity existed in this family although ancestors on both sides originated from the same area. In a second family, referred by Dr. Marie Joubert, some consanguinity was present (Figure 3). There followed a series of field trips and studies in the region which led to examination of an increasing number of patients identified by familial and genealogical studies and with the help of Dr. Denis Bergeron, then practicing neurology in Chicoutimi, and Dr. Larochelle, a pioneer pediatrician. The number of affected individuals of which we were aware rapidly increased. The earliest genetic studies revealed that the birthplaces of their ancestors were along the St. Lawrence River (Figure 4), but in later studies, birthplaces of grandparents were found along the Saguenay River and in the Lake St. Jean area itself (Figure 5).

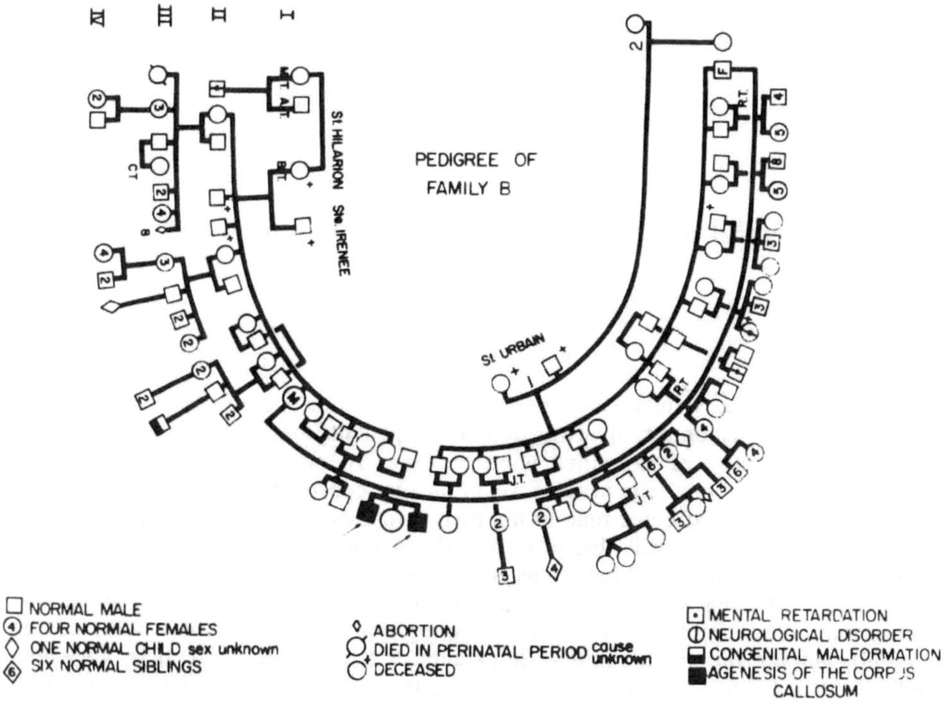

Figure 2. Pedigree of the first family in whom callosal agenesis with anterior horn cell disease was diagnosed.

An increasing number of patients were then examined both in their homes and institutions, and in Dr. Bergeron's clinic. These studies were continued by Richard Nagy, a graduate student in the Department of Neurogenetics at the Montreal Neurological Institute (Nagy 1982). By 1982, we were aware of 237 affected individuals in 144 sibships and 90 of these, members of 65 sibships, were examined by us. The neurological abnormalities in 54 patients are summarized in Table 2. A more detailed

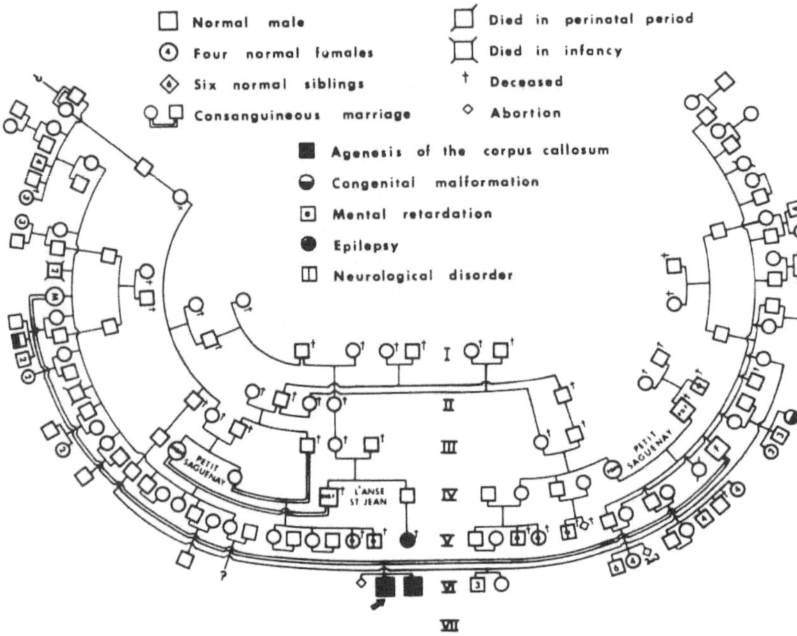

Figure 3. Pedigree of the second diagnosed family (Family BER).

description of the clinical symptoms is presented elsewhere in this volume (Andermann and Andermann). The laboratory investigations in 90 patients are outlined in Table 3.

Most of the patients we examined were between 10 and 20 years of age with occasional exceptions - one lady was 51 years of age, confined to a wheelchair, with marked contractures. An affected man in his early twenties who had four siblings with the full blown syndrome was still able to walk and use a small hand-saw, though he had quite marked atrophy of his small hand muscles. The number of patients with the syndrome born in the region varied between one and nine per year in this population basin of about 300,000 inhabitants. There were not many marriages with close consanguinity among the parents and grandparents of the patients, and there were no first cousin marriages at all. Among 34 families, there was one in which the parents were second cousins and in another, they were second cousins once removed. Parents were third cousins in 3 and distant cousins in 6, with a consanguinity rate of 32.4%. In maternal and paternal grandparents consanguinity rates were similar.

The average inbreeding coefficient calculated for eight affected families to one common ancestral pair was found to be F=0.002257. The average number of links of parents to the ancestral pair was 5 (Table 4). This corresponds to a mean degree of relationship between 3rd cousins and 3rd cousins once removed, but it may be assumed that there may have been other links between the parents of which we were not aware, and that the parents may have been somewhat more closely related. The degree of relationship was probably similar to that among other people from this area; thus, the degree of relationship is not increased in the parents of children with the Andermann syndrome. In a situation where a strong founder effect likely exists, more obvious or closer consanguinity need not be present.

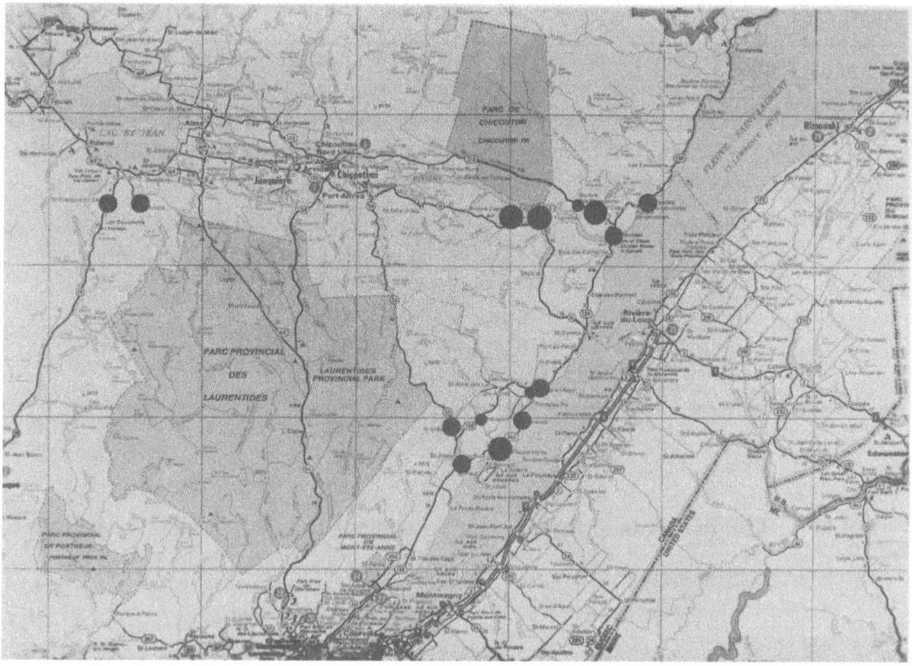

Figure 4. Places of origin of ancestors of patients with callosal agenesis illustrating the direction of migration to the Lake St-John area.

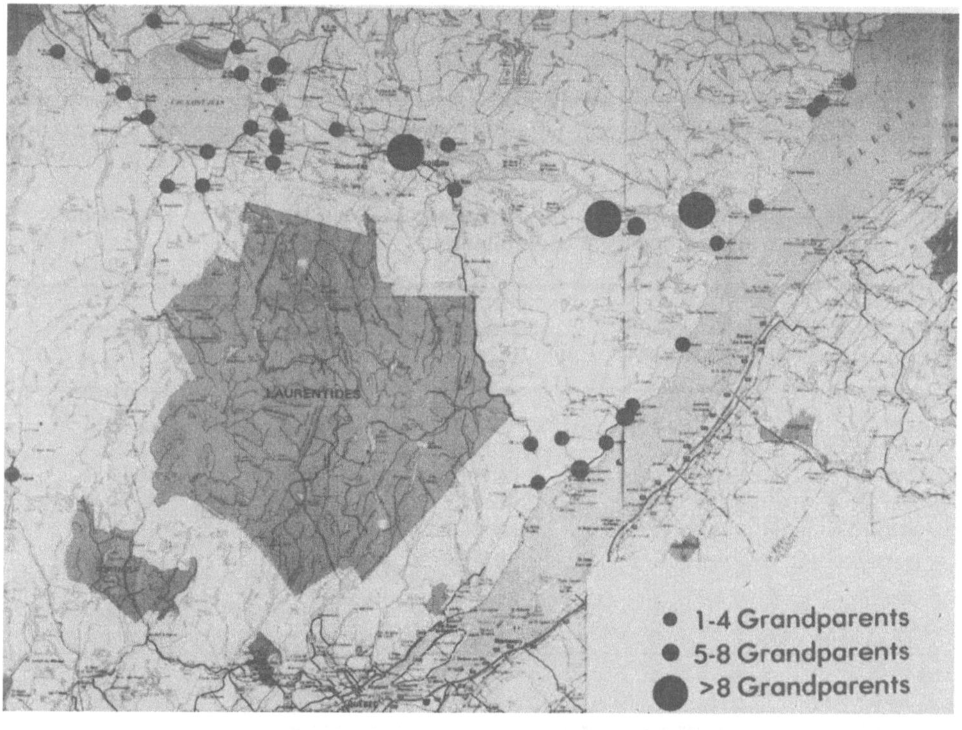

Figure 5. Places of origin of grandparents of the patients.

Table 2. Summary of neurological abnormalities found in 54 affected patients.

	No.	%
Areflexia	54	100.0
Hypotonia	54	100.0
Progressive muscle weakness and wasting	54	100.0
Asymmetrical ptosis	23	42.6
Cerebellar dysfunction	20	37.0
Nystagmus	16	29.6
Strabismus	13	24.1
Seizures	12	22.2
Sensory abnormalities	10	18.5

Table 3. Laboratory investigations in 90 patients with Andermann's syndrome.

	Examined Individuals (N=90)	
	No.	%
PEG/CT	57	63.3
Biopsy	35	38.9
EMG	15	16.7
EEG	26	28.9
Dermatoglyphics	33	36.7
Chromosome studies	9	10.0

Table 4. Test of significance for an autosomal recessive mode of inheritance assuming complete truncate ascertainment.

Sipship Size	No. of Sibships	No. of Affected Individuals		$\frac{(O-E)^2}{E}$
		Observed	Expected	
1	4	4	4.000	0.000
2	12	15	13.714	0.121
3	11	14	14.270	0.005
4	5	6	7.314	0.236
5	3	6	4.917	0.239
6	2	5	3.650	0.500
7	3	6	6.059	0.001
8	1	1	2.223	0.673
11	1	4	2.871	0.444
12	2	4	6.196	0.778
Total	**44**	**65**	**65.214**	**2.997**

$\chi^2_9 = 2.997 \quad p > 0.95$

Four different types of genetic analysis using various assumptions regarding the mode of ascertainment were then carried out. One has to correct for referral bias since families in which more than one child is affected tend to be more striking and thus referred more readily. Assuming complete truncate ascertainment for autosomal recessive mode of inheritance, there was excellent correlation between the observed and the expected findings (Figure 9). Using the proband method, segregation analysis, assuming the complete method of ascertainment, resulted in a figure of 25%. Multiple incomplete ascertainment showed 24.1% to be affected.

Genealogical studies carried out in our department, based on a list of couples and their children, tracing the consanguineous loops as far as possible, allowed us to trace 45 affected individuals in a pedigree to a couple who married in Quebec City in 1657 (Figure 6).

The incidence of agenesis of the corpus callosum and sensory motor neuropathy in this population is 1/173 with a gene frequency of 1/48 and a carrier frequency of 1/25. This compares with the incidence of Tay-Sachs disease in the Ashkenazi Jewish population, a classic example of a genetic disease aggregate in a group with an original small founding population living in an area where transportation was difficult and consanguinity frequent. In Montreal, the incidence of Tay-Sachs in Ashkenazi Jews was 1/3600 with a gene frequency of 1/60 and a carrier frequency of 1/29.

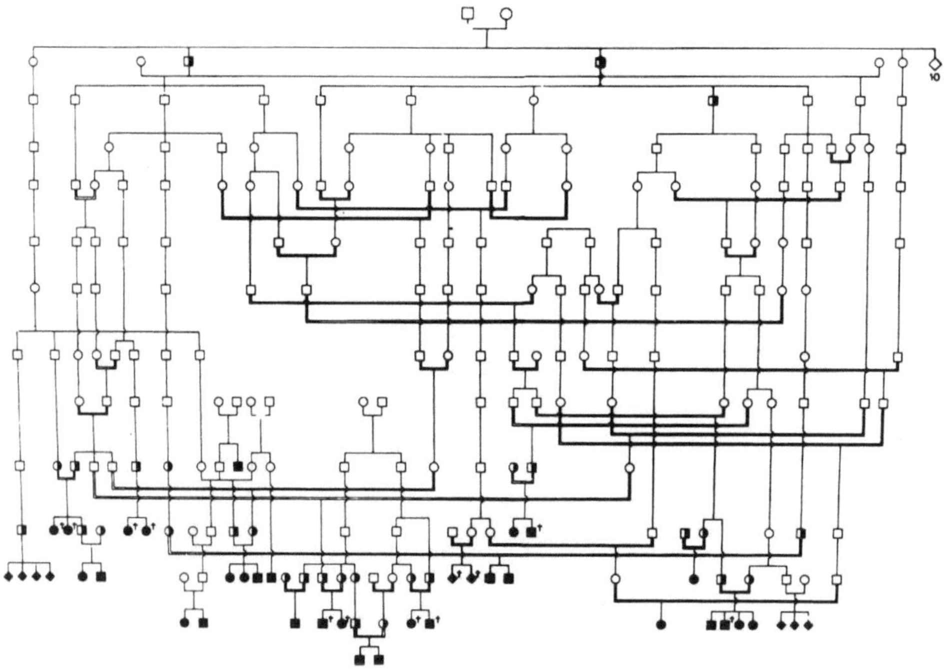

Figure 6. Extended pedigree showing consanguinity (bold lines) and a common ancestral couple.

CONCLUSION

Genetic studies have confirmed the autosomal recessive nature of the disease and led to early diagnosis and accurate counseling. Because of early diagnosis, sibships with multiple affected children are no longer seen. Unfortunately, prenatal diagnosis of this syndrome is complicated by the presence of the callosum in a percentage of affected children and thus is not sufficiently reliable.

Extensive genealogical studies attempting to map the entire population of the area have been developed at the University of Quebec in Chicoutimi by SOREP under the guidance of Dr. Gerard Bouchard. Linkage studies carried out with Dr. Guy Rouleau should permit more effective prenatal diagnosis and should, in due course, lead to prevention of this terrible disease.

REFERENCES

Mathieu, J., Bedard, F., Prevost, C. and Langevin, P., 1990, Neuropathie sensitivo-motrice héréditaire avec ou sans agénésie du corps calleux: Etude radiologique et clinique de 64 cas, *J. Can. Sci. Neurol.* 17:103-108.

Nagy, J.R., 1982, Familial agenesis of the corpus callosum with sensorimotor neuronopathy: genetic and epidemiological studies, Unpublished Master's thesis, McGill University, Montreal.

A NEW SYNDROME:

FAMILIAL FRONTO-NASAL DERMOID CYSTS

WITH AGENESIS OF THE CORPUS CALLOSUM

Ivan Jacobson[1] and Malcolm A. Jeeves[2]

[1]Department of Surgical Neurology
Dundee University
Dundee, Scotland, DD1 9ND
[2]Psychological Laboratory
University of St. Andrews
St.Andrews, Fife
Scotland, KY16 9JU

INTRODUCTION

In 1971 we documented a family in which members of two generations were found to have fronto-nasal dermoid cysts. At that stage of their clinical follow-up we had detected that one member of the family had agenesis of the corpus callosum (Plewes and Jacobson, 1971). Familial nasal dermoid cysts had been described in seven families up until then (De Wet and Cort, 1966; Furniss, 1938; Musabeck and Darst, 1967; New and Erich, 1937; Stupka, 1938). A further large family has since been reported (Mulbauar and Dittmar, 1976) with nasal dermoids, widened noses and bilaminal septa in a high proportion of the members

Baarsma (1980) mentioned a further nine cases of hereditary nasal dermoids but did not elaborate any further. However, no other familial fronto-nasal dermoid cysts have been published since our original paper. Long-term follow up and further investigations on three generations of this family have brought to light an additional consistent abnormality in all - namely agenesis of the corpus callosum. The family pedigree is illustrated in Figure 1.

CASE REPORTS

First Generation

Case 1

D.E.C. (d.o.b. 29.09.28) This first affected individual presented in 1961 with focal motor epilepsy. A nasal cyst had been removed some years previously. His facial appearance was normal (Figure 2). The only significant abnormality was anosmia. Nasal bones were radiologically normal. A carotid angiogram revealed a subfrontal mass which at

Callosal Agenesis, Edited by M. Lassonde and
M.A. Jeeves, Plenum Press, New York, 1994

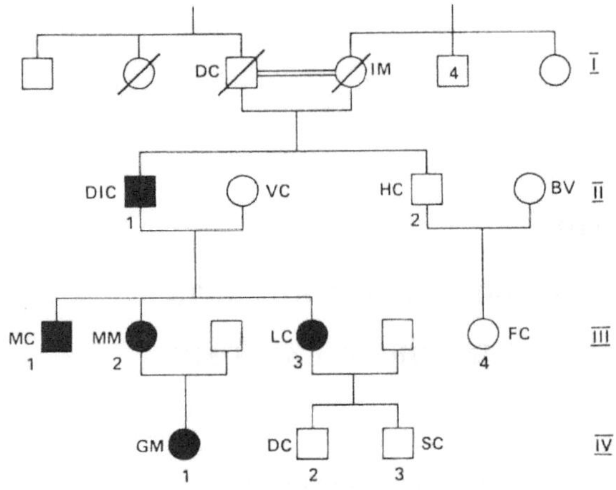

Figure 1. Pedigree of affected kindred.

Figure 2. D.E.C. facial appearance.

operation was found to be a large simple dermoid cyst. Epileptiform seizures have been a persistent post-operative problem. Close relatives claim to have observed changes in personality.

C.T. and M.R.I. scans (Figures 3, 4 and 5) show a low-density subfrontal abnormality with agenesis of the corpus callosum. The ventricles were noted to be significantly dilated. The following three cases are the offspring of D.E.C.

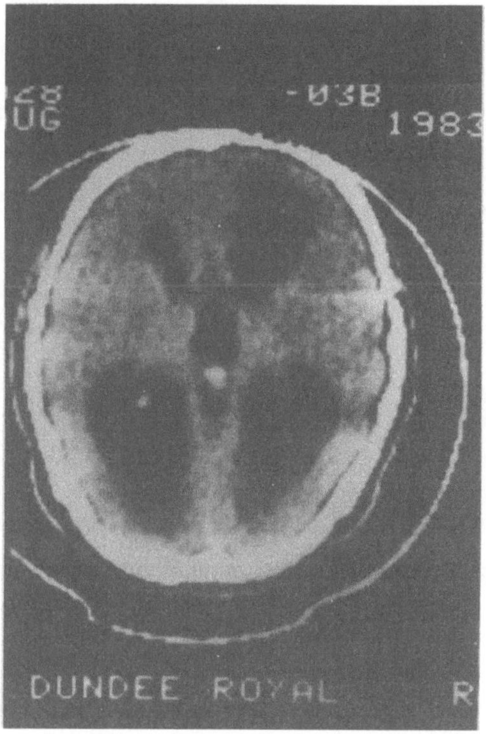

Figure 3. D.E.C. C.T. Scan showing subfrontal lesion with dilated lateral and third ventricles.

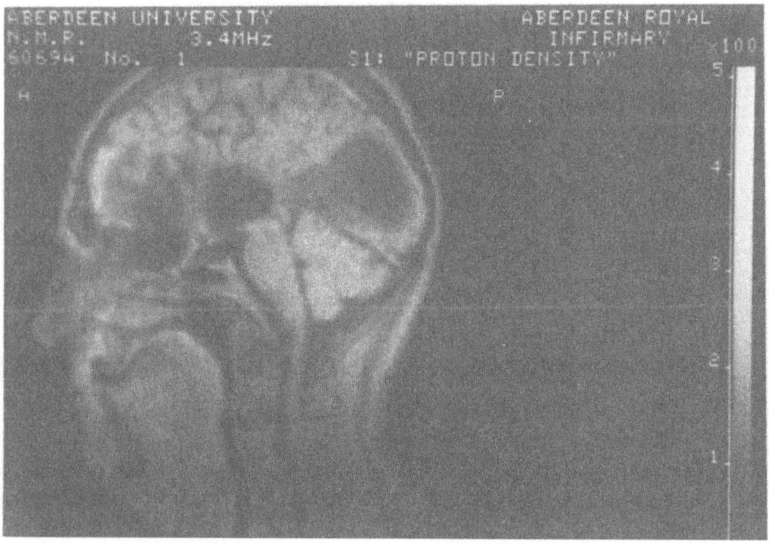

Figure 4. D.E.C. M.R.I. scan demonstrating agenesis of corpus callosum, subfrontal lesion and dilatation of lateral ventricles.

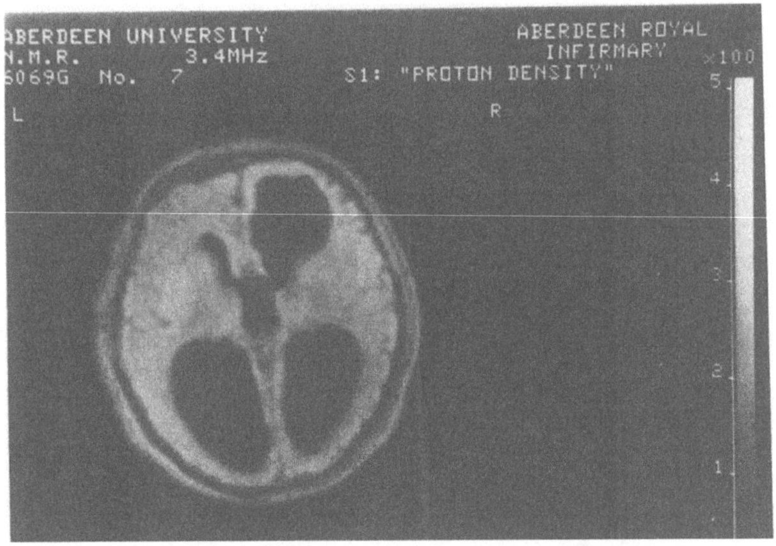

Figure 5. D.E.C. M.R.I. scan demonstrating agenesis of corpus callosum, subfrontal lesion and dilatation of lateral ventricles.

Second Generation

Case 2

M.C. (d.o.b. 21.05.53) The eldest and only male child presented at the age of five with nasal obstruction. His facial appearance showed extensive nasal abnormality (Figure 6). Radiology revealed an extremely widened nasal septum (Figure 7). He re-presented some years later with headache. He was noted to have a widened nose with mid-nasal puncta discharging yellow material. A transnasal resection of the widened

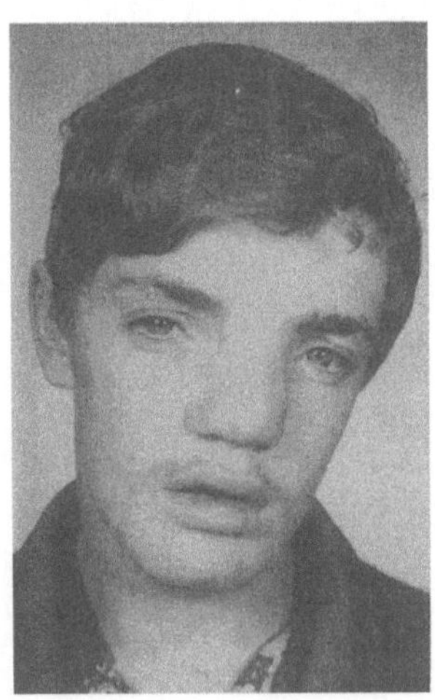

Figure 6. M.C. Facial appearance demonstrating extensive nasal abnormality.

septum confirmed a nasal dermoid extending upwards into the subfrontal region. Post-operatively meningitis developed. He responded well to antibiotics.

C.T. and M.R.I. scans (Figure 8) reveal a low density subfrontal mass and agenesis of the corpus callosum. The ventricles were noted to be dilated.

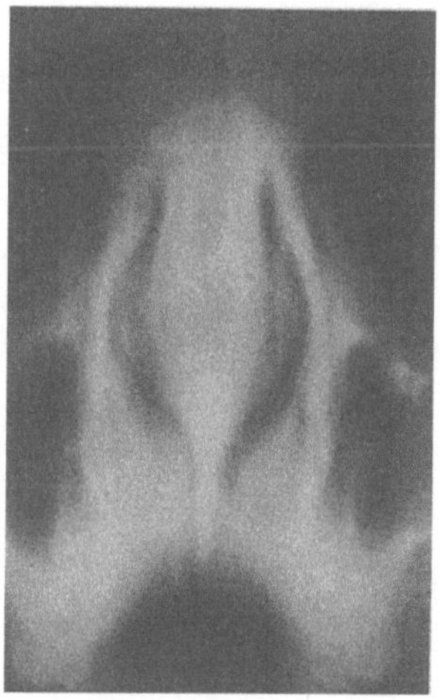

Figure 7. M.C. Nasal Tomogram showing widening of septum extending into subfrontal region.

Case 3

M.M. (d.o.b. 18.09.57) The oldest daughter of D.E.C. presented at the age of twelve with a personality change. She too had a nasal discharge. She was noted to have hyperteleriorism, a widened nasal bridge and dermoid sinus below the glabella. She had poor recent memory, a performance IQ on the Wechsler scale of 87 and a verbal IQ of 67. Radiological investigations confirmed a nasal dermoid extending subfrontally (Figure 9). The nasal and frontal dermoid cysts were excised transcranially. Post-operatively she was anosmic but otherwise well. Her facial appearance is shown (Figure 10).

C.T. and M.R.I. scans (Figure 11) revealed agenesis of the corpus callosum with ventriculomegaly. No subfrontal abnormality was detectable.

At the age of 21 she delivered a female child (G.M. Figure 1 IV 2).

Case 4

L.C. (d.o.b. 19.05.60). The youngest child of D.E.C. was referred with headaches and progressive decline in educational performance at the age of ten. No nasal stigmata were present. She was not anosmic. Her performance IQ was 87 and verbal IQ 75. A detailed neuropsychological investigation was commenced as summarized below. Carotid angiography and pneumo-encephalography excluded a subfrontal lesion but agenesis of the corpus callosum was noted. Some ten years later she had a grand mal seizure and was then noted to be anosmic. C.T. scan revealed a subfrontal mass and confirmed the callosal agenesis.

A subfrontal dermoid cyst was excised via a frontal craniotomy. Follow up C.T. and M.R.I. scans (Figures 12 and 13) confirm agenesis of the corpus callosum. No recurrence of the dermoid has occurred.

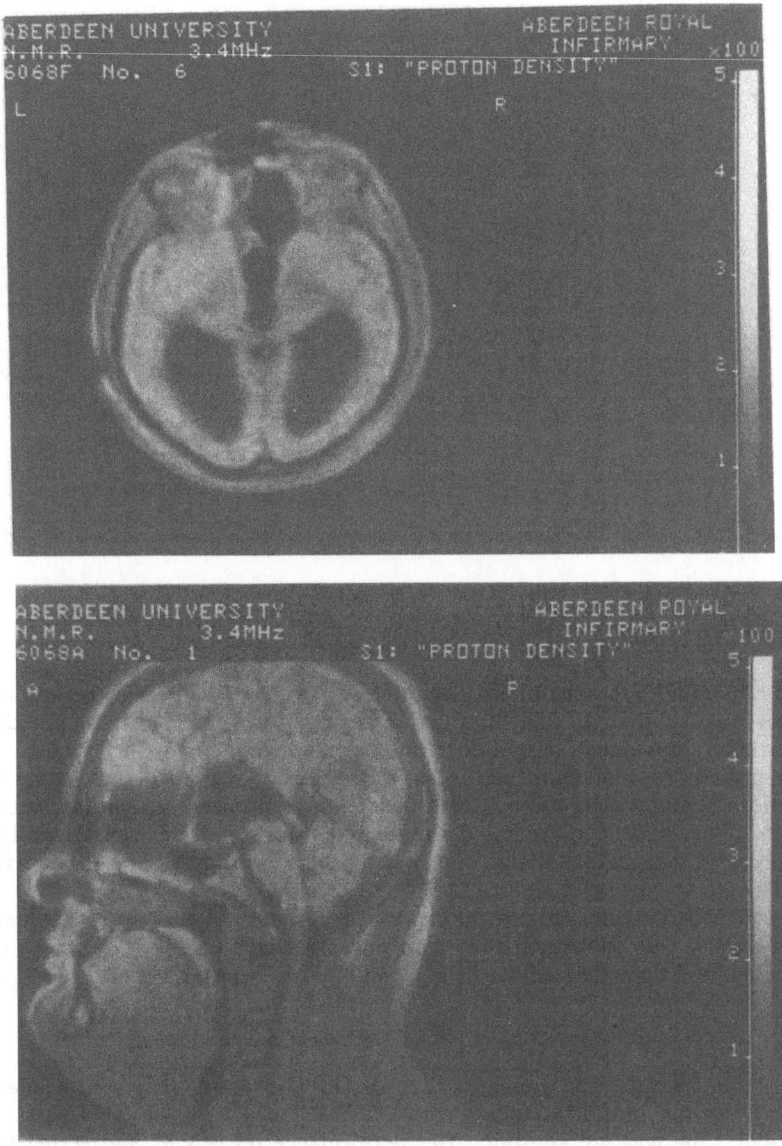

Figure 8. M.C. M.R.I. scans demonstrating subfrontal lesion, agenesis of corpus callosum and dilatation of lateral ventricles.

She has had two uneventful pregnancies with the delivery of two boys (S.C. and D.C.). She is currently asymptomatic. The following three cases are the third generation of the family. G.M. is affected.

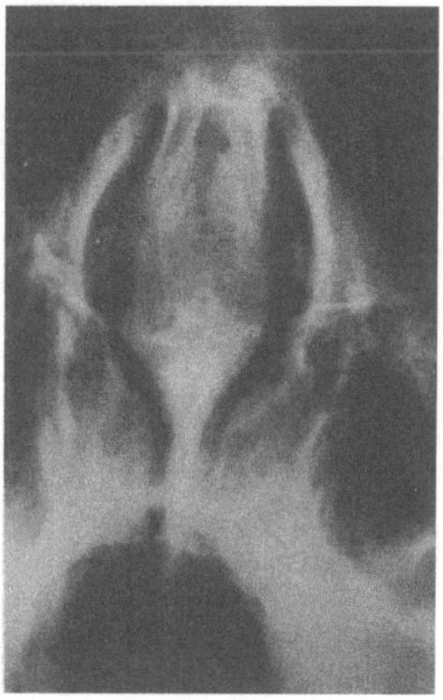

Figure 9. M.M. Nasal tomogram demonstrating septal mass extending into cribriform region.

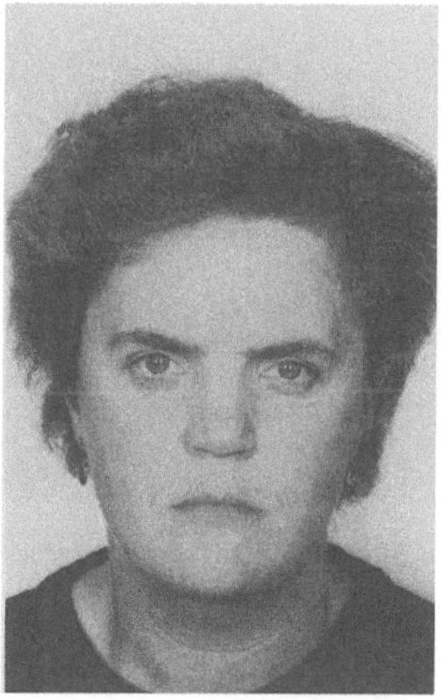

Figure 10. M.M. Facial appearance.

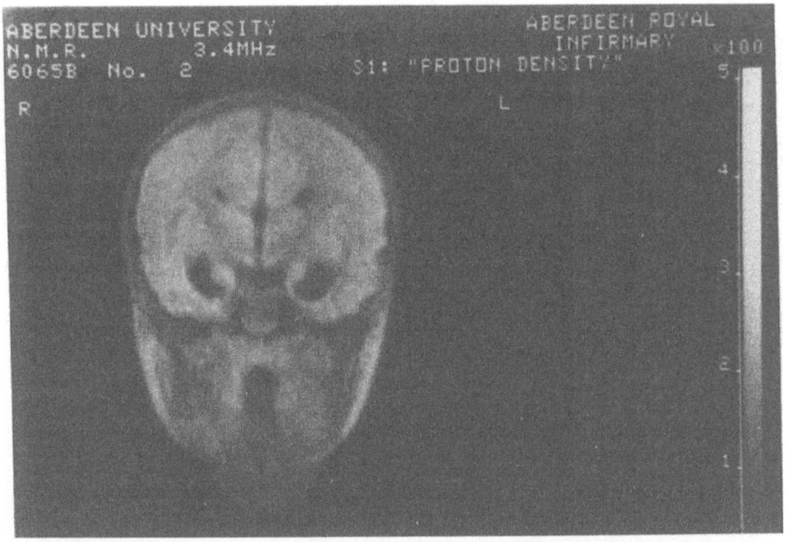

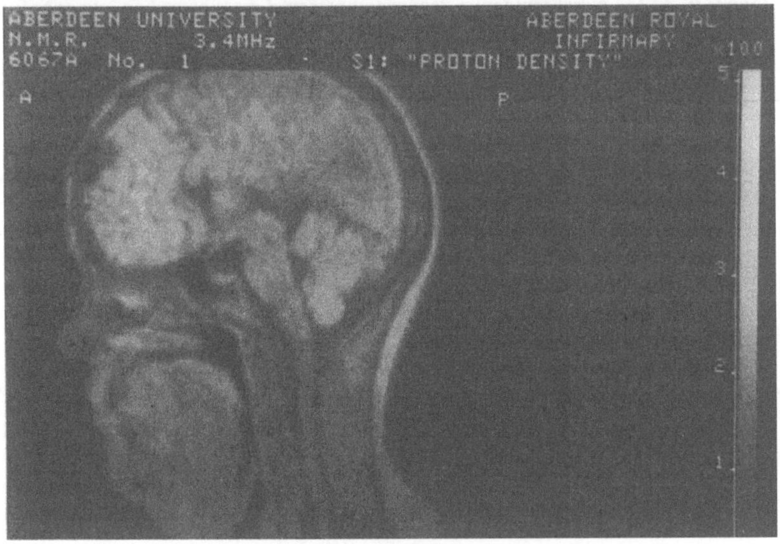

Figure 11. M.M. M.R.I. scans showing agenesis of corpus callosum.

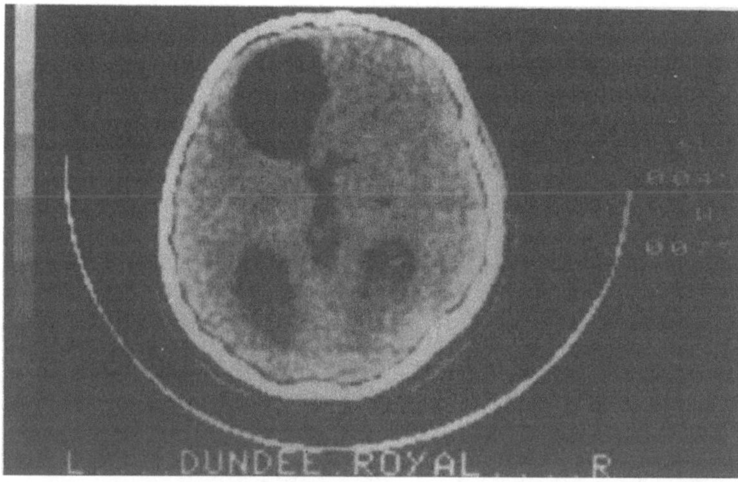

Figure 12. L.C. C.T. scan showing subfrontal cystic lesion and agenesis of corpus callosum.

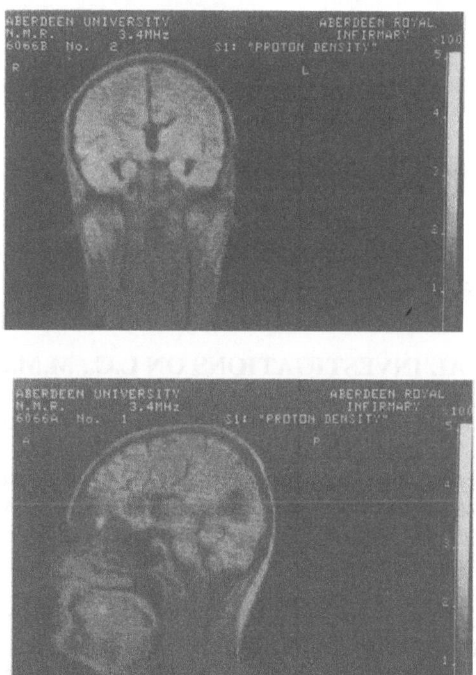

Figure 13. L.C. M.R.I. scans showing agenesis of corpus callosum.

Third Generation

Case 5

G.M. (d.o.b. 12.02.79). G.M. presented at the age of seven because close relatives believed they could detect personality changes similar to those noted in her mother and maternal aunt. She was anosmic but otherwise neurologically normal. Nasal bones were normal. C.T. scan revealed a subfrontal dermoid cyst with agenesis of the corpus callosum (Figure 14). The cyst was removed via a right frontal craniotomy. Recovery was uneventful.

Follow up C.T. and M.R.I. scans confirm agenesis with no subfrontal lesion. The nasal region was normal.

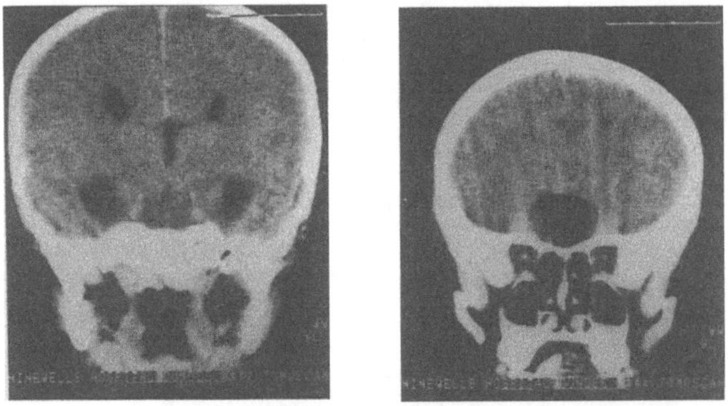

Figure 14. G.M. C.T. scans showing agenesis of corpus callosum and subfrontal midline dermoid cyst.

Cases 6 and 7

D.C. (d.o.b. 29.03.85). **S.C.** (d.o.b. 13.02.87). These are the children of L.C. with unrelated fathers. They are currently normal with normal C.T. and brain ultrasound.

NEUROPSYCHOLOGICAL INVESTIGATIONS ON L.C., M.M., G.M. AND D.I.C.

L.C. was intensively studied between 1971 and 1975. At that time M.M. was also studied being regarded as an ideal clinical control for L.C. *on the assumption* (in the absence of imaging evidence) that her corpus callosum was present. However, the results of the behavioral testing revealed that her performance on a variety of tasks was closely similar to that of her sister and other acallosals (Jeeves, 1965, 1969). Thus on a task of tactile cross-localization both L.C. and M.M. made more errors in the crossed condition when compared with the uncrossed condition than any of the age and mental age matched controls. These and other results led Reynolds (unpublished Ph.D. 1975) to comment (page 55) "the similar performance of the acallosal girl (L.C.) and the sibling (M.M.) supports the hypothesis that the sibling is also acallosal". This judgment was further confirmed when their interhemispheric transfer times were measured. This required them to make a manual response when a point source of light appeared randomly either to the right or left of a central fixation point. Whereas normals, whether children or adults, show an interhemispheric transfer time of 2 to 3 msecs, the acallosal L.C. and her sister M.M. produced times ranging from 10 to 30 msecs, depending on whether response was two-handed or single-handed and upon whether the light fell on the nasal or temporal retina. Again Reynolds (1975) wrote (page 111) "We conclude that the sibling is acallosal also, as all the pathway differences for the sibling as well as the acallosal are increased over those for the normals".

In the 1980s L.C.'s bimanual co-ordination was intensively studied (see Silver and Jeeves, this volume). The task used was similar to one devised by Preilowski (1972) for use with the Californian commissurotomy patients. Subjects were required to draw lines by rotating cranking handles in specific orientations on an X-Y plotter in which the X (horizontal) and Y (vertical) components of the movement were under the control of the right and left hands respectively. When the acallosals were allowed to see the lines that they were drawing they could learn to do the task, though slower than normals. However, when, after becoming very highly practiced, visual feedback was withdrawn, the acallosals performed like anterior commissurotomy patients. (Jeeves et al., 1988a; Jeeves et al., 1988b)

L.C.'s ability to transfer visuo-spatial information was investigated following a report by Meerwaaldt that an 8-year-old acallosal girl, when required to do the task by touch alone, was markedly inferior with her right hand. L.C.'s performance was very similar to that of Meerwaldt's acallosal girl. (Jeeves and Silver, 1988)

In 1986, L.C.'s language skills were studied in detail (Jeeves and Temple, 1988) revealing specific difficulties with phonology. These problems were subsequently extended in studies of further acallosals (Temple et al., 1989, 1990). It may be that these undetected and hitherto unreported difficulties in handling some aspects of language have contributed to the educational problems encountered by L.C. and some other members of the families reported here. Whilst none of the difficulties alone are markedly disabling in everyday living, the combined and cumulative effect of the deficits taken together may well have lowered overall competence to a level that created the unidentified educational difficulties referred to from time to time.

It is particularly worth noting, in retrospect, that the simplest of the behavioral tasks used, namely the measurement of interhemispheric transfer times when responding to unstructured visual input, which is totally non-invasive and easy to administer, gave a clear indication of callosal agenesis in L.C.'s sibling M.M. in 1972, a diagnosis not confirmed by imaging techniques until 1989.

Most recently (September 1990) it has been possible to administer this test to G.M. and D.I.C. Both showed significantly lengthened interhemispheric transmission times, G.M.'s was 10 msecs and D.I.C.'s was 43 msecs.

DISCUSSION

Three well-recognized congenital abnormalities are present in this family. All the affected members at this stage have agenesis of the corpus callosum and frontal dermoid cysts. Three in addition have nasal dermoids communicating through the cribriform plate with the intracranial dermoid. Of these three, the first case had no external stigmata of the nasal dermoid but his two oldest children have very obvious abnormalities in their nasal configuration and these are still in evidence. The other two cases without nasal dermoids show minor or no features suggestive of facial dysmorphism.

The clinical presentation has been very similar in all, apart from the age of presentation and leading nasal obstructive symptoms in D.C., M.C. and M.M. The age of presentation has spanned from 7 to 26 years but a careful analysis and early referral with appropriate investigations would probably have led to the diagnosis in all except L.C. She was investigated extensively at the age of 9 with a presumed diagnosis of a frontal cyst but by the techniques available at that time only agenesis was confirmed. It is possible that her dermoid cyst, which subsequently declared itself at a much later stage, would have been visualized had C.T. or M.R.I. scanning been available then.

The clinical features seem in this family to be virtual mirrors of each other - anosmia, nasal symptoms, loosely defined early personality disorders, poor schooling and intellectual development, headache and seizures. These symptoms, of course, reflect the anatomical and pathological features common to all of them and do not require any further detailed elaboration except to point out that, apart from L.C., a detailed examination of the significance of the agenesis of the other members of the family has not been completed.

Median nasal sinuses with associated dermoid cysts of the nasal septal area are relatively rare congenital abnormalities, representing between 3.7% and 12.6% of all dermoids of the head and neck (New and Erich, 1937; Crawford and Webster, 1952, Ryan, 1948). Nasal dermoids enlarge slowly, tend not to cause space-occupying symptoms and although many have been reported in the first and second decades of life a fair number have presented in older age (McClean et al., 1964). These cysts have a tendency to be associated with extensions through and into the skull but intracranial complications have not been widely discussed (McLean et al., 1964; Littlewood, 1961)

In our original report on this family (Plewes and Jacobson, 1971), we were able to find a further seven families with 19 affected individuals but no series then were described with the combined frontal and nasal dermoids of a familial nature (De Wet and Cort, 1966; Furniss, 1938; Musabeck and Darst, 1967; New and Erich, 1937; Stupka, 1938) although Baarsma (1980) comments that familial nasal dermoids have been described in nine cases with an autosomal dominant inheritance pattern but he did not elaborate further on these patients. In particular he made no comment as to the association of a frontal dermoid in any of these families. A further series of 13 nasal dermoids was described by Sessions (1982) Four of these had intracranial extensions but no familial tendency was noted in that series. To date we have been unable to trace a family similarly affected to the one we describe in the respect of their frontal and nasal association.

The sex ratio of nasal dermoids has been claimed to have a male preponderance of 2:1 (1982). This however has not been confirmed in other series. Crawford and Webster (1952) had 14 patients of whom 10 were males and Griffiths (1976) described 17 patients, 12 of whom were female. In our small series, two were male and three were female.

Two broad theories have been proposed to explain the embryopathogenesis of nasal dermoid cysts - the anomaly is either due to a deep-seated cranial developmental error or may be due to a cutaneous malformation. The variations in the family we describe of nasal-frontal lesions with a consistent anomaly of the callosum (agenesis) in all of them lends very strong support to a central developmental error in common with nasal gliomas and encephaloceles (Hughes et al., 1980). Nasal dermoids associated with other congenital abnormalities are rare. Combinations with medial nasal grooves, nasal aplasias, spina bifida and cheilognathopalatoschisis have been discovered (Taylor et al., 1966). To date, however, no association of callosal agenesis has been documented apart from the one we are now describing.

Neither the incidence nor the prevalence of agenesis of the corpus callosum can be assessed from current evidence. The proportion of symptomless cases is not known, even though this is claimed to be fair (Unterharnscheidt et al., 1968). In a series of pediatric encephalograms over a 15-year period, 45 cases out of a total of 6,450 studies were found (Grogono, 1968). However only one case was found in a series of 19,000 autopsies. Some 2% of mentally retarded patients have been claimed to have callosal agenesis (Andermann, 1981). The majority of reported cases are sporadic but a number of familial incidences have been recorded (Andermann et al, 1972, 1975; Menkes et al., 1964; Naiman and Fraser, 1955; Jolly, 1955; Henneaux, 1956; Ziegler, 1958; Shapiro and Cohen, 1973; Guazzi et al., 1974). Up-to-date reviews appear elsewhere in this volume.

To date, three different modes of inheritance have been reported. The family described by Menkes et al. (1964) has a clear X linked inheritance. One other family, probably X linked and recessive, has been recorded (Kaplan, 1983). Autosomal recessive transmission may account for the heterogeneous syndromes described by Lachiewicz et al. (1985) and Pineda et al. (1984).

Dominant inheritance in a father and son was reported by Lynn et al. (1980). The family we describe certainly suggests a dominant but non sex-linked syndrome. Recognizable syndromes (see elsewhere in this volume) with agenesis of the corpus callosum have been widely documented - Aicardi syndrome (Aicardi et al., 1965), Andermann syndrome (Andermann et al., 1975), Shapiro syndrome (Shapiro and Cohen, 1973), Trisomy 8 and acro-callosal syndrome (Schinzel, 1982). These five syndromes usually have the finding of agenesis. Our family is the sixth such group.

Callosal agenesis is a frequent but not constant feature in a number of other syndromes - Dandy-Walker, Aperts, Lissencephaly, Trisomy 18 and the oral facial digital syndrome (see Geoffroy, this volume).

The association of agenesis with a few cases of basal encephalocele (Manelfe et al., 1978), median facial cleft and frontonasal dysplasia (François et al., 1973; Fontaine et al, 1983) is of interest in considering that a similar embryological link must be present in our family.

There are substantial differences in this family in comparison with any of the others published to date. The vast majority were discovered in extremely young patients with a number of other anomalies showing no consistent pattern except perhaps in their presentation of early seizures, mental retardation and in some a high early mortality. Some autosomal dominant syndromes have also been described with the occasional finding of agenesis of the corpus callosum - tuberous sclerosis, cystic fibrosis, median facial cleft syndrome, Aperts syndrome, reno-facial dysplasia. The syndrome of Opitz and Kaveggia (1974) with hyperteleriorism and agenesis has minimum but some common features with our family. Attempts to systematically group callosal agenesis and other cerebral malformations have not been successful (Parrish et al., 1979).

Chromosomal abnormalities have not been detected in this family. In contradistinction, a number of chromosomal abnormalities seem to be associated with callosal agenesis. In all instances of Trisomy 8 investigated to date partial or total agenesis has been found (Zellweger et al., 1977; Riccardi, 1977). A lower incidence has been reported in Trisomy 18 and 13 (Sumi, 1970). Only occasional cases in a large number of other chromosomal anomalies have been found (Schinzel, 1984; Warkany et al., 1966).

It is possible that the personality changes observed by close relatives in two of the family were due in part to the expansion of their frontal cysts. The learning difficulties common to all the younger two generations could be ascribed to the callosal agenesis, especially since very recently published studies have now identified specific language competence deficits associated with callosal agenesis (Jeeves and Temple, 1987; Temple et al, 1989, 1990). IQ assessment is so far available on only two of the cases. In both instances, this indicates a discrepancy between verbal and performance IQ outside the norm reported by Sauerwein (this volume). To date we have only studied one of our cases (L.C.) extensively in an attempt to assess the significance of acallosal state on bimanual co-ordination and other tasks (Jeeves et al., 1988). We are therefore unable to comment as to whether the other affected members would show similar results. Early indications are that this may well be the case.

CONCLUSION

We have reported a unique family of three generations, all of whom have agenesis of the corpus callosum and a midline subfrontal dermoid cyst. Three have in addition nasal dermoids. The two youngest members are unaffected. The syndrome seems to be inherited as a dominant non sex-linked characteristic with a normal chromosomal pattern. A molecular genetic approach used in the analysis of this syndrome is as yet inconclusive.

REFERENCES

Aicardi, J., Defebre, J., and Lerique-Koechlind., 1965, A new syndrome - Spasms in flexion callosal agenesis and ocular abnormalities, *Encephalogr. Clin. Neurophysiol.* 19:609-610.

Andermann, E., 1981, Agenesis of the corpus callosum, in: "Handbook of Clinical Neurology," Vol. 4, P.J. Vinken and G.W. Bruyn, eds., North Holland, Amsterdam.

Andermann, E., Andermann, F., and Joubert, M., 1972, Familial agenesis of the corpus callosum with anterior horn cell disease, *Transact. Amer. Neurologic Ass.* 97:242-244.

Andermann, E., Andermann, F., and Joubert, M., 1975, Three familial midline malformations of the central nervous system, *Birth Defects* 11:269-293.

Baarsma, E.A., 1980, The median nasal sinus and dermoid cyst, *Archiv. Oto-Rhino-Laryng.* 226:107-113.

Crawford, J.K. and Webster, J.P., 1952, Congenital dermoid cysts of the nose, *Plast. Reconstr. Surg.* 9:235-260.

De Wet, H.C. and Cort, A., 1966, Dermoicde siste van dee neusgebied, *S. Afr. Med. J.* 40:557-560.

Fontaine, G., Walbaum, R., Poupard, D.B., et al., 1983, La dysplasie fronto-nasale, *J. Geriet. Hum.* 31:351-365.

Francois, J.E., Eggermont, E., and Evans, N., 1973, Agenesis sof the corpus callosum in the median facial cleft syndrome and associated ocular malformations, *Am. J. Ophthalmol.* 76:241-245.

Furniss, F.W., 1938, Mesial nasal dermoid cysts of the nose occurring in twin boys, *J. Laryng.* 53:314-315.

Griffith, B.H., 1976, Frontonasal tumours: Their diagnosis and management, *Plast. Reconstruct. Surg.* 57:692-698.

Grogono, J.L., 1968, Children with agenesis of the corpus callosum, *Dev. Med. Child Neurol.* 10:613-616.

Guazzi, G.C., Stoppoloni, G., Ventruto, V., and Iorio, G., 1974, Immature neuronale corticale avec agenesie des grandes commissures interhemispheriques et hypoplasie des voies optico piramidales ches trois enfants issus d'une même famille, *Acta Neurol. (Napoli)* 29:6.

Henneaux, J., 1956, Malformations vertébrales et division palatine chez deux soeurs, associées chez l'une d'elles, à une agénésie du corps calleux, *Schweiz Arch. Neurol. Psychiatr.* 77:433-441.

Hughes, G.B., Sharpina, G., Hunt, W., and Tucker, H.M., 1980, Management of the congenital midline mass: A review, *Head Neck Surg.* 2:222-233.

Jeeves, M.A., 1965, Psychological studies of three cases of congenital agenesis of the corpus callosum, *in:* "Functions of the Corpus Callosum," CIBA Foundation Study Groups, Vol. 20, E.G. Ettlinger, ed., London: Churchill.

Jeeves, M.A., 1969, A comparison of interhemispheric transmission times in acallosals and normals, *Psychonom. Sci.* 16:245-246.

Jeeves, M.A., 1990, Agenesis of the corpus callosum, *in:* "Handbook of Neuropsychology," F. Boller and J. Grafman, eds., Elsevier, Holland.

Jeeves, M.A. and Silver, P.H., 1988, Interhemispheric transfer of spatial tactile information in callosal agenesis and partial commissurotomy, *Cortex* 24:601-604.

Jeeves, M.A., Silver, P.H., and Jacobson, I., 1988, Bimanual co-ordination in callosal agenesis and partial commissurotomy, *Neuropsychologia* 28:262-269.

Jeeves, M.A., Silver, P.H., and Milne, A.B., 1988, Role of the corpus callosum in the development of a bimanual motor skill, *Dev. Neuropsychol.* 4(4):305-323.

Jeeves, M.A. and Temple, C.M., 1987, A further study of language function in callosal agenesis, *Brain Lang.* 32:325-335.

Jolly, D.H., 1955, (Quoted by Naiman, J. and Fraser, F.C., 1955), *Arch. Neurol. Psychiatr.* 74:182-185.

Kaplan, P., 1983, X-linked recessisve inheritance of agenesis of the corpus callosum, *J. Med. Geriet.* 20:122-124.

Lachiewica, A.M., Kogan, S.F., Levitt, S.B., and Weiner, R.L., 1985, Concurrent agenesis of the corpus callosum and ureterocoeles in siblings, *Pediatr.* 75:904-907.

Littlewood, A.H.M., 1961, Congenital nasal dermoid cysts and fistules, *Br. J. Plast. Surg.* 14:169-179.

Lynn, R.B., Buchanan, D.C., Fenichel, G.M., and Freemon, F.R., 1980, Agenesis of the corpus callosum, *Arch. Neurol.* 37:444-445.

Manelfe, C., Starlingjardin, D., Touibi, S., Bonafe, A., and David, J., 1978, Transphenoidal encephalocoele associated with agenesis of corpus callosum: Value of metrizamide computed asternography, *J. Comput. Assist. Tomogr.* 2:221-235.

Matson, D.D. and Ingraham, F.D., 1951, Intracranial complications of congenital dermal sinuses, *Paediatr.* 8:463-474.

McLean, G.E., Broadbent, T.R., and Woolf, R.M., 1964, Dermoid cysts of the nose, *Am. Surg.* 30:3.

Menkes, J.H., Phillipart, M., and Clark, D.B., 1964, Heriditary partial agenesis of the corpus callosum, *Arch. Neurol.* 11:198-208.

Mulbauar, W. and Dittmar, W., 1976, Heriditary median dermoid cysts of the nose, *Br. J. Plast. Surg.* 29:334-340.

Musebeck, K. and Karst, H., 1967, Zur familiären Häufung medianer dermoid Fisteln der Nase, *HNO* 15:326-328.

Naiman, J. and Fraser, F.C., 1955, Agenesis of the corpus callosum: A report of two cases in siblings, *Arch. Neurol. Psychiatr.* 74:182-185.

New, G.B. and Erich, J.B., 1937, Dermoid cysts of the head and neck, *Surg. Gyn. Obstet.* 65:48-55.
Opitz, J.M. and Kaveggia, E.G., 1974, Studies of malformation Syndromes of Man 33: The F.G. syndrome an X-linked recessive syndrome of multiple congenital anomalies and mental retardation, *Z. Kinderheilkd.* 117:1-18.
Parrish, M.L., Roessmann, U., and Levinsohn, M.V., 1979, Agenesis of the corpus callosum, A study of the frequency of associated malformations, *Ann. Neurol.* 6:349-354.
Plewes, J.L. and Jacobson, I., 1971, Familial fronto-nasal dermoid cysts, *J. Neurosurg.* 34:683-686.
Pineda, M., Gonzales, A., Fabregues, I., and Fernandez-Alverez, E., 1984, Familial agenesis of the corpus callosum with hypothermia and apneic spells, *Neuropediatrics* 15:63-67.
Preilowski, B.F.B., 1972, Possible contributions of the anterior forebrain commissures to bilateral motor co-ordination, *Neuropsychol*ogia 10:267-277.
Reynolds, D.M., 1975, Hemispheric differences and interhemispheric relations with special reference to the functions of the corpus callosum. Unpublished Ph.D. thesis, University of St. Andrews, St. Andrews, U.K.
Riccardi, S.M., 1977, Trisomy 8: An international study of 70 patients, *Birth Defects* 13:171-184.
Ryan, A.J., 1948, Dermoid cyst of the bridge of the nose, *Conn. St. Med. J.* 12:218-221.
Schinzel, A., 1982, Four patients including two sisters with the acrocallosal syndrome, *Hum. Genet.* 62:382.
Schinzel, A., 1984, "Catalogue of Unbalanced Chromosome Aberrations in Man," W. de Gruyter, Berlin.
Sessions, R.B., 1982, Nasal dermal sinuses new concepts and explanations, *Laryngoscope* 92 *(Suppl.)* 29:1-30.
Shapira, Y. and Cohen, T., 1973, Agenesis of the corpus callosum in two sisters, *J. Med. Genet.* 10:266-269.
Stupka, W., 1938, Die Missbildungen und Anomalien der Nase und des Nasenrachenraumes, Springer, Berlin.
Sumi, S.M., 1970, Brain Malformations in the Trisomy 18 Syndrome, *Brain* 93:821-830.
Taylor, B.W., Erich, J.B., and Dockerty, M.B., 1966, Dermoids of the head and neck, *Minn. Med.* 49:1535-1549.
Temple, C.M., Jeeves, M.A., and Vilarroya, O., 1989, Ten pen men: Rhyming skills in two children with callosal agenesis, *Brain Lang.* 37:548-564.
Temple, C.M., Jeeves, M.A., and Vilarroya, O., 1990, Reading in callosal agenesis, *Brain Lang.* 39:235-253.
Unterharnscheidt, F., Jachnik, D., and Gött, H.., 1968, Der Balkenmangel, *Monogr. Neurol. Psychiat. (Berlin)* 128:1-232.
Warkany, J., Passange, E., and Smith, L.B., 1966, Congenital malformations in autosomal trisomy syndromes, *Amer. J. Dis. Child.* 112:502-517.
Zellweger, H., Ionasescu, V., Shochet, J.R.S., Simpson, J., and Waziri, M., 1977, Chromosomal aneuploides excluding Downs Syndrome, *in:* "Handbook Clinical Neurology: Congenital Malformations of the Skull and Brain," Vol. 31, Part 11, P.J. Vinken and G.W. Bruyn, eds., North Holland, Amsterdam.
Ziegler, E., 1958, Bösartige familiäre frühinfantile Krampfkrankheit, teilweise verbunden mit familiärer Balkenaplasie, *Helv. Pediatr. Acta* 13:169-184.

OTHER SYNDROMES FREQUENTLY ASSOCIATED WITH CALLOSAL AGENESIS

Guy Geoffroy

Service de Neurologie, Département de Pédiatrie
Hôpital Sainte-Justine pour les Enfants
3175, Chemin Côte Ste. Catherine
Montréal, Qué., H3T 1C5
Canada

INTRODUCTION

More than 50 disorders associated with agenesis of the corpus callosum (ACC) have been reported in medical literature. In most cases, the association appears to be a chance occurrence. Some of the cases have chromosomal abnormalities, others have central nervous malformation or metabolic disease and a few have peroxysomal disease or apparently unrelated systemic malformations (see Tables 1 and 2). A number of cases without concomitant pathology have also been described. Clinicians are at present considering the possibility that inherited metabolic disease may be responsible for cerebral malformations (Dobyns, 1989; Kolodny, 1989). In this context, we recently examined a brother and sister similarly afflicted with a complex neurological syndrome and an unusual metabolic abnormality of lipids. Both had agenesis of the cerebellar vermis.

Callosal agenesis is a constant feature of five syndromes (Acrocallosal, Aicardi, Andermann, Menkes and Shapiro) and a frequent feature of Apert Syndrome, Chiari malformation, Dandy-Walker malformation, hydrocephaly, holoprosencephaly, median facial cleft syndrome and neurofibromatosis type 1. Aicardi and Andermann (this volume) have already presented the syndromes which they first described. The present chapter will concentrate on the review of Acrocallosal, Menkes and Shapiro syndromes. Furthermore, the clinical features of 83 cases of callosal agenesis that have been seen at our hospital will be reported.

ACROCALLOSAL SYNDROME (ACS)

In 1980, Schinzel and Schmid reported two unrelated cases with macrocephaly, unusual facial appearance, polydactyly (mainly hallux duplication), mental retardation and ACC. Six additional cases, unrelated and produced by healthy non-consanguineous

Callosal Agenesis, Edited by M. Lassonde and
M.A. Jeeves, Plenum Press, New York, 1994

parents, were reported by Nelson and Thompson (1982), Sanchis et al. (1985) and Legius et al. (1985). Toriello et al. (1986) described another case with frontal "dysplasia", ACC and Dandy-Walker malformation.

Table 1. Syndromes and abnormalities associated with agenesis of the corpus callosum.

Syndrome/Abnormality	Source
Acrocallosal syndrome	Schinzel and Schmid, 1980
Agyria	Josephy, 1944
Aicardi syndrome	Aicardi et al., 1969
Andermann syndrome	Andermann et al., 1972
Apert syndrome	De Leon et al., 1989
Apert syndrome with limbic malformations	Jeret et al., 1989
"Cri du chat" (deletion of short arm of chromosome 5), G syndrome	Rethoré, 1982
Dandy-Walker malformation	Hart et al., 1972
F.G. syndrome (or Opitz-Kaveggia)	Opitz and Kaveggia, 1974
Focal dermal hypoplasia	Baughman and Worcester, 1970
Hydrocephaly	Harner, 1977
Median cleft face syndrome	De Meyer, 1967
Menkes syndrome	Menkes et al., 1964
Monosomy 21 (partial)	Holbeck et al., 1974
Multiple basal-cell nevi syndrome ("Gorlin")	Gorlin et al., 1965
Neu syndrome	Neu et al., 1971
Oro-facial-digital syndrome	Gorlin and Psaume, 1962
Oro-facial-digital syndrome type I with polycystic kidneys	Connacher et al., 1987
Rubinstein-Taybi syndrome	Smith, 1976
Shapiro syndrome	Shapiro and Plum, 1967
Taybi-Linder syndrome	Taybi and Linder, 1967
Trisomy 13-15	Warkany, 1971
Trisomy 18	Passarge et al., 1966
Zellweger syndrome	Bowen et al., 1964

At first the syndrome was thought to be sporadic, but it soon became apparent that it was transmitted through autosomal recessive inheritance (Schinzel and Kaufmann, 1986; Salgado et al., 1989; Temtamy and McGuid, 1989). More recently, other systemic malformations have been included in the syndrome such as abnormal pulmonic valve with pulmonary hypertension, hypothalamic dysfunction (diabetes insipidus and central hypothyroidism), imperforated anus with recto-vaginal fistula and Dandy-Walker malformation (Moeschler et al., 1989), dextro position of aorta and ostium secundum, anteriorly placed anus, malignant hyperthermia (Philip et al., 1988) and tetralogy of Fallot, multiple cyst of right kidney, liver and pancreas, absence of left kidney and right testis (Casamassima et al., 1989). To date, approximately 20 cases have been reported in the literature. The clinical features of the syndrome are listed in Table 3.

SHAPIRO SYNDROME

The Shapiro syndrome or EHH (Episodic Hyperhidrosis, Hypothermia and ACC) was first described by Shapiro and Plum (1967), who observed a rare disorder of central thermoregulation consisting of episodes of excessive sweating with profound

Table 2. Other abnormalities associated with agenesis of the corpus callosum.

Abnormality	Source
Asphyxiating thoracic dysplasia, hepatic ductal hypoplasia and Dandy-Walker syndrome	Trabelsi et al., 1990
Combined immunodeficiency, bilateral cataract and hypopigmentation in two brothers	Vici et al., 1988
Congenital oculomotor apraxia	Vercelleto-Frior et al., 1986
Craniosynostosis, mental retardation, camptodactyly, hypogonodism and distinctive facies	Lin and Gettig, 1990
Double urinary collection system and trigonocephaly	Ben Ari et al., 1989
Enlarged parietal foramina	Barredo and Geoffroy, 1967
Facial anomalies, Robin sequence and other anomalities	Toriello et al., 1986
Holoprosencephaly	Zingesser et al., 1966
Holoprosencephaly and Chiari malformation	Byrd, 1988
Holoprosencephaly, hydrocephaly, octadactyly and cardial malformation	Grote et al., 1984
Hypohidrotic ectodermal dysplasia and primary hypothyroidism	Fryns et al., 1989
Interhemispheral neuroepithelial (glio-ependymal) cysts and neocortical maldevelopment	Barth et al., 1984
Macrocephaly in siblings	Young et al., 1985
Median cranioencephalic dysraphias and olfacto-genital dysplasia	De Morsier, 1962
Nasodermal cysts	Jacobson, this volume
Neurofibromatosis	Atlas et al., 1988
Non ketotic hyperglycinaemia	Dobyns, 1989; Wariyar et al., 1990;
Schizophrenia	Lewis et al., 1988
Sphenopharyngeal encephalocele	Mauersberger and Diaz, 1985
Trigonoencephaly	Larsen and Osborn, 1982
Ulrich-Turner syndrome	Kimura et al., 1990
No abnormalities	Lynn et al., 1980

Table 3. Clinical features of the acrocallosal syndrome (20 cases reported, autosomal inheritance).

Constant	Frequent	Infrequent
Agenesis of the corpus callosum Macrocephaly Mental retardation Polydactyly or syndactyly Unusual facial appearance such as: - broad and prominent forehead - downward obliquity - epicanthal folds - highly arched or cleft palate - hypertelorism - large anterior fontanelle - small nose	Funduscopic abnormality (reduced pigmentation, optic atrophy)	Congenital heart defects Hypogenitalism Hypothalamic dysfunction Visceral abnormalities

hypothermia. Neuroradiological investigation of the patients revealed agenesis of the corpus callosum. In the early 1980s, 12 additional cases (8 males and 4 females) weredescribed (Lewitt et al., 1983). A few new cases have since been reported (Mooradian et al., 1984). During a typical episode, the patient shows fatigue, lethargy, withdrawal, slowness of comprehension, mutism, amnesia and confusion. The neurological examination reveals ataxia, tremor, dysarthria and hyperreflexia (Table 4).

Table 4. Clinical description of the Shapiro syndrome (EHH).

Age of onset:	6 months - 25 years
Duration of episode:	30 minutes - few weeks
Frequency of episodes:	4 - 5 per day ; each 5-10 years
Suspected cause:	endocrine or hypothalamic dysfunction (in 7/12 cases) such as: - amenorrhea - diabetes insipidus - hyponatremia etc. - polydipsia - precocious puberty - testicular atrophy
EEG:	abnormal in all patients
Neuro-radiology:	agenesis of the corpus callosum, no destructive lesion
Epilepsy:	in 4/12 cases
Improved with:	phenobarbital (Lewitt et al., 1983: case 1) chlorpromazine (Lewitt et al.,1983: case 2) oxybutynin (Lewitt, 1988) cyproheptadine (Arroyo et al., 1990)

The pathogenesis of this spectacular syndrome is unknown. Is it the result of an abnormality in neurotransmission? Or is it rather an expression of the so-called "diencephalic" epilepsy? Interestingly, one patient who had many daily episodes of severe sweating and hypothermia was given 5 mg of oxybutynin (ditropan) twice a day for urinary emergency. Within a few hours, excessive sweating and hypothermia were abolished and did not reappear in the following six months (LeWitt, 1988). Oxybutynin is an antimuscarinic compound which does not possess antinicotinic effects on autonomic ganglia. The beneficial effect on the symptoms of Shapiro syndrome remains to be explained. Two other cases with EHH, reported by Arroyo et al. (1990), responded well to cyproheptadine, a serotonin antagonist. Considering the relationship between serotonin and the principal symptoms of the syndrome (hypothermia, bradycardia, increased blood pressure) on one hand, and the good response to cyproheptadine on the other, the authors postulated a dysfunction of serotonergic central pathways. Finally, a post-mortem examination performed on one case (Noel, 1973) has revealed severe neural loss and fibrillary gliosis in the hypothalamic infundibular nuclei, particularly in the arcuate nucleus and the premamillary formations.

The relation of the clinical picture to ACC is enigmatic in itself. Hypothermia has never been reported in callosotomized patients or in cases with Marchiafava-Bignami disease.

MENKES SYNDROME

This affliction should not be confused with the "kinky hair disease" or "trichopoliodystrophy" also described by Menkes.

Table 5. Syndromes and abnormalities associated with callosal agenesis in 40 of 83 patients studied at the Hospital Sainte-Justine in Montreal.

Syndrome/Abnormality	Frequency
Andermann syndrome	9
Chromosomal abnormality	6
Medial facial cleft	3
Arnold Chiari syndrome	2
Dandy-Walker malformation	2
Pierre-Marie-Robin syndrome	2
Acrocallosal syndrome	1
Aicardi syndrome	1
Apert syndrome	1
Aplasia cutis congenita	1
Hallerman-Streiff syndrome	1
Hemangiomatosis	1
Neurofibromatosis	1
Smith-Lemli-Opitz syndrome	1
Zellweger syndrome	1
No abnormalities	7

Table 6. Clinical features (besides those associated with the syndromes listed in Table 5) in the callosal agenesis patients studied at the Hospital Sainte-Justine in Montreal.

Feature	Frequency
Epilepsy	19
Mental retardation (moderate to severe)	19
Cardiac malformations	10
Hydrocephaly	10
Cystic formation (brain)	9
Dysmorphic features	9
Renal and genital malformations	8
Cortical dysplasia	6
Cleft lip palate	5
Blindness and/or deafness	4
Endocrine dysfunction	3
Anal imperforation	2
Porencephaly	1
Spina bifida	1

The syndrome observed by Menkes et al. (1964) is a familiar form of ACC that is transmitted through an X-linked inheritance, affecting only male probands. All of the five infants examined by Menkes et al. (1964) had convulsions in the first days of life which were intractable. They subsequently showed severe mental retardation, most of them failing to achieve milestones. Three of the five patients died at the age of 19, 17 and 14 months, respectively. The remaining two patients had severe epileptogenic encephalopathy accompanied by profound mental retardation. Pathological examination revealed ACC, polymicrogyria, heterotopias of the grey matter and absence of the longitudinal callosal bundle (Probst bundle). Chemical analysis performed on the brain of one patient showed a reduction in lipid content.

PERSONAL STUDIES

Eighty-three children with callosal agenesis were admitted to the Hôpital Sainte-Justine in Montreal between January 1985 and December 1990. Forty were females and 43 were males. In approximately one half of the cases this malformation was associated with 15 definite clinical syndromes (Table 5), mainly Andermann syndrome and chromosomal abnormalities. Seven children (8%) were clinically "normal". Nineteen (22%) were severely to moderately retarded, 19 had epilepsy, 10 (12%) hydrocephalus and 10 cardiac malformations (Table 6).

CONCLUSION

More than 50 disorders associated with callosal agenesis have been reported to date. In most cases the association appears to be a chance occurrence. In other cases, the malformation is a constant or frequent feature of a syndrome. In our series of 83 patients, callosal agenesis was associated with a definite clinical syndrome in 50 percent of the cases, 22 percent having Andermann syndrome. The pathogenesis of the various syndromes that include ACC is unknown. Several observations point to an inherited metabolic disorder.

With today's sophisticated sonographic methods of investigation, antenatal diagnosis of ACC will be made more frequently. Many of these children will probably have normal outcome. Children with ACC accompanied by developmental delay and/or seizure disorder, however, should be screened for metabolic diseases.

REFERENCES

Aicardi, J., Lefevre, J., and Lerique-Koechlin, A., 1965, A new syndrome: spasm in flexion, callosal agenesis, ocular abnormalities, *Electroencephalogr. Clin. Neurophysiol.* 19:609-610.

Andermann, F., Andermann E., Joubert, M., Karpati, G., Carpenter, S., and Melanson, D., 1972, Familial agenesis of the corpus callosum with anterior horn cell disease: a syndrome of mental retardation, areflexia and paraparesis, *Trans. Am. Neurol.* Assoc. 97:242-244.

Arroyo, H.A., Di Blasi, A.M., and Grinszpan, G.J., 1990, A syndrome of hyperhidrosis, hypothermia and bradycardia possibly due to central monoaminergic dysfunction, *Neurolo*gy 40:556-557.

Atlas, S.W., Zimmerman, R.A., Bruce, D., et al., 1988, Neurofibromatosis and agenesis of the corpus callosum in identical twins: MR diagnosis, *AJNR* 9:598-601.

Barredo, F. and Geoffroy, G., 1967, L'association des lacunes pariétales à l'agénésie du corps calleux, *Can. Med. Assoc. J.* 96:1531-1533.

Barth, P.J., Uylings, H.B., and Stam, F.C., 1984, Interhemispheral neuroepithelial (glioependymal) cysts associated with agenesis of the corpus callosum and neocortical maldevelopment. A case study, *Childs Brain* 11:312-319.

Baughman, F.A. and Worcester, D.D., 1970, Agenesis of the corpus callosum in a case of focal dermal hypoplasia, *Mt. Sinai J. Med.* 37:702-709.

Ben Ari, J., Shuper, A., Mismonni, A., Rosen, O., Grunbaum, M., and Merlob, P., 1989, Agenesis of the corpus callosum associated with double urinary collecting system, trigonocephaly and other minor anomalies: a new association, *Eur. J. Pediatr.* 148:787-788.

Bowen, P., Lee, C.S.N., and Zellweger, H., 1964, A familial syndrome of multiple congenital defects, *Bul. Johns Hopk. Hosp.* 114:402-414.

Byrd, S.E., 1988, Disorders of midline structures: holoprosencephaly, absence of corpus callosum, and Chiari malformation, *Semin. Ultrasound CT MR* 9:201-2115.

Casamassima, A.C., Beneck, D., Gewitz, M.H., et al., 1989, Acrocallosal syndrome: additional manifestations, *Am. J. Med. Genet.* 32:311-317.

Connacher, A.A., Forsyth, D.D., and Stewart, W.K., 1987, Orofaciodigital syndrome type I associated with polycystic kidneys and agenesis of the corpus callosum, *J. Med. Genet.* 24:116-118.

De Leon, G.A., De Leon, G., Grover, W.D., Zaeri, N., and Alburger, P., 1989, Agenesis of the corpus callosum in Apert syndrome?, *Arch. Neurol.* 46:479.

De Meyer, W., 1967, The median cleft face syndrome. Differential diagnosis of cranium bifidum occultum, hypertelorism and median cleft nose, lip and palate, *Neurolo*gy 17:961-971.

De Morsier, G., 1962, Median cranioencephalic dysraphias and olfacto-genital dysplasia, *World Neurol.* 3:485.

Dobyns, W.B., 1989, Agenesis of the corpus callosum and gyral malformations are frequent manifestations of non ketotic hyperglycenemia, *Neurology* 39:817-820.

Fryns, J.P., Chrzanowska, K., and Van den Berghe, H., 1989, Hypohidrotic ectodermal dysplasia, primary hypothyroidism and agenesis of the corpus callosum, *J. Med. Genet.* 26:520-521.

Gorlin, R.J. and Psaume, J., 1962, Orodigitofacial dysostosis: a new syndrome. A study of 22 cases, *J. Pediatr.* 61:520-530.

Gorlin, R.J., Vickers, R.A., Keln, E., and Williamson, J.J., 1965, The multiple basal-cell nevi syndrome. An analysis of a syndrome consisting of multiple nevoid basal-cell carcinoma, jaw cysts, skeletal anomalies, medullo-blastoma and hyporesponsiveness to parathormone, *Cancer* 18:89-104.

Grote, W., Rehder, H., Weisner, D., and Wiedemann, H.R., 1984, Prenatal diagnosis of a probable hereditary syndrome with holoprosencephaly, hydrocephaly, octodactyly and cardiac malformations, *Eur. J. Pediatr.* 143:155- 57.

Hart, M.N., Malamud, N., and Ellis, W.G., 1972, The Dandy-Walker syndrome. A clinicopathological study based on 28 cases, *Neurology* 22:771-780.

Harner, R.N., 1977, Agenesis of the corpus callosum and associated defects, in: "Scientific Approaches to Clinical Neurology," Vol. I., E. S. Goldensohn and S. H. Appel, eds, Lea & Febiger, Philadelphia.

Holbeck, S., Friedrich, U., Grostrom, K., and Petersen, G.B., 1974, Monosomy for the centromeric and juxta-centromeric region of chromosome 21, *Humangenetik* 24:191-195.

Jeret, J.S., Serur, D., and Wisniewski, K., 1989, Agenesis of the corpus callosum and limbic malformations in Apert syndrome, *Arch. Neurol.* 46:10.

Josephy, H., 1944, Congenital agyria and defect of corpus callosum, *J. Neuropath. Exp. Neurol.* 3:63-68.

Kimura, M., Nakajima, M., and Yoshino, K., 1990, Ulrich-Turner syndrome with agenesis of the corpus callosum, *Am. J. Med. Genet.* 37:227-228.

Kolodny, E.H., 1989, Agenesis of corpus callosum: a marker for inherited metabolic diseases?, *Neurology* 39:847-848.

Larsen, P.D. and Osborn, A.G., 1982, Computed tomographic evaluations of corpus callosum agenesis and associated malformations, *CT* 6:225-230.

Legius, E., Fryns, J.P., Casaer, P., Boel, M., and Eggermont, E., 1985, Schinzel acrocallosal syndrome: a variant example of the Greig syndrome?, *Am. Genet. (Paris)* 28:239-240.

Lewis, S.W., Reveley, M.A., David, A.S., and Ron, M.A., 1988, Agenesis of the corpus callosum and schizophrenia: a case report, *Psychol. Med.* 18:341-347.

LeWitt, P.A., 1988, Hyperhidrosis and hypothermia responsive to oxybutynin, *Neurology* 38:506.

LeWitt, P.A., Newman, R.P., Greenberg, H.S., Rocher, L.L., Calne, D.B., and Ehrenkranz, J.R.L., 1983, Episodic hyperhidrosis, hypothermia and agenesis of the corpus callosum, *Neurology* 33:1122-1129.

Lin, A.E. and Gettig, E., 1990, Craniosynostosis, agenesis of the corpus callosum, severe mental retardation, distinctive faces, camptodactyly and hypogonadism, *Am. J. Med. Genet.* 35:582-585.

Lynn, R.B., Buchanan, D.C., Fernichel, G.M., and Freemon, F.R., 1980, Agenesis of the corpus callosum, *Arch. Neurol.* 37:444-445.

Mauersberger, W. and Diaz, L.A., 1985, Sphenopharyngeal encephalocele associated with agenesis of the corpus callosum, *Neurochirurgia (Stuttg.)* 28:238-242.

Menkes, J., Philippart, M., and Clark, D., 1964, Hereditary partial agenesis of the corpus callosum, *Arch. Neurol.* 11:198-208.

Moeschler, J.B., Pober, B.R., Holmes, L.B., and Graham jr., J.M., 1989, Acrocallosal syndrome: new findings, *Am. J. Med. Genet.* 32:306-310.

Mooradian, A.D., Morley, G.K., McGeachie, R., Lundgren, S., and Morley, J.E., 1984, Spontaneous periodic hypothermia, *Neurology* 34:79-82.

Nelson, M.M. and Thomson, A.J., 1982, The acrocallosal syndrome, *Am. J. Med. Genet.* 12:195-199.

Neu, R.L., Kadjii, T., Gardner, L.I., Nagyfy, S.F., and King, S., 1971, A lethal syndrome of microcephaly with multiple congenital anomalies in three siblings, *Pediatrics* 47:610-612.

Noel, P., Hubert, J.P., Ectors, M., Franken, L., and Flament-Durant, J., 1973, Agenesis of the corpus callosum associated with relapsing hypothermia, *Brain* 96:359.

Opitz, J.M. and Kaveggia, E.G., 1974, Studies of malformation syndrome of man XXXIII. The FG syndrome: an X-linked recessive syndrome of multiple congenital anomalies and mental retardation, *Z. Kinderheilkd.* 117:1-18.

Passarge, E., Trus, C.W., Sueoka, W.T., Baumgartner, N.R., and Keer, K.R., 1966, Malformations of the central nervous system in trisomy 18 syndrome, *J. Pediatr.* 69:771-778.

Philip, N., Apicella, N., Lassman, I., Ayme, S., Mattei, J.F., and Giraud, F., 1988, The acrocallosal syndrome, *Eur. J. Pediatr.* 147:206-208.

Rethoré, M.O., 1982, Chromosome 5: monosomy: partial deletion of short arm, *in*: "Handbook of Clinical Neurology: Neurogenetic Directory," Vol. 43, Part II, P. J. Vinken and G. W. Bruyns, eds., Elsevier, North Holland , Amsterdam.

Salgado, L.J., Ali, C.A., and Castilla: E.E.: 1989: Acrocallosal syndrome in a girl born to consanguineous parents, *Am. J. Med. Genet.* 32:298-300.

Sanchis, A., Cervero, L., Martinez, A., and Valverde, C., 1985, Duplication of hands and feet, multiple joint dislocations, absence of the corpus callosum and hypsarrhythmia: acrocallosal syndrome?, *Am. J. Med. Genet.* 20:123-130.

Schinzel, A. and Kaufmann, U., 1986, The acrocallosal syndrome in sisters, *Clin. Genet.* 30:399-405.

Schinzel, A. and Schmid, W., 1980, Hallux duplication, postarial polydactyly, absence of the corpus callosum, severe mental retardation and additional abnormalities in two unrelated patients: a new syndrome?, *Am. J. Med. Genet.* 6:241-249.

Shapiro, W.R. and Plum, F., 1967, Paroxysmal hypothermia and mutism accompanying agenesis of the corpus callosum, *Neurology* 17:317.

Smith, D.W., 1976, Rubinstein-Taybi syndrome, *in:* "Recognizable Patterns of Human Malformation," D.W. Smith, ed., W.B. Saunders Co., New York.

Taybi, H. and Linder., D., 1967, Congenital familial dwarfism with cephaloskeletal dysplasia, *Radiology* 89:275-281.

Temtamy, S.A. and McGuid, N.A., 1989, Hypogenitalism in the acrocallosal syndrome, *Am. J. Med. Genet.* 32:301-305.

Toriello, H.V., Radecki, L.L., Sharda, J., Looyenga, D., and Mann, R., 1986, Frontonasal "dysplasia", cerebral anomalies and polydactyly: report of a new syndrome and discussion from a developmental field perspective, *Am. J. Med. Genet. (Suppl.)* 2:89-96.

Trabelsi, M., Hammon-Jeddi, A., Kammoun, A., Bennaceur, B., and Gharbi, H.A., 1990, Dystrophie thoracique asphyxiante de Jeune associée à une hypoplasie ductulaire hépathique, une agénésie du corps calleux et un syndrome de Dandy-Walker, *Pediatrie* 45:35-38.

Vercelletto-Frior, M., Lavenant, F., Bricout, J.H., Venisse, T., Vercelletto, P., and Feve, J.R., 1986, Apraxie oculo-motrice congénitale avec agénésie du corps calleux et atrophie sous-tentorielle, *Rev. Neurol. (Paris)* 142:140-144.

Vici, C.D., Sabetta, G., Gambarara, M., et al., 1988, Agenesis of the corpus callosum combined with immunodeficiency, bilateral cataract and hypopigmentation in two brothers, *Am. J. Med. Genet.* 29:1-18.

Wariyar, U.K., Welch, R.J., Milligan, D.W., and Perry, R.H., 1990, Sonographic and pathological features of callosal hypoplasia in non-ketotic hyperglycinaemia, *Arch. Dis. Child.* 65:670-671.

Warkany, J., 1971, "Congenital Malformations," III: Year Book Medical Publishers, Chicago.

Young, I.D., Trounce, J.Q., Levene, M.I., Fitzsimmonds, J.S., and Moore, J.R., 1985, Agenesis of the corpus callosum and macrocephaly in siblings, *Clin. Genet.* 28:225-230.

Zingesser, L.H., Schlechter, M.M., and Medina, A., 1966, Angiographic and pneumoencephalographic features of holoprosencephaly, *Am. J. Roentgenol.* 97:561-574.

ANTENATAL SONOGRAPHIC FINDINGS OF AGENESIS OF CORPUS CALLOSUM

Patrizia Vergani and Nicola Strobelt

Dept. of Obstetrics Gynecology
Ospedale San Gerardo di Monza
Via Solferino, 16, Monza
Italia

INTRODUCTION

The corpus callosum consists of predominantly white fibers connecting the cerebral hemispheres. It develops in the upper portion of the lamina terminalis, in the cephalic extremity of the neural tube near the closure point of the rostral neuropore. Among the three telencephalic commissures, the anterior commissure, the fornix and the corpus callosum, the latter is the last structure that undergoes development. The anterior commissure appears in the second month of gestation, the fornix develops in the third month and while the fornix is partially regressing, the first outlines of the corpus callosum can be identified during the fourth month.

The corpus callosum develops along an antero-posterior axis and acquires its final shape during the sixth month. A degeneration of the white matter located between the corpus callosum and the fornix occurs at the same time, thus forming the cavum septi pellucidi and the Verga's ventricle (Romero et al., 1988).

NORMAL CORPUS CALLOSUM

The normal corpus callosum may be routinely visualized by sonography after the 20th week of gestation.

On a midline sagittal scan, it appears as a hypoechogenic structure over the cavum septi pellucidi and the Verga's ventricle. Several structures can be observed in the same plane: third ventricle, cerebral peduncles, pons, medulla, cerebellum and roof of the fourth ventricle (Figure 1) (Vergani et al., 1988; Babcock, 1984).

In an axial scan the genu of the corpus callosum can be defined by the pulsation of the pericallosal artery. On coronal views, the corpus callosum appears as a hypoechoic structure forming anteriorly, the roof of the lateral ventricles and of the cavum septi pellucidi and posteriorly, the roof of the lateral and Verga's ventricles (Figure 2).

Callosal Agenesis, Edited by M. Lassonde and
M.A. Jeeves, Plenum Press, New York, 1994

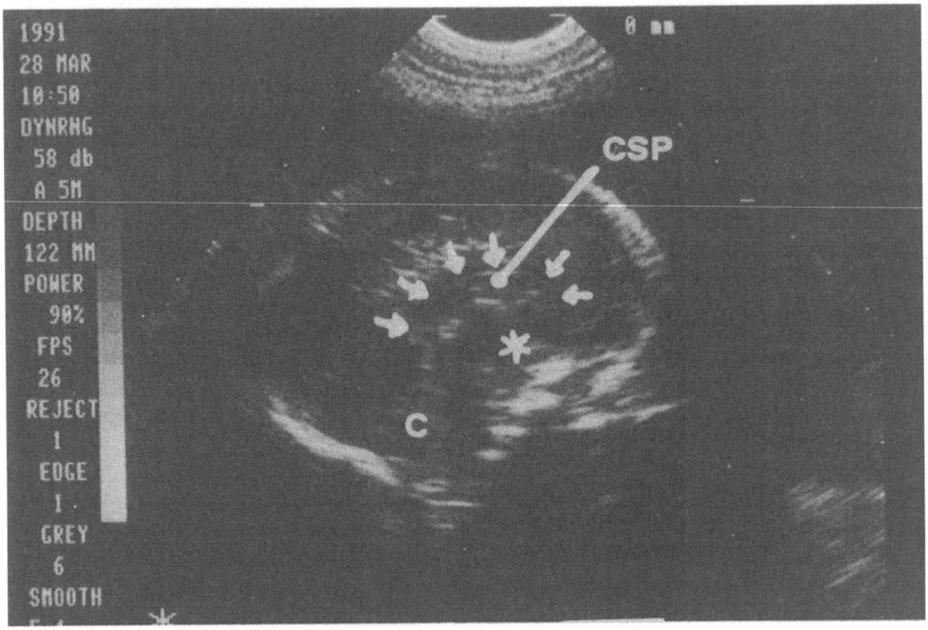

Figure 1. Median sagittal scan of the fetal head showing the corpus callosum (arrows); CSP=Cavum septi pellucidi; *=Third ventricle; C=Cerebellum.

CALLOSAL AGENESIS

Agenesis of the corpus callosum can be isolated or associated with various conditions such as Dandy-Walker malformation, polymicrogyria, interhemispheric cysts, porencephalic paramedian cysts and Arnold-Chiari malformation (Atlas et al., 1986; Byrd et al., 1978; Parrish et al., 1979; Romero et al., 1988; Young et al., 1985).Ultrasound scanning allows a prenatal diagnosis of callosal agenesis per se. Differential diagnosis of callosal agenesis must include isolated hydrocephalus, holoprosencephaly, arachnoid or porencephalic cysts and lipoma of the corpus callosum (Romero et al., 1988).

Axial view

Agenesis of the corpus callosum can be suspected on an axial scan of the fetal head by observing ventriculomegaly with lateral displacement of the lateral ventricles. The latter structures are then characterized by a disproportion between their pointed frontal horns and their enlarged occipital horns (Babcock, 1984).

In cases in which the posterior horns and atria of the lateral ventricles are primarily involved, the sonographic image of the frontal parts may resemble the shape of bull horns (Figure 3A).

In other cases, still on the axial plane, the frontal horns are dilated and the third ventricle is elongated and superiorly displaced (Babcock, 1984; Gebarsky et al., 1984; Vergani et al., 1988). The end-result of this arrangement may suggest the image of a leaping frog (Figure 3B). The elevation of the third ventricle is a very specific sign of callosal agenesis but is observed in only 40% of fetuses (Romero et al., 1988).

When performed, serial examinations reveal a lack of progression of ventriculomegaly. Regression may occasionally be observed. Head circumference is generally not increased (Chervenak et al., 1985; Comstock et al., 1985).

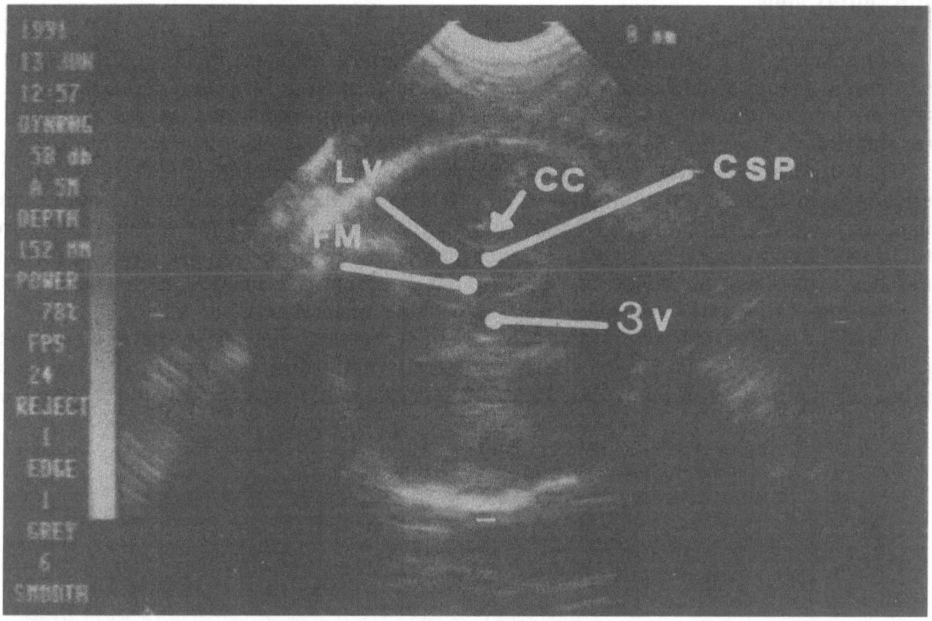

Figure 2. Coronal scan of the fetal head. CC=Corpus callosum; CSP=Cavum septi pellucidi; LV=Lateral ventricle; FM=Foramen of Monro; 3V=Third ventricle.

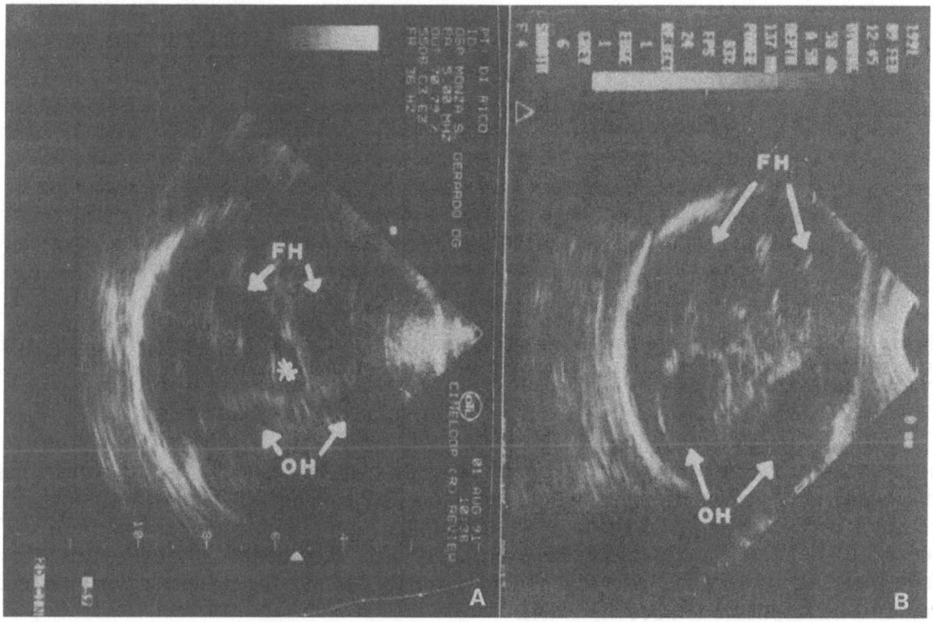

Figure 3. Two cases of agenesis of the corpus callosum (axial plane). Note the enlargement and elevation of the third ventricle in Figure. 3A and the disproportion between the pointed frontal horns (FH) and the enlarged occipital horns (OH) in Figure. 3B.

Midsagittal view

The absence of a normally sized and placed corpus callosum in the midline sagittal and coronal planes would confirm the sonographic suspicion; however, these two planes (sagittal and coronal) are not easily achieved when the fetus is in a cephalic presentation.

On a midline sagittal scan, the corpus callosum is not interposed between the pericallosal cisterns and the cingulate gyri. When not directly taken on the midline, this plane may yield a false image of the corpus callosum and of all the median structures previously mentioned (Figure 1). In such cases, visualization of the caudate nucleus indicates a parasagittal plane, passing through the lateral ventricles, and what is seen as the corpus callosum would only be the upper wall of the lateral ventricle.

A sagittal plane may also allow to distinguish between partial or complete callosal agenesis (Figure 4). Hypogenesis of the corpus callosum, otherwise referred to as partial agenesis (see Rauch and Jinkins, this volume) may affect the anterior or

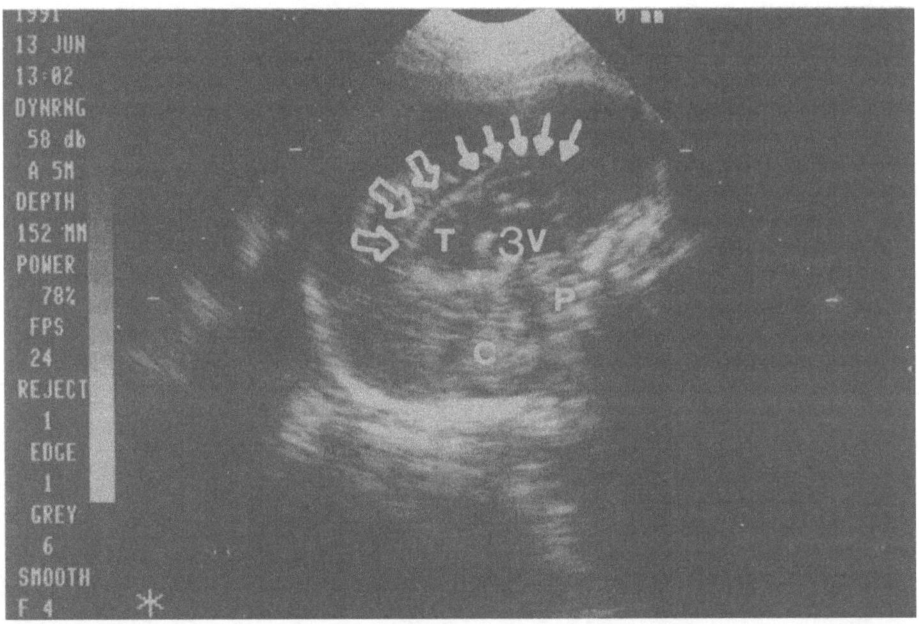

Figure 4. Partial agenesis of the corpus callosum. Note the anterior part normally developed (full arrows) and the missing posterior part (empty arrows). T=Thalamus; 3V=Third ventricle; P=Pons; C=Cerebellum.

posterior portion. A pseudoporencephalic etiology may be suspected in the former case. Hypogenesis of the anterior part would then be a consequence of local destruction of the cerebral parenchyma by a vascular or infectious factor that occurred in utero. The second case, posterior hypogenesis, is probably a developmental anomaly caused by a failure of migration of the cells that should normally connect the posterior portions of the cerebral hemispheres (Babcock, 1984).

Coronal view

On a coronal (frontal) plane, the lateral ventricles appear to be laterally displaced, showing sharply angled lateral peaks of their bodies and frontal horns (Babcock, 1984).

The third ventricle may be particularly well visualized on this plane, allowing for the evaluation of its eventual dilatation as well as its displacement in the superior axis (Vergani et al., 1988).

Upper displacement of the third ventricle can be associated with partial or complete agenesis of the falx cerebri and with a dilatation of the interhemispheric space. Using color-flow doppler in prenatal ultrasonography, an anterior cerebral artery running over the third ventricle without the physiological interposition of the corpus callosum can be found.

CONCLUSION

Since agenesis of the corpus callosum is associated with other central nervous system anomalies in 80% of cases, in utero identification of this defect should prompt a careful examination of the fetal anatomy. The diagnosis of isolated agenesis of corpus callosum should help avoid unnecessary intervention (Romero et al., 1988; Vergani et al., 1988)..

Prognosis is in fact correlated with the presence and severity of associated anomalies. Unfortunately, ultrasonography may not detect all minor structural abnormalities which may be related to postnatal neurological deficiencies.

The use of ultrasonography in detecting ACC in utero should be followed by further studies based on a correlation between prenatal and neonatal sonographic screening.

REFERENCES

Atlas, S.W., Zimmerman, R.A., Bilaniuk, L.T., et al., 1986, Corpus callosum and limbic system: Neuroanatomic MR evaluation of developmental anomalies, *Radiology* 160:355.

Babcock, D.S., 1984, The normal, absent and abnormal corpus callosum: Sonographic findings, *Radiology* 151:449.

Byrd, S.E., Harwood-Nash, D.C., and Fitz, C.R., 1978, Absence of the corpus callosum: Computed tomographic evaluation in infants and children, *J. Can. Assoc. Radiol.* 29:108.

Chervenak, F.A., Hobbins, J.C., Wertheimer, I. et al, 1985, The natural history of ventriculomegaly in a fetus without obstructive hydrocephalus, *Am. J. Obstet. Gynecol.* 152:574.

Comstock, C.H., Culp, D., Gonzalez, J. et al., 1985, Agenesis of the corpus callosum in the fetus : Its evolution and significance, *J. Ultrasound. Med.* 4:613.

Gebarski, S.S., Gebarski, K.S., Bowerman R. et al., 1984, Agenesis of the corpus callosum: Sonographic features, *Radiololgy* 151:443.

Parrish, M.L., Roessmann, U., and Levinsohn, M.W., 1979, Agenesis of the corpus callosum: A study of the frequency of associated malformations, *Ann. Neurol.* 6:349.

Romero, R., Pilu, G., Jeanty, A. et al., 1988, Agenesis of corpus callosum, *in*: "Prenatal Diagnosis of Congenital Anomalies," R. Romero et al., eds, Appleton & Lange, East Norwalk, CT.

Vergani, P., Ghidini, A., Mariani, S. et al., 1988, Antenatal sonographic findings of agenesis of corpus callosum, *Am. J. Perinatol.* 5:105.

Young, I.D., Trounce, J.Q., Levene, M.I. et al., 1985, Agenesis of the corpus callosum and macrocephaly in siblings, *Clin. Genet.* 28:225.

CALLOSAL AGENESIS: POSTNATAL SONOGRAPHIC FINDINGS

Giovanni Cioni,[1] Laura Bartalena,[2]
Enrico Biagioni,[1] and Antonio Boldrini[2]

[1]Institute of Child Neurology and Psychiatry
University of Pisa and Stella Maris Foundation
Via dei Giacinti, 2
I-56018 Calambrone (Pisa), Italy
[2]Neonatal Intensive Care Unit, University of Pisa, Italy

INTRODUCTION

In the last decade, cranial ultrasound scanning (US) has become a routine examination in neonatal intensive care units (NICUs). This technique allows a non-invasive quick visualization of newborns' brains, even in very low birth weight infants admitted to intensive care. The examination is conducted directly at the cot or in the incubator by means of a transportable real-time US-apparatus, thus making it unnecessary to move the baby. A 5 MHz (or 7.5 MHz) transducer is applied over the anterior fontanel which remains open during the first several months of life. Coronal and sagittal scans are usually performed, allowing a good visualization of supratentorial and subtentorial structures. Only the regions located immediately below cranial bones are not well seen. The most hypoechoic structures (i.e. cerebro-spinal fluid) appear black and the most hyperechoic ones (i.e. bones) appear white, whereas tissues with intermediate echogenicity are represented by different shades of grey.

In most NICUs, US is carried out within the first days of life in all neonates who present some risk for intracranial abnormalities. The examinations are generally repeated at different intervals depending on the gestational age of the infants and on previous US findings, until discharge from the unit. Subsequently, infants with an abnormal scan generally have additional examinations before fontanel closure.

The diagnostic and prognostic value of this technique for haemorrhagic and ischemic lesions has been proven by several studies (Stewart et al., 1987; De Vries et al., 1988; De Vries et al., 1989; Cioni et al., 1992). US findings correlate well with the results of other more invasive imaging techniques, such as computed tomography (CT) and magnetic resonance imaging (MRI), and also with autopsy. US appears also highly predictive of the neurologic outcome of the baby.

US is thus of remarkable value for the early diagnosis of cerebral malformations. Most of these are easily detectable by means of this technique (e.g. hydrocephalus, cerebellar abnormalities etc.) whereas other alterations (i.e. grey matter heterotopias, dysgyrias etc.) are often not recognizable by US. Callosal agenesis (CA) belongs to the

Callosal Agenesis, Edited by M. Lassonde and
M.A. Jeeves, Plenum Press, New York, 1994

first group. In the following paragraphs the normal appearance of corpus callosum (CC) at US and the sonographic features of complete and partial CA will be described.

NORMAL US APPEARANCE OF CORPUS CALLOSUM

The CC can be easily identified during routine cranial sonographic scanning. On coronal scans (Figure 1A), it is recognizable as a hypoechoic band that lies in the floor of the interhemispheric fissure, forming the roof of the lateral ventricles. These cavities are located next to the median line and are separated by the septum pellucidum which is attached to the inferior surface of the CC. They normally show a triangular shape; the superior and the medial borders appear convex. The pericallosal cistern with the pericallosal vessels can be identified over the CC as an echogenic structure. Cingulate gyri are detectable on the two sides of the interhemispheric fissure.

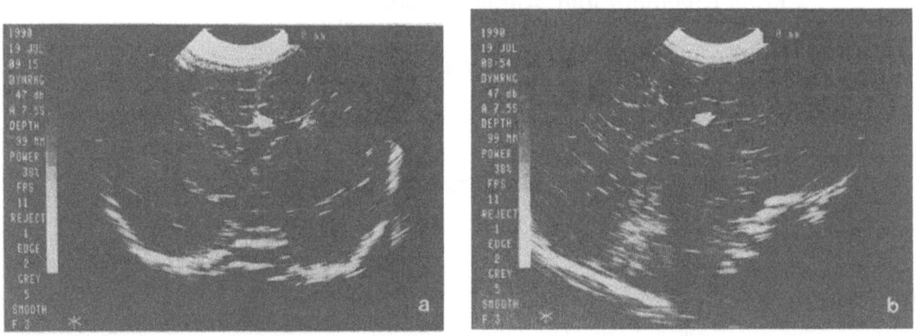

Figure 1A. Coronal US scan of a normal newborn, showing the CC as a hypoechoic band (arrow) in the floor of the interhemispheric fissure. **B.** US midline sagittal scan of the same infant; the CC appears as a broad arched anechoic band (arrow) below the interhemispheric fissure.

On midline sagittal scans (Figure 1B), the CC is recognizable as a broad arched anechoic band in the floor of the interhemispheric fissure. At the anterior end of the CC, it is possible to identify the genu, forming a semicircular sweep downward, and the rostrum, the narrow connection to the lamina terminalis. Posteriorly, the genu is continuous with the body of the CC which arches upward and backward and then ends in the posterior extremity, where the splenium is detectable as a thick, rounded structure. This portion of the CC is related to the falx cerebri and to the inferior sagittal sinus. The free margin of the tentorium and the straight sinus can be identified behind the splenium. Over the CC, the echogenic pericallosal cistern with its vessels and, on each side of the fissure, the cingulate gyri are recognizable. The septum pellucidum is attached to the inferior concave surface of the CC.

US FINDINGS IN CALLOSAL AGENESIS (CA)

In general, CA is easily recognizable by means of US-scan and the US features of this pathological condition have been reported in a few studies (Skeffington, 1982; Cadier et al., 1982; Gebarsky et al., 1984; Babcock, 1984; Baarsma et al., 1987). In the NICU of the University of Pisa, 11 cases of CA have been detected between 1985 and 1991. The main clinical and neuroradiological data of these patients are reported in Table 1. The sonographic findings of these patients will be compared to other cases reported in the literature.

Table 1. Neuroimaging and clinical findings in 11 acallosal infants.

Case	Sex	US	CT-MRI	Other malformations	Follow-up
1	F	CA arach.cyst	CT, MRI: CA arach.cyst heterotopia	-	6y: mild R hem. epil.
2	M	CA	CT, MRI: CA	urinary	5y: dyst.c.p.
3	M	CA	CT: CA	-	3y: normal
4	F	CA	-	-	?
5	M	CA	CT, MRI: CA	urinary cardiac	3y: ment.ret.
6	M	CA	CT, MRI: CA microcephaly	-	3y: dyst.c.p. ment.ret.
7	M	part.CA lob.holopr.	MRI: part.CA lob.holopr.	-	3y: normal
8	F	part.CA	CT: part.CA	urinary	3y: ment.ret.
9	M	CA	MRI: CA	-	deceased
10	M	CA	CT: CA	-	6m: normal
11	M	CA	CT: CA	musculo- skeletal	?

Abbreviations: arach.: arachnoid; CA: callosal agenesis; dyst.c.p.: dystonic cerebral palsy; epil.: epilepsy; lob.holopr.: lobar holoprosencephaly; ment.ret.: mental retardation; part.: partial; R hem.: right hemiplegia.

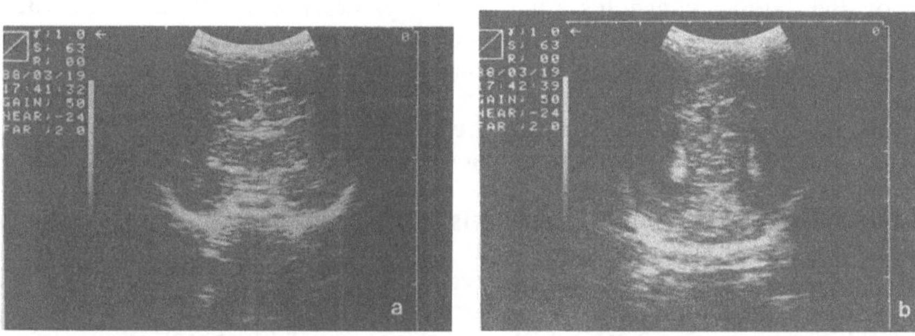

Figure 2A. Coronal anterior US scan of a newborn with CA; the frontal horns of the lateral ventricles are markedly separated; their medial margins have a concave shape ("bull's horns"). **B.** Coronal posterior US scan of the same infant; colpocephalic aspect of the lateral ventricles which appear parallel.

First of all, the US picture of CA is characterized by the failure to detect the CC in the floor of the interhemispheric fissure. The absence of that broad hypoechoic band, which normally represents the callosal body, is easily noted in coronal (Figure 2A and B) and sagittal scans (Figure 3A). The direct recognition of callosal absence is also possible by means of MRI midsagittal scans (Figure 3B), but not by CT.

Other indirect findings, also very important for the sonographic diagnosis of CA, were well known before the introduction of US. Davidoff and Dyke (1934) were the first to describe the neuroradiological picture of CA in three patients examined by means of pneumoencephalography. Subsequently, their observations have been applied to CT (Wing and Osborn, 1977; Guibert-Trainer et al., 1982; Kendall, 1983) and to MRI (Davidson et al., 1985; Atlas et al., 1986; Barkovich, 1990). The abnormalities are the following: 1) Marked separation of frontal horns and bodies of lateral ventricles; 2) Angular shape of the dorsal margin of the lateral ventricles; 3) Concave shape of the medial border of the lateral ventricles; 4) Dilatation of the occipital horns; 5) Elongation of the foramina of Monro; 6) Dilatation and dorsal extension of the third ventricle; 7) Radial arrangement of the medial cerebral sulci, which extend downwards through the zone normally occupied by the CC.

The occurrence in CA of these abnormalities of the ventricular system and medial sulci has been explained by means of anatomical studies. In the absence of the CC, lateral ventricles tend to separate from each other. The caudate heads and lentiform nuclei keep the size of the frontal horns relatively normal, whereas the posterior portions of the ventricles extend superiorly into the underdeveloped white matter. This last phenomenon gives rise to an enlargement of the fornices and of the occipital and posterior temporal horns. This condition is known as "colpocephaly". The lateral ventricles are deformed by the Probst's bundle, formed by callosal fibers which failed to cross the midline and have a course parallel to the interhemispheric fissure, indenting the medial walls of the lateral ventricles. In the case of callosal absence, the third ventricle sometimes tends to be widened and to extend superiorly while the foramina of Monro appear enlarged and elongated. The cingulate sulcus does not form and the cingulate gyri remain everted; the sulci of the medial surface of the hemispheres then extend into the third ventricle with a radial arrangement.

These abnormalities are easily observable by means of US through coronal (Figure 2A and B) and sagittal scans (Figure 3A). The anterior and central portions of the lateral ventricles appear markedly separated on the coronal scan (Figure 2A). On the same view these parts of the ventricular cavities show sharply angled lateral peaks, giving a double horn appearance: this aspect, when associated with dilatation and dorsal extension of the third ventricle, is called "bull's head" (Cadier et al., 1982). In our cases and in others described in the literature, the particular concave shape of the medial border of the lateral ventricles (due to the protrusion of Probst's bundle), the "colpocephaly" and the relative narrowness of frontal horns are always observable. By contrast, the dilatation and the dorsal extension of the third ventricle are not constant features of the sonographic picture of CA; as reported by some authors (Skeffington, 1982; Babcock, 1984), the shape and the location of this cavity may appear completely normal in certain cases. This particular abnormality was observed in only four of our 11 cases. On sagittal scans (Figure 3A), the radial arrangement of the medial cerebral sulci around the roof of the third ventricle is easily detectable: this disposition replaces the normal concentric shape of pericallosal sulci and is specific of CA. In the case of partial CA, this radial arrangement is observable only where the CC has not developed. A final anatomical hallmark of CA is detectable only by means of US-scan (save arteriographic examinations): this feature is represented by the missing pericallosal artery, as described by Baarsma et al. (1987). This vessel usually circumvents the

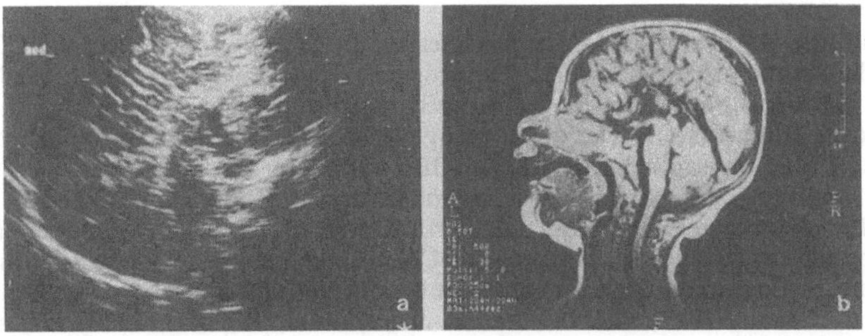

Figure 3A. US midline sagittal scan of a newborn with CA. The CC is absent; the medial cerebral sulci have a radial arrangement around the roof of the third ventricle. **B.** MRI of the same infant at 7 months; midsagittal scan confirming CA.

anterior extremity of the CC, forming a semicircular sweep around the genu and the rostrum. As confirmed by anatomical and arteriographic studies (Bull, 1967), in CA the pericallosal artery assumes a wandering course and is not seen pulsating in the pericallosal cistern on the US midsagittal scan plane.

The sonographic diagnosis of complete forms of CA is generally easy and may be reached by means of the different findings described above. The recognition of partial forms of CA, however, is often difficult. In these cases, the shape of the lateral ventricles may appear only slightly abnormal with a moderate posterior dilatation, the pericallosal artery may keep a normal course around the genu, which is more frequently preserved, and the third ventricle is usually normal. The radial arrangement of cerebral sulci in some portions of medial hemispheric borders is often the most significant sign of partial CA (Couture and Cadier, 1983).

The sonographic diagnosis of partial and complete forms of CA appears reliable. Most of the cases described in the literature had their diagnosis confirmed by CT (Skeffington, 1982; Cadier et al., 1982; Gebarsky et al., 1984; Babcock, 1984) or post-mortem analyses (Babcock, 1982; Gebarsky et al., 1984). Ten out of the 11 cases observed in our Clinic were submitted to CT and/or MRI, one baby having escaped our observation after the neonatal period. A CT-scan was carried out in four patients and an MRI in two cases. The other four babies underwent both examinations (see Table 1). CT and MRI controls were also carried out after the neonatal period except in two cases. In all patients the sonographic diagnosis of partial (2 cases) or complete (8 cases) CA was confirmed. One baby died during the third month of life and the autopsy was consistent with neonatal US findings.

US IN CA ASSOCIATED WITH OTHER CEREBRAL MALFORMATIONS

A very high incidence of associated CNS malformations has been reported in CA. Jellinger et al. (1981) examined 50 autopsy cases of CA and 44 of them proved to have other cerebral abnormalities. These consisted of hydrocephalus (22 cases), microcephaly (19 cases), dysgyrias (19 cases), rhinencephalic defects (16 cases), absence of forebrain commissures and other dysraphic changes, lobar hypoplasia and arachnoid cysts (6 cases) and other rarer malformations such as cerebellar and brainstem dysplasia, optic nerve and tract abnormalities, encephalocele and choroid plexus lesions. The authors also included as an associated malformation the presence of Probst's longitudinal bundle (detected in 32 cases). A wide range of facial and non-nervous system abnormalities (affecting the musculo-skeletal, genito-urinary, cardiovascular and gastrointestinal systems) were also reported.

Associated CNS malformations such as dysgyrias or grey matter heterotopias are not always revealed by US (see case 1). These abnormalities are usually detected by CT-scan and especially by MRI. Other malformations do not modify the typical sonographic picture of CA and can easily be revealed by US. Posterior fossa abnormalities such as a large cisterna magna (Gebarsky et al., 1984) and arachnoid cysts in the anterior temporal lobe (Couture and Cadier, 1983) belong to this class. A last group is constituted by malformations which, when associated with CA, give rise to complex US pictures. This class includes hydrocephalus (Babcock, 1984; Couture and Cadier, 1983), interhemispheric abnormalities such as cavum velum interpositum (Gebarsky et al., 1984), lipoma (Babcock, 1984; Couture and Cadier, 1983), arachnoid cyst (case 1) and also complex abnormalities such as septo-optic dysplasia (Gebarsky et al., 1984), encephalocele (Babcock, 1984) or mild forms of lobar holoprosencephaly (case 7). In these patients the detection of CA features may be difficult, particularly when the ventricular system is displaced and deformed by the associated malformations. Partial or complete CA can then be revealed by the presence of particular signs, such as the missing pericallosal artery and the radial arrangement of medial cerebral sulci in some portions of the interhemispheric fissure.

In our patients, associated CNS malformations were detected in two cases. One patient (case 1) with complete CA showed a large grey matter heterotopia in the centrum semiovale of the left hemisphere and a large interhemispheric arachnoid cyst located in the left parietal area and communicating with the subarachnoidal space. The latter malformation was recognized by means of US during the neonatal period (Figure 4A), whereas heterotopia was detected only by a subsequent MRI examination (Figure 4B).

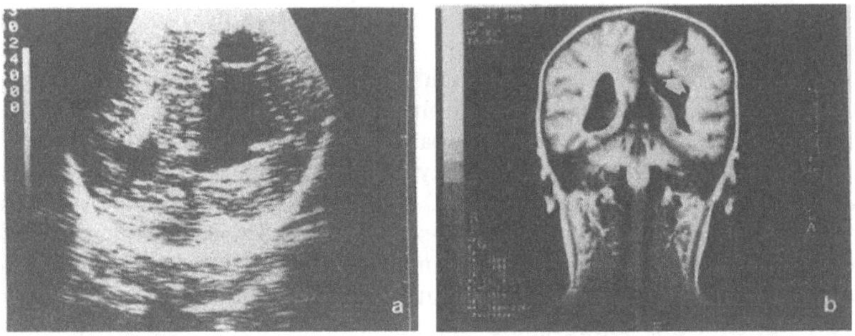

Figure 4A. Coronal posterior US scan of another newborn with CA; large paramedial anechoic image, colpocephalic aspect of the right ventricle. **B.** MRI of the same infant at the age of 6; coronal posterior scan showing a midline arachnoid cyst, and a grey matter heterotopia on the left hemisphere (arrow).

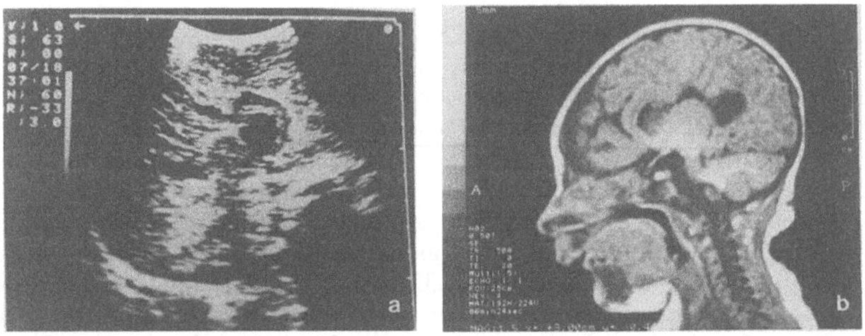

Figure 5A. US midsagittal scan of a newborn with partial CA associated with lobar holoprosencephaly; the CC is partially absent as also indicated by the radial arrangement of the medial cerebral sulci; a duct is connecting the posterior horns of the lateral ventricles. **B.** MRI of the same infant at 2 weeks; midsagittal scan confirming US findings.

The sonographic picture of the other patient (case 7) was characterized by a partial form of CA associated with a duct connecting the posterior portions of the lateral ventricles (Figure 5A). This complex abnormality was interpreted as a mild form of lobar holoprosencephaly. These findings were later confirmed by MRI (Figure 5B).

CONCLUSION

US is a very reliable tool for the detection of CA. Other neuroradiological techniques and postmortem findings have always confirmed the diagnosis, as reported in the literature and as shown in our cases. Nevertheless, an MRI examination is generally necessary to exclude the presence of associated malformations which are frequent and not always detectable by the sole means of sonographic scanning.

The large use of cranial US in the neonatal period may provide new data on the incidence of CA in the general population. The statistics reported in different studies vary from 1/19000 based on post-mortem examinations (Grogono, 1968) to 2% of mentally retarded patients (Jeret et al., 1987). In our clinic, 11 cases of partial or complete CA were detected among 1359 infants examined (0.81 %) in the period between January 1, 1985 and July 31, 1991.

These data were obtained by performing US on almost all premature infants and on fullterm neonates presenting CNS risk condition (i.e. perinatal asphyxia, low birth weight, malformations etc.). Some infants with CA may thus have escaped our observation. Comparison with the studies reported above is difficult. A more systematic use of cranial sonographic scanning in different NICUs may provide more reliable data about the real incidence of this malformation in the future.

Neonatal diagnosis of CA by means of US allows a follow-up of the neurological and cognitive development of these subjects from the first months of life. As already reported, two of our patients presented associated CNS malformations (cases 1 and 7). Four subjects (cases 2, 5, 8 and 11) were affected with non-nervous system abnormalities concerning cardio-vascular, muscolo-skeletal and urinary apparatus. Four infants (cases 5, 6, 9 and 11) showed cranial and facial dysmorphism (i.e. microcephaly, hypertelorism and micrognatia). One infant (case 9) died when he was three months old. He had inadequate sucking and hypertonia from the second week of life and opistotonus from the second month. MRI confirmed CA and excluded the presence of other abnormalities. All laboratory tests were normal, including karyotype and blood and urine metabolic screening. Autopsy confirmed the US and MRI findings but did not add any clue to the understanding of the pathogenesis. Two of the remaining 10 patients escaped our observation after the neonatal period (cases 4 and 11) and one at six months of age (case 10). This last patient had shown a normal neurological development until that time.

The other 7 infants have been followed during the first years: the age of their follow-up ranges from 3 to 6 years. Three of them (cases 1, 2 and 3) appear mentally normal, one presents a cognitive development at the lower limit of the normal range (the infant affected with a mild form of lobar holoprosencephaly - case 6), whereas the other four showed a developmental retardation. One of these infants has partial CA (case 8). Cases 2 and 5 present a dystonic form of cerebral palsy. Case 1, with a complex malformation including CA, grey matter heterotopia and interhemispheric arachnoid cyst, shows a mild right hemiplegia and focal epileptic seizures. These data confirm the different outcome of patients affected with CA.

Acallosal infants represent a high-risk population for mental retardation and other neurologic and psychiatric disorders. Neonatal diagnosis of CA by means of US will allow more detailed studies of the early development of these patients and facilitate the creation of specific intervention programs.

Acknowledgments

The authors wish to thank Dr. R. Canapicchi who performed most of the MRIs on our infants and P. Freer for reviewing the text. This study was supported by grants of the Italian Ministry of Health.

REFERENCES

Atlas, S.W., Zimmerman, R.A., Bilaniuk, L.T., et al., 1986, Corpus Callosum and Limbic System: neuroanatomic MR evaluation of developmental anomalies, *Radiology* 160(2):355-62.

Baarsma, R., Martijn, A., and Okken, A., 1987, The missing pericallosal artery on sonography: a sign of agenesis of the corpus callosum in neonatal brain?, *Neuroradiology* 29:47-9.

Babcock, D.S., 1984, The normal, absent and abnormal Corpus Callosum: sonographic findings, *Radiology* 151(2):449-53.

Barkovich, A.J., 1990, "Pediatric Neuroimaging," Raven Press, New York.

Bull, J., 1967, The corpus callosum, *Clin Radiol.* 18:2-18.

Cadier, L., Souyris, S., and Diard, F., 1982, Agénésie complète du corps calleux: aspect échographique, *J. Radiology (Paris)* 63:429-32.
Cioni, G., Bartalena, L., Biagioni, E., Boldrini, A, and Canapicchi, R., 1992, Neuroimaging and functional outcome of neonatal leukomalacia, *Behav. Brain Res.* 49:7-19.
Couture, A. and Cadier, L., 1983, "Echographie Cérébral par Voie Transfontanellaire," Vigot, Paris.
Davidoff, L.M. and Dyke, C.G., 1934, Agenesis of the corpus callosum. Its diagnosis by encephalography. Reports of three cases, *AJR* 32:1-10.
Davidson, H.D., Abraham, R., and Steiner, R.E., 1985, Agenesis of corpus callosum: magnetic resonance imaging, *Radiology* 155(2):371-3.
De Vries, L.S., Dubowitz, L.M.S., Pennock, J.M, and Bydder, G.M., 1989, Extensive cystic leukomalacia: correlation of cranial ultrasound, magnetic resonance imaging and clinical findings in sequential studies, *Clin. Radiol.* 40:158-66.
De Vries, L.S., Wigglesworth, J.S., and Dubowitz, L.M.S., 1988, The natural evolution of periventricular leukomalacia during the neonatal period and infancy: correlation of imaging and postmortem findings, *Early Hum. Dev.* 17:205-19.
Gebarsky, S.S., Stiennon, Gebarsky, K., Bowermann, R.A,. and Silver, T.M., 1984, Agenesis of the corpus callosum: sonographic features, *Radiology* 151:443-8.
Grogono, J.L., 1968, Children with agenesis of the corpus callosum, *Dev. Med. Child Neurol.* 10:613-6.
Guibert-Trainer, F., Piton, J., Billerey, J., and Caill, J.M., 1982, Agenesis of the corpus callosum, *J. Neuroradiol.* 9:135-60.
Jellinger, K., Gross, H., Kaltenback, E., and Grisold, W., 1981, Holoprosencephaly and Agenesis of the corpus callosum: frequency of associated malformations, *Acta Neuropathol.* (Berlin) 55:1-10.
Jeret, J.S., Serur, D., Wisniewski, K.E, and Lubin, R.A., 1987, Clinicopathological findings associated with agenesis of the corpus callosum, *Brain Dev.* 9(3):255-64.
Kendall, B.E., 1983, Dysgenesis of the corpus callosum, *Neuroradiology* 25:239-56.
Skeffington, F.S., 1982, Agenesis of the corpus callosum: neonatal ultrasound appearance, *Arch. Dis. Child* 57(9):713-4.
Stewart, A.L., Reynolds, E.O.R., Hope, P.L., et al., 1987, Probability of neurodevelopmental disorders estimated from ultrasound appearance of brains of very preterm infants, *Dev. Med. Child Neurol.* 29:3-11.
Wing, S.D. and Osborn, A.G., 1977, Normal and pathological anatomy of the corpus callosum by computed tomography, *J. Comput. Tomogr.* 1:183-92.

CT FINDINGS IN CALLOSAL AGENESIS

Denis Melanson, Armando Salazar,
and Raquel del Carpio-O'Donovan

Department of Radiology
The Montreal Neurological Institute and Hospital
Mc Gill University
3801 University St.
Montreal, Que., H3A 2B4
Canada

INTRODUCTION

Absence of the corpus callosum (ACC) represents a complex malformation of neural development, which may occur in isolation or in combination with other cerebral anomalies (Barkovich, 1991). Most cases are sporadic but familial forms do exist. Abnormalities in the limbic formations and the anterior commissure may accompany deficiencies in the formation of the corpus callosum, as they all relate to a common ancestor, the lamina terminalis: crossing of fiber tracts proceeds from the anterior commissure to the hippocampal structures, and then to the corpus callosum itself.

The enigmas of the development and function of the corpus callosum still persist (Larsen and Osborn, 1982), but better in vivo imaging has improved our understanding of this complex anomaly. Although Magnetic Resonance (MR) has become the examination of choice in the investigation of cerebral abnormalities, such as those of the corpus callosum, Computed Tomography (CT) has retained a major role and still can achieve the diagnosis of callosal agenesis.

CT FINDINGS IN CALLOSAL AGENESIS

The standard protocols for head examination consist of 10mm transverse (axial) sections from the base of the skull to the vertex. These will suffice, in most instances, to demonstrate callosal agenesis and many of possible associated anomalies. When the anomaly is suspected, additional sections can be obtained. These sections consist of further axial slices, thinner, with overlap, and of coronal sections. The latter technique is only possible if the patient can maintain the difficult position of cervical hyper-extension. If not, the former technique can provide data for coronal and sagittal reformatting. Newer CT technology can provide all of these possibilities in a shorter period of time, thus reducing the patient's discomfort associated with position.

Callosal agenesis results in several anatomical features that are distinct from the normal configuration. Axons that usually cross the interhemispheric fissure within the

Callosal Agenesis, Edited by M. Lassonde and
M.A. Jeeves, Plenum Press, New York, 1994

corpus callosum, instead form fiber bundles that run medially to the medial walls of the lateral ventricles, parallel to the fissure. These bundles are known as the longitudinal callosal bundles, or the bundles of Probst. Another characteristic feature of callosal absence is eversion of the cingulate gyrus. Normally, the crossing callosal fibers produce a superomedial displacement of the cingulate gyri, the end result being a cingulate gyrus that is oriented differently to the interhemispheric fissure. This normal inversion results in the formation of the cingulate sulcus. When the corpus callosum is absent, the cingulate gyri remain everted, and the cingulate sulci remain unformed. As a result, the course of the mesial interhemispheric sulci is uninterrupted from the third ventricle to the periphery, in a radial manner (Pomeranz, 1989; Barkovic, 1991; Mori, 1992).

Computed tomography in the axial plane demonstrates a "Batwing" appearance of the frontal horns (Figure 1 and Figure 2). The frontal horns also appear narrow, unless there is dilatation secondary to hydrocephalus or associated with other anomalies. The bodies of the lateral ventricles and the frontal horns have sharply angled lateral beaks: this is better defined in the coronal sections (Figure 2). Because of the long callosal bundles running longitudinally, the medial walls of the bodies of the lateral ventricles are concave.

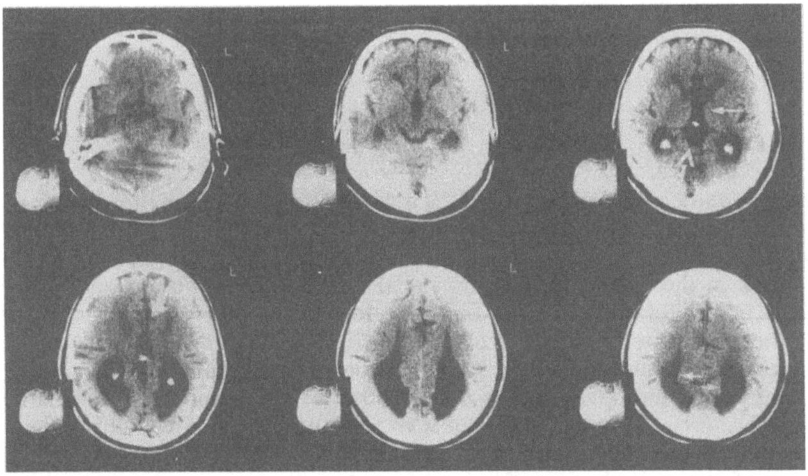

Figure 1. Callosal Agenesis: sequential transverse scans (thickness 10 mm). Wide third ventricle (←); Prominent interhemispheric fissure (↓); Colpocephaly (← →); Prominent quadrigeminal cistern (→).

Interposition of the third ventricle between the bodies of the lateral ventricles is an important finding: usually the third ventricle is also wider than normal (Figure 2). The height of the interposed third ventricle may vary greatly, sometimes extending upward between the interhemispheric fissure to the vertex. Although arachnoid cysts can occur in the interhemispheric fissure in association with agenesis of the corpus callosum, most commonly this collection of fluid is lined by ependymal cells communicating with the third ventricle, or with one or both of the lateral ventricles (Barkovich, 1991). If the falx cerebri is hypoplastic or locally absent, large collections of cerebrospinal fluid at that level may result from the dilatation of the fissure itself (Mori, 1992)

A prominent suprapineal recess is usually present; at times, it may be dilated and may extend upwards towards the vertex (Larsen and Osborn, 1982). Conversely, the superior (quadrigeminal) cistern may act similarly together with, or at the expense of, the recess (Figure 1).

The occipital horns may show relative dilatation when compared to the width of the bodies (Mori, 1992). In fact, when the corpus callosum is absent, the posterior bodies, trigones, occipital horns and posterior temporal horns of the lateral ventricles are bordered only by loose and deficient white matter. The ventricles therefore expand superiorly and laterally, with a resulting dilatation of the trigones and occipital horns. This configuration of the ventricles has been termed colpocephaly (Figure 1). It can also be encountered when the corpus callosum is hypoplastic, or partially absent (hypogenetic), when the splenium is absent. CT axial views are however far from optimal in this situation. Good reformatted sagittal views may better reveal the anomaly.

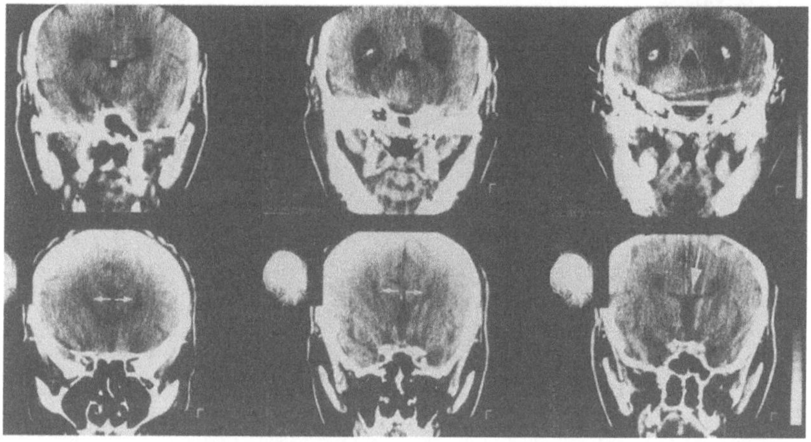

Figure 2. Callosal Agenesis: sequential coronal scans (thickness 8 mm). High position of third ventricle (↓); Batwing appearance of frontal horns (← →).

Finally, there can also be absence or hypoplasia of the fornix with complete absence of the corpus callosum but CT cannot show it accurately.

ASSOCIATED ANOMALIES

Agenesis of the corpus callosum has been reported with a wide variety of cerebral anomalies (Barkovich and Norman, 1988; Zuluaga and del Carpio-O'Donovan, 1991). For instance, ACC can be associated with migrational abnormalities, Arnold-Chiari and Dandy-Walker malformations, schizencephaly and holoprosencephaly. Interhemispheric lipoma is also a known association that may coexist with the corpus callosum so that the latter is dysgenetic, but not absent (Figure 3).

In septo-optic dysplasia, ACC may occur in addition to optic atrophy, absence of the septum pellucidum and hypopituitarism (Manelfe and Rocchiccioli, 1979). It can also be present in Aicardi's syndrome (Aicardi, Chevrié and Rousselle, 1969).

Agenesis of the corpus callosum has also been associated with cerebellar vermian hypoplasia (Joubert, Eisenreing, Robb and Andermann, 1969) resembling the Dandy-Walker syndrome. An extensive report (62 patients) of Andermann syndrome was made by Filteau et al. (1991). They concluded that there is an association between cerebellar anomalies and the schizophrenia-like syndrome in these patients, but ruled out an implication of the callosal defects (Mathieu et al., 1990; Filteau et al., 1991).

Most of these anomalies can be suspected by CT, on axial or coronal images. When more subtle anomalies are present, such as migration disorders (heterotopias and cortical dysplasias), CT has limitations, and may overlook them (Figure 4).

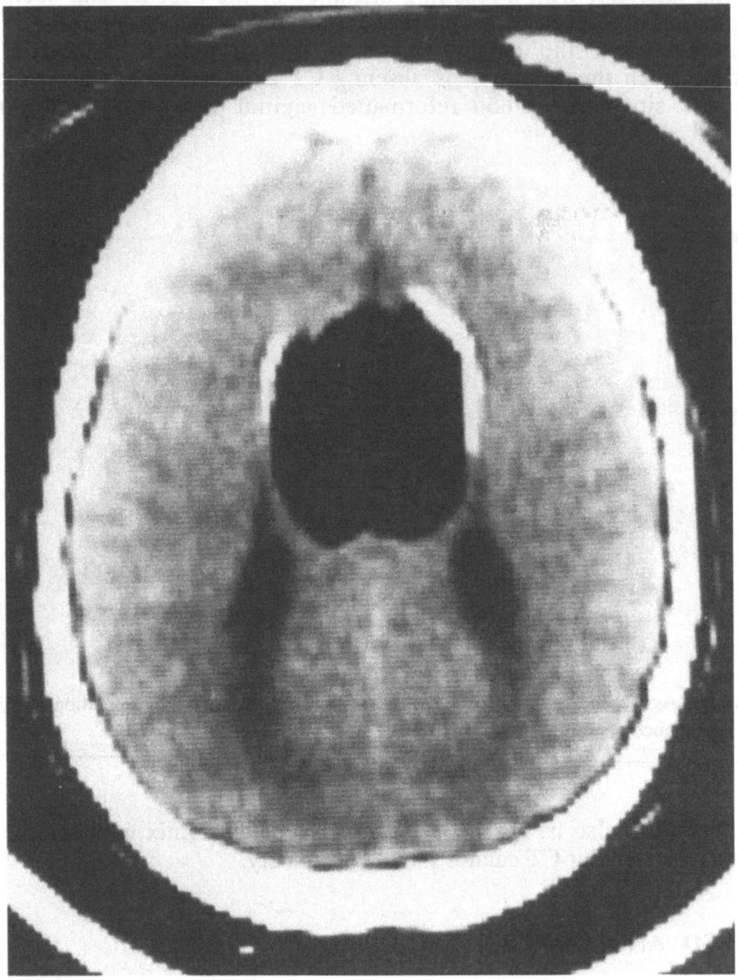

Figure 3. Classical appearance of lipoma associated with callosal dysplasia and cortical dysplasias; CT has limitations and may overlook cortical dysplasias (Figure 4).

SUMMARY OF FINDINGS

The classical CT findings of callosal agenesis can be summarized as follows (Figure 1 and Figure 2):

1. Separation of the lateral ventricles and absence of the septum pellucidum
2. Batwing and concave appearance of the frontal horns, due to indentation by Probst's bundles and eversion of cingulate gyri

3. Widening and sometimes elongation of the foramina of Monro
4. Widening and ascent of the third ventricle, to a variable degree, sometimes extending in the interhemispheric fissure
5. Dilatation of the trigones and occipital horns
6. Wide superior cistern, due to the absence of splenium
7. Wide interhemispheric fissure, due to hypoplasia of falx

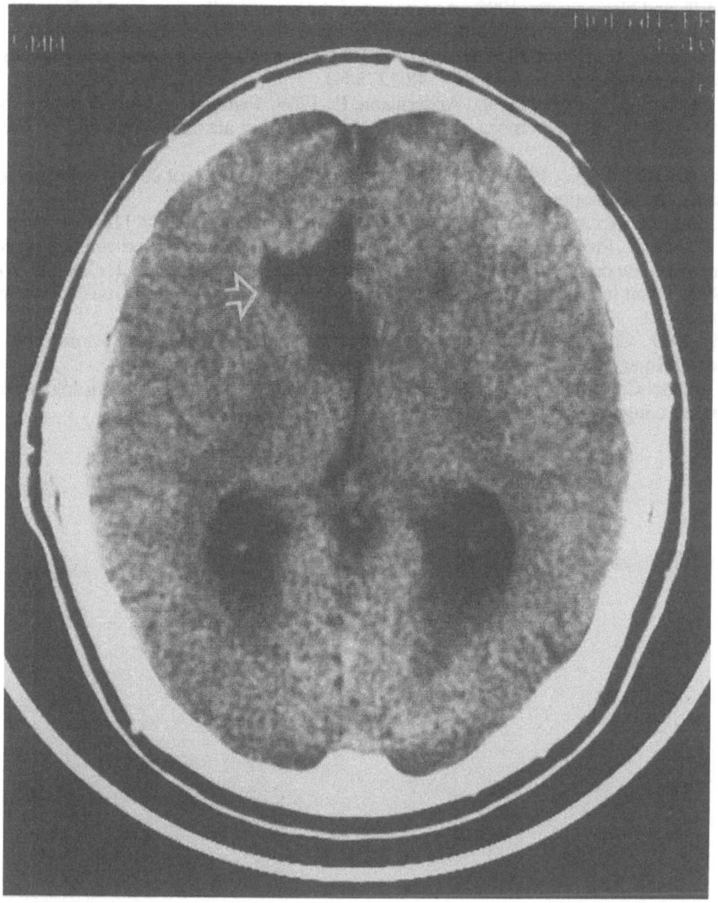

Figure 4. Callosal Agenesis: with associated cortical dysplasia, right mesial frontal (→).

CONCLUSION

Callosal agenesis has interested brain scientists for many years, but more so recently with the advent of more sophisticated clinical tests and also new imaging modalities. What pneumo-encephalography could already show decades ago, Computed Tomography (CT) and more recently Magnetic Resonance (MR) are now disclosing with more accuracy and with much less discomfort to the patient. Whereas MR appears superior for all varieties of the anomaly, CT still allows diagnosis of the classical agenesis. Newer improvements in scanning time and reformatting capabilities now offer better possibilities then ever. But as MR becomes more and more available, it will replace CT as the examination of choice for all suspected cerebral dysplasias.

REFERENCES

Aicardi, J., Chevrie, and J.J., Roussell, F., 1969, Le syndrome agénésie calleuse, spasmes en flexion, lacunes chorio-rétiniennes, *Arch. Fr. Pediatr.* 26:1103-1120.

Barkovich, AJ. 1991, Brain development: normal and abnormal, *in:* "Magnetic Resonance Imaging of the Brain and Spine," S.W. Atlas, ed., Raven Press, New-York.

Barkovich, A.T. and Norman, D., 1988, Anomalies of the corpus callosum: correlation with further anomalies of the brain, *AJR* 151:171-9.

Filteau, M.J., Poucher, E., Bouchard, R.H., et al., 1991, Corpus callosum agenesis and psychosis in Andermann syndrome, *Arch. Neurol.* 48:1275-80.

Joubert, M., Eisenreing, J., Robb, P., and Andermann, F., 1969, Familial agenesis of the cerebellar vermis. A syndrome of episodic hyperpnea, abnormal eye movements, ataxia and retardation, *Neurology* 19:813-25.

Larsen, P.D. and Osborn A.G., 1982, Computed tomographic evaluation of corpus callosum agenesis and associated malformations, *J. Comput. Assist. Tomogr.* 6:225-230.

Manelfe, C., and Rocchiccioli, P., CT of septo-optic dysplasia, *AJR* 133:1157-1160.

Mathieu, J., Bédard, F., Prévost, C., and Langevin, P., 1990, Neuropathie sensori-motrice héréditaire avec ou sans agénésie du corps calleux: étude radiologique et clinique de 64 cas, *J. Can. Sci. Neurol.* 17:103-8.

Mori, K., 1992, Giant interhemispheric cysts associated with agenesis of the corpus callosum, *J. Neurosurg.* 76(2):224-30.

Pomeranz, S., 1989, Congenital pediatric anomalies of the brain and spine, *in:* "Craniospinal Magnetic Resonance Imaging," L. Bralow, ed., W.B. Saunders, Philadelphia.

Zuluaga, A. and del Carpio-O'Donovan, R., 1991, Cuerpo calloso: desarrollo embriológico, mielinización y anomalías congenitas, *Rev. Mex. Radiol.* 45 *(Suppl. 2)*:121-6.

MAGNETIC RESONANCE IMAGING OF CORPUS CALLOSUM DYSGENESIS

Ronald A. Rauch and J. Randy Jinkins

Neuroradiology Section
Department of Radiology
The University of Texas Health Science Center
San Antonio, Texas 78284-7800, U.S.A.

INTRODUCTION

The normal corpus callosum is probably best viewed in the midline sagittal section (Figure 1). This plane can be achieved by neuropathologists at brain cutting or, noninvasively, by Magnetic Resonance Imaging (abbreviated MRI or simply MR). By comparison, using Computed Tomography (CT) it is generally physically impossible to produce direct midsagittal images. Ultrasound, another commonly used imaging modality, can image patients in the midsagittal plane but requires a sonographic "window" through the skull. This is only available naturally in infants with fontanels which are still open. MR, on the other hand, allows high resolution and high contrast images in multiple directly acquired planes in patients of all ages. For this reason, MR is by far the most superior tool currently available for evaluation of the corpus callosum in its normal and pathologic state. While it is true that neuropathologists are able to obtain very high resolution images of the brain, such studies are not available until after the patient's demise. Obviously at that time it is too late to administer neuropsychological tests to these patients to investigate their callosal function. As will be seen, MR not only allows the identification of abnormal development of the corpus callosum in the living patient, but allows accurate delineation of the extent of callosal abnormality as well. Another advantage over neuropathology is that MR can perform direct acquisitions in the axial, coronal, and sagittal planes. Once the neuropathologist has made the standard cuts in the axial (or coronal) plane, it is often difficult to piece the brain back together to produce precise sagittal images in the case of an incidental finding such as callosal dysgenesis.

The normal anatomy of the corpus callosum as it is seen on midsagittal MR images is shown in Figure 1A. It is generally artificially divided into four sections (shown in the drawing, Figure 1B): the genu, the body, the splenium and the rostrum, listed in the order in which they form embryologically. Other related structures which are normally visible on MR and which are included in the discussion are the anterior commissure, the columns of the fornix, and the cingulate gyrus. The third ventricle is also shown in its normal position beneath the corpus callosum.

Callosal Agenesis, Edited by M. Lassonde and
M.A. Jeeves, Plenum Press, New York, 1994

While most individuals have a corpus callosum which looks very similar to Figure 1, there are a number who reveal anomalies of callosal formation. These anomalies in the past were often lumped into the single category of "agenesis of the corpus callosum". We believe that these developmental anomalies can better be thought of as a continuing spectrum of anomalies. We have designated these anomalies dysgenesis of the corpus callosum. This difference between the term dysgenesis and agenesis is more than simple semantic quibbling, since the MR appearance of the corpus callosum allows callosal anomalies to be divided into one of three distinct radiological categories: agenesis, hypogenesis, and hypoplasia (Jinkins et al., 1989).

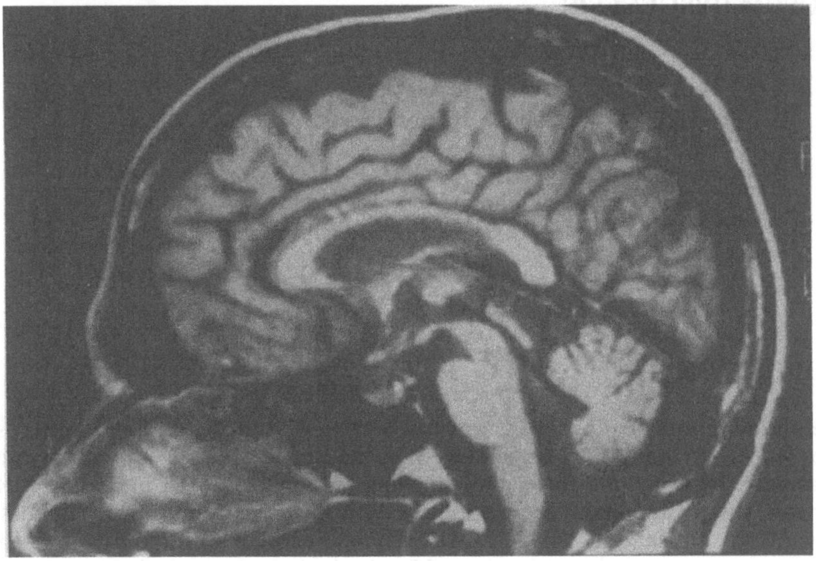

Figure 1A. MR, midsagittal plane, showing normal corpus callosum.

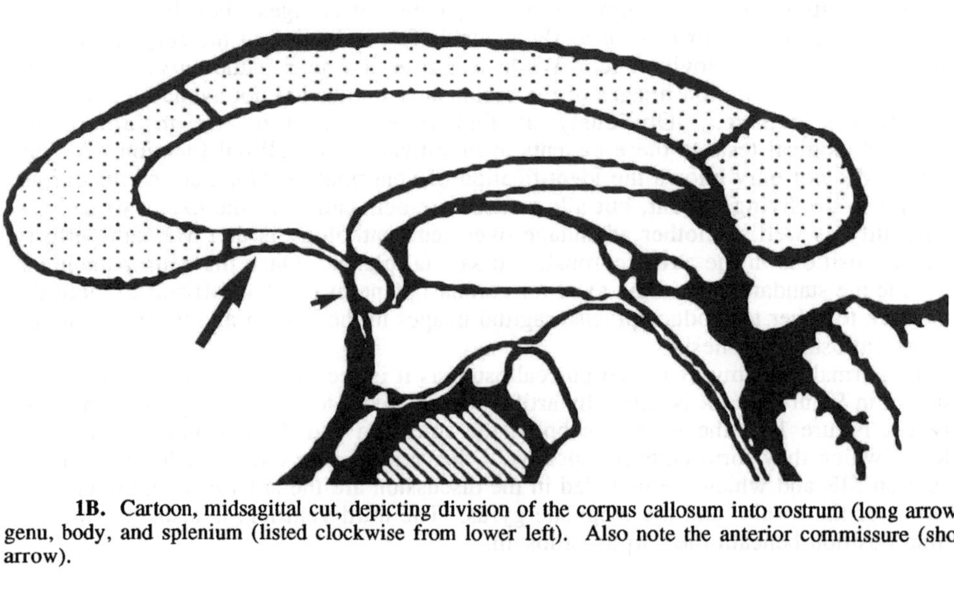

1B. Cartoon, midsagittal cut, depicting division of the corpus callosum into rostrum (long arrow), genu, body, and splenium (listed clockwise from lower left). Also note the anterior commissure (short arrow).

Each of these anomalies has a characteristic appearance in an MR image of the corpus callosum. Each category is also likely to exhibit a unique neuropsychological profile, based on the degree of corpus callosum anomaly as well as the presence and degree of other anomalies within the brain. Attempting to understand the neuropsychological and neurological implications of this anatomic spectrum of callosal dysgenesis is likely to be hampered if these differences are ignored and all are simply grouped together into one category.

AGENESIS OF THE CORPUS CALLOSUM

Agenesis of the corpus callosum simply means that there is no identifiable corpus callosum present. MR allows direct visualization of the normal corpus callosum, and thus also readily shows its absence in agenesis (Figures 2 and 3). MR also enables the identification of other anatomical anomalies which are frequently associated with the failure of development of the corpus callosum. The third ventricle can be seen to balloon upwards and it is difficult or impossible to separate the third ventricle from the interhemispheric fissure on MR. The cingulate gyrus and cingulate sulcus appear to be deficient, with the midline sulci radiating directly into the region of the third ventricle. On the coronal images (Figure 3) the cingulate gyrus is seen to be present, but rotated to face downward. This malposition is thought to be due to the lack of characteristic physical forces normally exerted by the corpus callosum at the time the cingulate gyri and sulci are forming. The fornices also appear to be absent, but as with the cingulate gyrus they are just misplaced. Because they are fairly small structures, and because they are widely separate in agenesis of the corpus callosum, they are generally difficult to identify on MR in patients with callosal agenesis. The fornices, however, can usually be identified at brain cutting by neuropathologists.

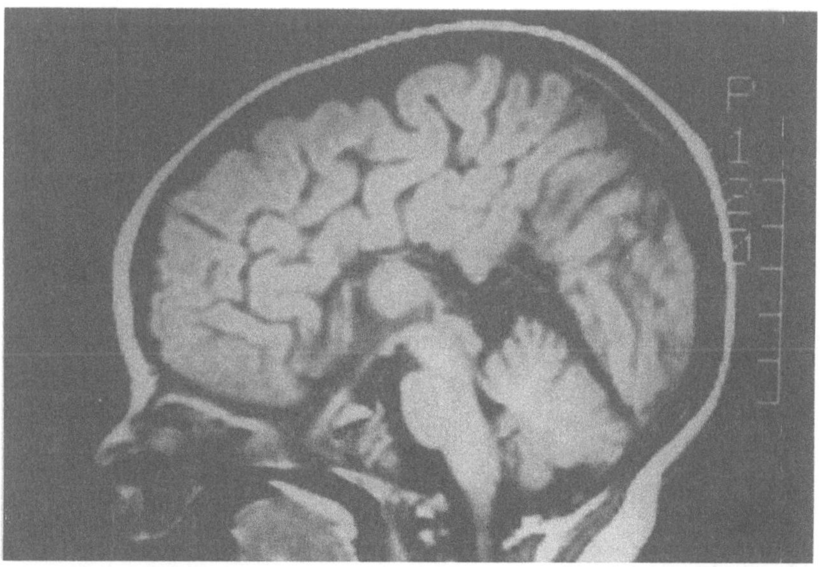

Figure 2. MR, midsagittal plane, showing corpus callosum agenesis, with high riding 3rd ventricle, and apparent absence of the cingulate gyrus. Also note that no anterior commissure can be identified.

Patients with agenesis of the corpus callosum typically have white matter tracts of varying size located just medial to the lateral ventricles. They were first described by Probst and were named after this early investigator. These Probst bundles, which are not found in individuals with a normal corpus callosum, are best seen on coronal MR images (Figures 3 and 4) but can be seen on axial images as well if they are large. The Probst bundles appear to contain the axonal fibers which were intended to have crossed the midline in the corpus callosum during fetal brain development. However, because the corpus callosum is absent, the axons follow a course paralleling the lateral ventricles. If a count of the axons present is made, the Probst bundles do not contain nearly sufficient axons to account for the total which would have typically formed the corpus callosum. Presumably a minority of the axons persist to form the Probst bundles, and the remainder disappear. This may be similar (but greater in magnitude) to the normal process of axonal diminution which occurs during development of the normal corpus callosum. In laboratory animals up to 70% of the original axons disappear from the corpus callosum during the process of brain maturation (LaMantia and Rakic, 1990a).

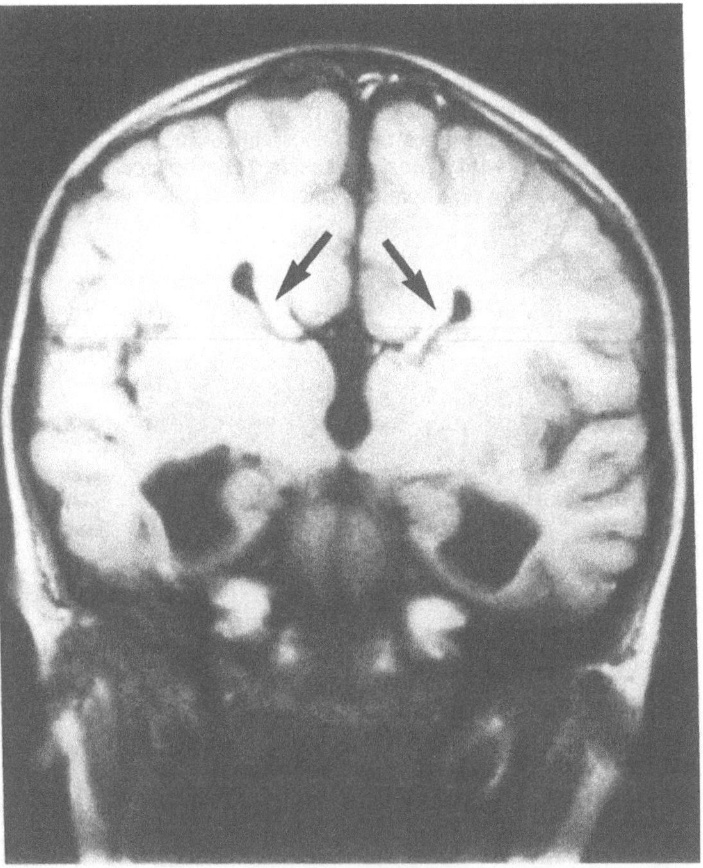

Figure 3. MR, coronal plane, showing callosal agenesis and Probst bundles (arrows). Note the abnormal position of the cingulate gyrus as well.

Agenesis of the corpus callosum is frequently associated with some degree of enlargement of the occipital and temporal horns of the lateral ventricles. This anomaly, called colpocephaly, represents a passive dilatation of the ventricles due to the loss of support normally provided by axons which become part of the corpus callosum (Atlas et al., 1986). This concept of axonal involution in callosal agenesis is consistent with the explanation suggested above for the discrepancy in the number of axons present in typical Probst bundles as compared to the total number of fibers seen in a normal corpus callosum. In addition, we have observed a tendency to find the degree of colpocephaly to be inversely related to the size of the Probst bundles, further suggesting that colpocephaly is the result of a relative paucity of axons around the lateral ventricles.

An alternate explanation for colpocephaly is that the axonal drop-out is the result of cortical maldevelopment. As will be discussed later, cortical migrational anomalies may cause fewer axons to be present in the corpus callosum. This is true both in the midline (where one typically views the corpus callosum) and laterally (where the axon loss could cause colpocephaly) This is supported by the findings of Ludwin and Norman, (1985).of an association between colpocephaly and polymicrogyria (a cortical migrational disorder). Since the migrational anomalies are often hard to diagnose on MR, colpocephaly, especially marked colpocephaly, might suggest that there are additional cerebral abnormalities present beyond agenesis of the corpus callosum.

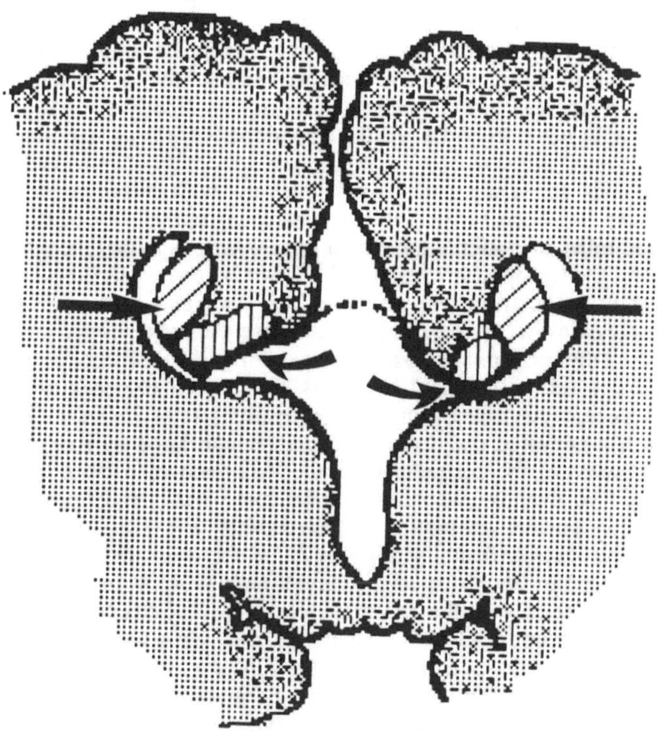

Figure 4. Drawing in coronal plane illustrating Probst bundles, shown by diagonal lines (straight arrow) and cingulate gyri, shown by vertical lines (curved arrow).

HYPOGENESIS OF THE CORPUS CALLOSUM

The embryological development of the corpus callosum begins with the formation of the lamina reuniens during the seventh week of fetal life. This structure acts as the precursor to both the anterior commissure and the corpus callosum (Barkovich et al., 1988). By the twelfth week of fetal development, the first axonal fibers can be found in the genu, the first part of the callosum to form. Development of the corpus callosum then progresses caudally until the splenium has formed by the eighteenth week of gestation. The rostrum, the most anterior portion of the corpus callosum, forms last, between weeks eighteen and twenty. Various insults to the developing fetus may result in absence of structures which were to form at the time of the insult. Thus an injury around the seventh week could damage the lamina reuniens, preventing the formation of not only the corpus callosum but also the anterior commissure. An injury between weeks seven and twelve would not interfere with formation of the anterior commissure but might result in total absence (i.e. agenesis) of the corpus callosum. Injuries between weeks twelve and twenty should not prevent formation of the genu of the corpus callosum but might result in lack of formation of other portions of the corpus callosum. This would cause the corpus callosum to be only partially present, a condition we term hypogenesis (Jinkins et al., 1989).

The MR in hypogenesis of the corpus callosum shows a corpus callosum which has a genu but is lacking other portions. Generally this involves the rostrum, the splenium and may involve a part of the body (Figures 5 and 6). We have chosen the term "hypogenesis" deliberately, even though most of medicine classifies these lesions as "partial agenesis" of the corpus callosum. The term hypogenesis is more correct and should denote that there is some corpus callosum present. Further, hypogenesis makes more semantic sense than partial agenesis since agenesis implies that there is no corpus callosum present at all, yet these patients clearly have a portion of their corpus callosum intact. Alas, the term "partial agenesis" is firmly rooted in the medical literature and it will not likely be easily abandoned without a concerted effort at correction.

Based on the MR appearance of several patients with hypogenesis of the corpus callosum, this entity has been further divided into interruptive and obstructive forms of hypogenesis. A number of patients appear to have a corpus callosum which is deficient in the portions which develop after fetal week 12 but before week 20. Thus they have a genu but are missing the rostrum, the splenium, and perhaps a portion of the body of the corpus callosum (Figure 5). It is believed that these patients simply suffered an insult after the genu of the corpus callosum had formed, resulting in an interruptive hypogenesis. This is felt to be identical in its mechanism to cases of agenesis of the corpus callosum except that the timing of the insult is later in fetal life. In contrast, some patients with a curtailed corpus callosum appear to have a mass which obstructed the caudal development of the corpus callosum. This mass (usually a lipoma) is seen to lie in the path of the corpus callosum and in some cases may envelop the corpus callosum as if the corpus callosum collided with the obstructing mass during embryogenesis. The rostrum usually appears to be intact in these subjects (Figure 6), suggesting that it was not an insult to the developing brain which interrupted callosal development, but rather it was the mass which simply obstructed the growth of the callosum. This preservation of the rostrum of the corpus callosum in obstructive hypogenesis should result in sparing of callosal function dependent upon the rostrum. This would potentially mean that neuropsychological studies might be able to separate these subjects from subjects with interruptive hypogenesis in whom the rostrum of the corpus callosum was absent. In both interruptive and obstructive forms of hypogenesis, Probst bundles can generally be seen although they are frequently smaller than seen in agenesis. This is not surprising if one believes that Probst bundles form from "callosal" axons which were unable to find a pathway across the midline. Since a portion of the callosum is present in hypogenesis, many neurons are able to cross the corpus callosum normally. Therefore, a smaller number of uncrossed axons remain (compared to callosal agenesis patients) to form the segmental Probst bundles seen in hypogenesis.

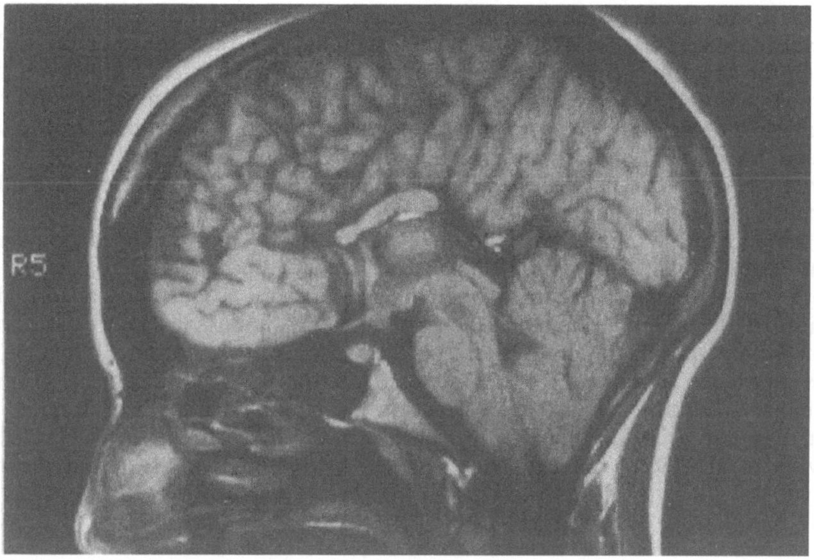

Figure 5. MR, midsagittal plane, showing interruptive type of hypogenesis of corpus callosum. Note that the splenium and rostrum are missing, as is a portion of the body.

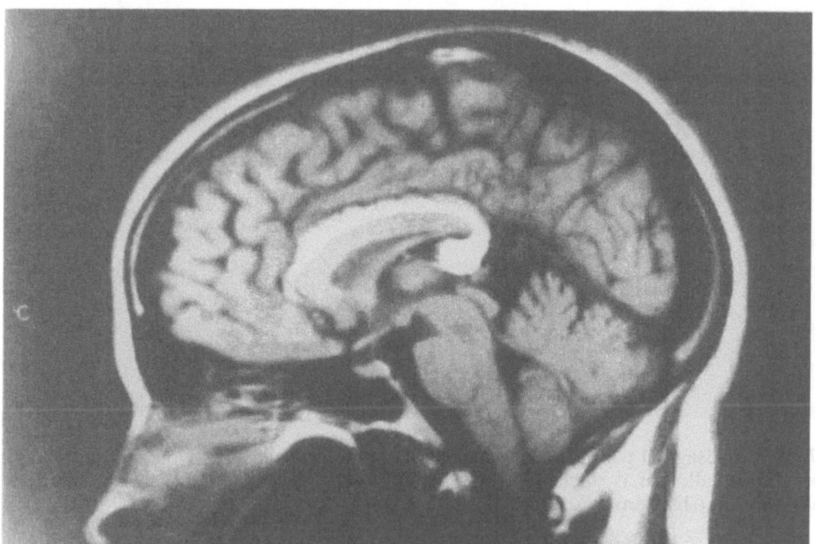

Figure 6. MR, midsagittal plane, showing obstructive hypogenesis. The lipoma appears as a very bright structure enveloping the corpus callosum on this MR image. The splenium is absent but the rostrum is present.

HYPOPLASIA OF THE CORPUS CALLOSUM

Agenesis and hypogenesis result when an insult occurs during the formation of the corpus callosum which yields a corpus callosum in which portions are frankly missing. However, if an insult to the developing fetus occurs after the corpus callosum has completely formed (that is, after 20 weeks gestational age), the entire corpus callosum, including the rostrum, genu, body, and splenium should be present. However, because the corpus callosum is not yet fully mature, the number of axons available to cross the midline in the corpus callosum may be reduced and the corpus callosum may not reach its normal adult thickness or cross-sectional area (Jinkins et al., 1989, Barkovich, 1990). We have termed this thinning of the corpus callosum hypoplasia of the corpus callosum (Figures 7, 8 and 9). In general, the mechanism responsible for such a thinning of the corpus callosum involves an injury which damages the developing cerebral cortex, the site of origin of the axons which will potentially cross in the corpus callosum. In the fetus, these cortical anomalies are usually produced by abnormalities in migration of the neurons from their origin in the germinal matrix to their intended location within the six ordered layers of the cortex. Interruption of this migration process produces a number of anomalies including: polymicrogyria (Figure 9), lissencephaly, schizencephaly, and various forms of gray matter heterotopias. These may or may not be evident on MR (Osborn et al., 1988, Titelbaum et al., 1989, and Barkovich et al., 1988) but can be diagnosed by histological means at the time of autopsy.

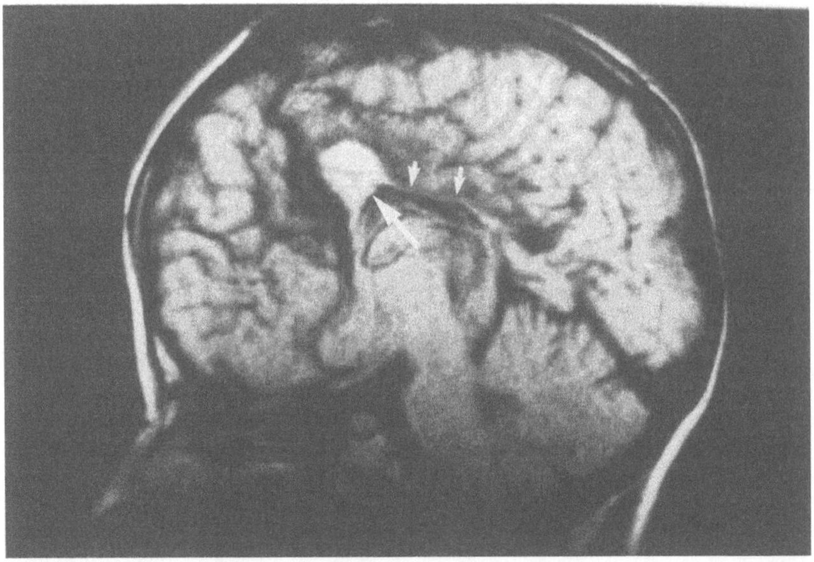

Figure 7. MR, midsagittal plane, showing hypoplasia of corpus callosum. Note that in the region of the genu, the corpus callosum is thickened (long arrow), while the rest of the corpus callosum appears abnormally thinned (short arrows).

The MR appearance of hypoplasia of the corpus callosum can be somewhat varied. Hypoplasia may either be focal or diffuse, depending on the extent of the cortical dysgenesis. In some cases of hypoplasia of the corpus callosum, irregular focal thickening as well as focal thinning has been observed (Figure 7). This suggests that there has been some migrational anomaly affecting the axons which are crossing in the

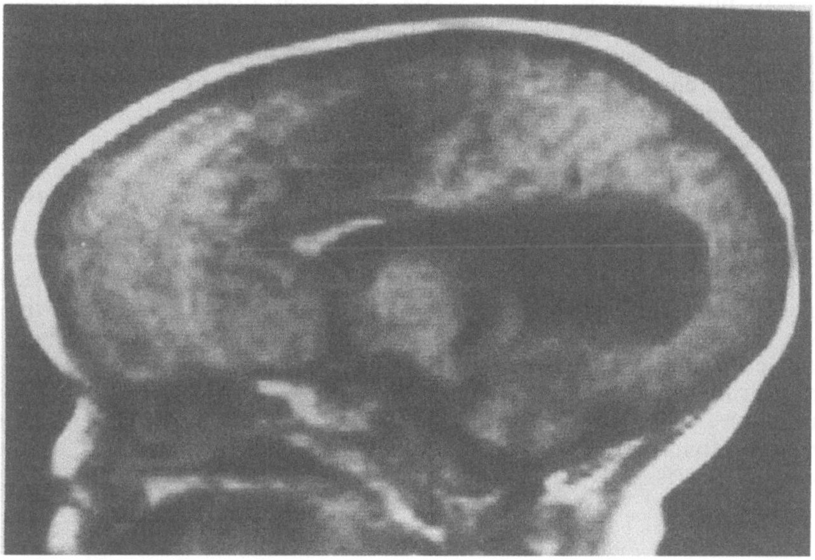

Figure 8. MR, midsagittal plane, showing hypoplasia of corpus callosum. Note that the corpus callosum is thinned. It also appears to be too short. However, this patient was twisted in the scanner, so that while the front of the brain is shown in the midline, the back is not. This gives the false impression that there is hypogenesis of the callosum as well (see Figure 9).

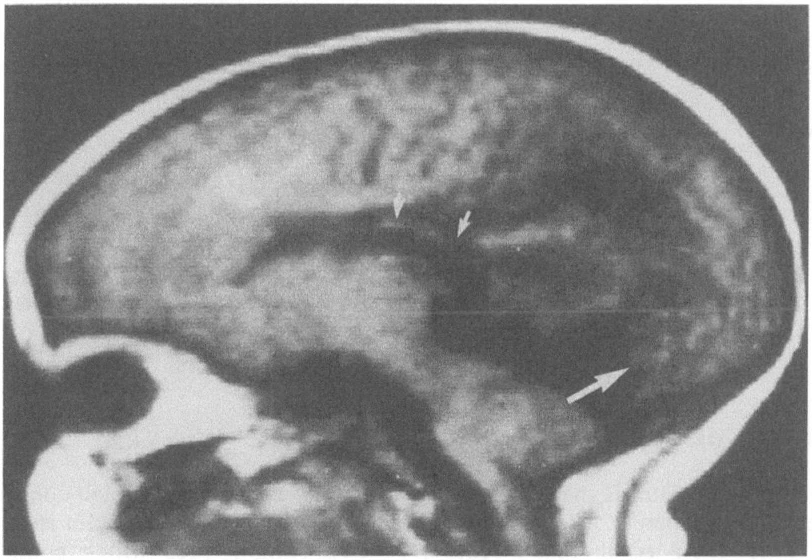

Figure 9. Same patient as in Figure 8, slightly to left of prior image. Now the posterior half of the callosum is seen (short arrow), hypoplastic, but not hypogenetic. Also note the suggestion of polymicrogyria present in the occipital lobe (long arrow). This diagnosis is difficult from the MR and is generally still best made by histopathological means.

corpus callosum, causing too many to be found at some sites and too few at others. Alternatively, this thickening could represent abnormal proliferation of the myelin or other glial elements which are also found within the corpus callosum.

Because the insult which results in pure hypoplasia of the corpus callosum appears to occur after the corpus callosum is formed (i.e. after 20 weeks), these patients should not have the degree of disconnection syndrome manifested by patients with agenesis and to a lesser extent in patients with hypogenesis. Instead one might expect to see neuropsychological findings which reflect the malformation of the cortex. The importance of hypoplasia, in relation to callosal agenesis or hypogenesis, is that hypoplasia not be confused with these other two entities, even though it can superficially resemble hypogenesis. This is especially true on MR images which are not obtained in a true sagittal plane (Figures 8 and 9). Finally, it should be realized that agenesis, hypogenesis, and cortical anomalies resulting in callosal hypoplasia are all thought to arise as a result of some insult to the developing brain. Therefore one may find a mixture of callosal and cortical anomalies in the same patient, giving rise to a corpus callosum which is both too short and too thin (hypogenesis and hypoplasia of the corpus callosum, Figure 10). If the cortical abnormalities are not identified and taken into consideration, abnormalities in neuropsychological function could be erroneously attributed to the callosal dysgenesis alone.

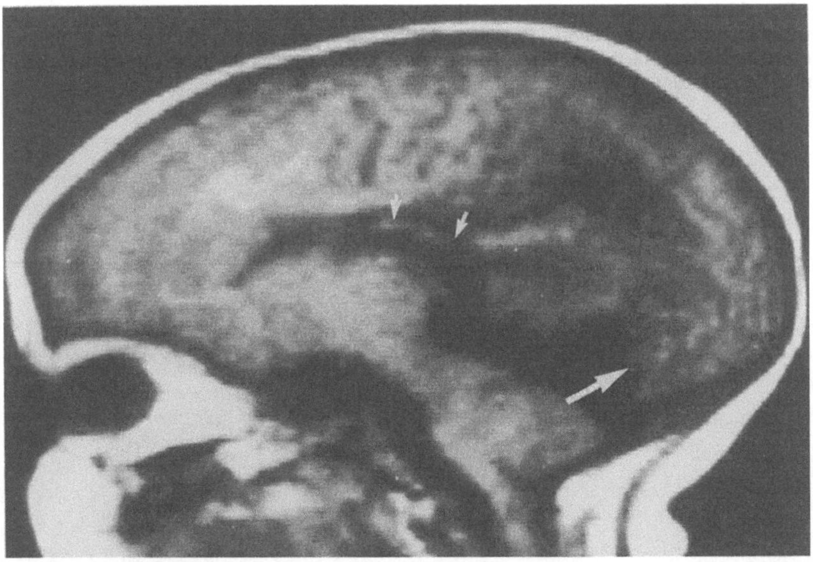

Figure 10. MR, midsagittal plane, showing hypoplasia and hypogenesis of the corpus callosum. The corpus callosum is lacking a splenium (long arrow) and is too thin in its body (short arrow). Note also the polymicrogyria and the Chiari II malformation.

OTHER COMMISSURES

The corpus callosum is the largest but not the only commissure which connects the two cerebral hemispheres. The posterior commissure mainly connects brain stem nuclei and so is not included in this discussion. Two commissures which connect the archicortical regions of the cerebral hemispheres in animals are the hippocampal commissure (also called the psalterium) and the basal telencephalic commissure (BTC). The latter has not been definitively described in man and was only recently described in any primate (LaMntia and Rakic, 1990b). The BTC travels with the anterior commissure, and probably cannot be reliably separated from the anterior commissure by current MR techniques. It is small and has been reported to be only 1% to 2% of the size of the anterior commissure in primates. The psalterium is also quite small, being approximately 10% of the size of the anterior commissure, and is normally quite difficult to see on MR. It is normally present in humans and is located just beneath the corpus callosum where the columns of the fornix approach the corpus callosum. Arising from the same precursor as the corpus callosum and anterior commissure, the psalterium is frequently absent in cases of callosal agenesis. Loeser and Alvord (1968) reported that the psalterium was never enlarged in cases of callosal agenesis. However, Barkovich (1990) reported several cases of agenesis of the corpus callosum with apparent associated hyperplasia of the psalterium. Since the psalterium carries axons from the hippocampus and the corpus callosum carries axons from neocortical regions of the brain, it would seem unlikely that the psalterium would be enlarged by aberrant axons which were intended to pass through the corpus callosum. It is possible that the psalterium is enlarged by a process similar to that which caused the focal enlargement of the corpus callosum in some of our cases of callosal hypoplasia.

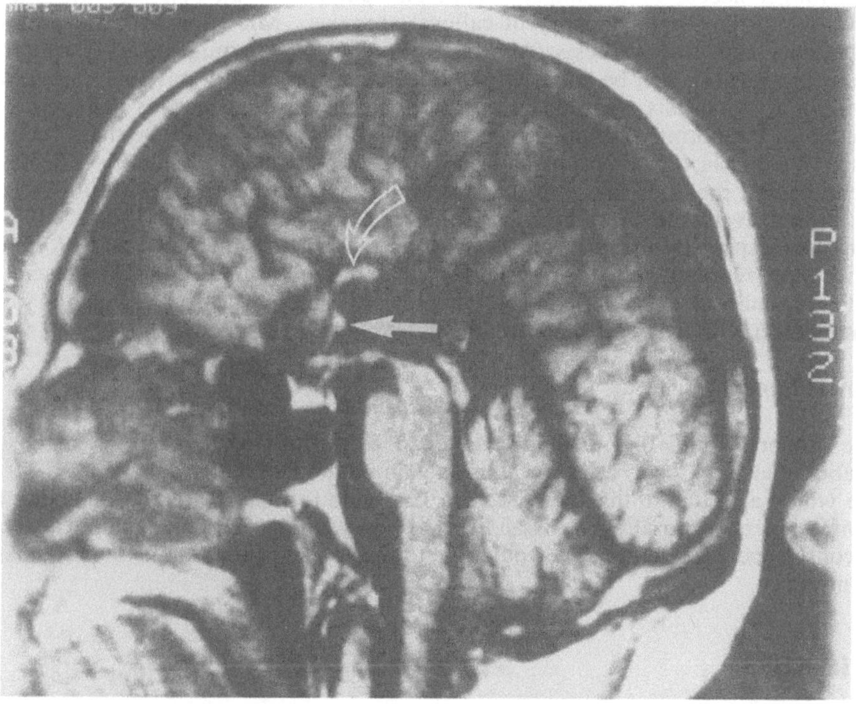

Figure 11. MR, midsagittal plane, in marked callosal hypogenesis (curved arrow), showing that the anterior commissure is enlarged (straight arrow).

The anterior commissure and corpus callosum arise from the same embryological precursor. Both carry fibers from the neocortical regions of the brain. It should therefore not be too surprising that the anterior commissure may appear abnormal in some cases of dysgenesis of the corpus callosum. The anterior commissure has been reported to be too small or absent in approximately 10% of cases of agenesis of the corpus callosum, and alternatively it appears enlarged in approximately 10% of cases of agenesis of the corpus callosum (Atlas et al, 1986, Loeser and Alvord 1968, and Barkovich 1990). In cases where the anterior commissure is absent, one might speculate that the injury to the developing brain occurred prior to the development of the lamina reuniens at seven weeks gestation (Figure 2A). The cases with an enlarged anterior commissure in subjects with callosal agenesis could be due to re-routing of some of the neuronal axons through the anterior commissure, thus increasing its size (Figure 11). Alternatively, the enlargement could be due to some dysplastic process similar to that already suggested to occasionally occur in callosal hypoplasia. The effects that either an absent anterior commissure or an enlarged anterior commissure might have on neuropsychological testing are unknown. However, with MR the opportunity now exists to examine the anterior commissure. This has the possibility of allowing for correlation of neuropsychological function data with anatomical data on the anterior commissure of living patients.

CONCLUSION

With its ability to produce directly acquired high contrast images in multiple planes, MR is particularly well suited to evaluation of the normal and abnormal corpus callosum. It is no longer necessary to simply accept the lumping of all patients with callosal dysgenesis into the single category of callosal agenesis. MR allows the separation of those patients with true agenesis from other patients with hypogenesis. Further, the degree of hypogenesis can be accurately defined, and it is usually possible to determine whether the hypogenesis was due to an injury to the forming corpus callosum or to obstruction of the normal formation of the corpus callosum by a midline mass such as a lipoma. Hypoplasia of the corpus callosum, on the other hand, generally reflects that there has been a previous insult to the developing cortical neurons, resulting in fewer axons being available for interhemispheric decussation. The state of other commissures, especially the anterior commissure, may also be definable on MR. Use of MR will allow researchers to more accurately determine the precise defect present in a subject with callosal dysgenesis. This should allow more specific correlation of the anatomic defect with data collected from neuropsychological studies.

Acknowledgments

The authors would like to thank Ms. Joanne Murray for her help in manuscript preparation and Mr. Cono Farias for assistance in photographic reproduction.

We would also like to thank Dr. Maryse Lassonde for her efforts in organizing both the conference on callosal agenesis and this volume summarizing the findings reported at the conference. In addition, we would like to thank her for inviting us to participate in this conference.

REFERENCES

Atlas, S.W., Zimmerman, R.A., Bilaniuk, L.T., et al., 1986, Corpus callosum and limbic system: Neuroanatomic MR evaluation of developmental anomalies, *Radiology* 160:355-362.

Barkovich, A.J., 1990, Apparent atypical callosal dysgenesis, *AJNR* 11:333-339.

Barkovich, A.J. and Norman, D., 1988, Anomalies of the corpus callosum: correlation with further anomalies of the brain, *AJNR* 9:493-501.

Barkovich, A.J., Chuang, S.H., and Norman, D., 1988, MR of neuronal migrational anomalies, *AJR* 150:179-187.

Curnes, J.T., Laster, D.W., Koubek, T.D., Moody, D.M., Ball, M.R., and Witcofski, R.L., 1986, MRI of corpus callosum syndromes, *AJNR* 7:617-622.

Jinkins, J.R., Whitmore, A.R., and Bradley, W.G., 1989, MR imaging of callosal and corticocallosal dysgenesis, *AJNR* 10:339-344.

LaMantia, A.S. and Rakic, P., 1990a, Axon overproduction and elimination in the corpus callosum of the developing rhesus monkey, *J. Neurosci.* 10:2156-75.
LaMantia, A.S. and Rakic, P., 1990b, Cytological and quantitative characteristics of four cerebral commissures in the rhesus monkey, *J. Comp. Neurol.* 291:520-537.
Loeser, J.D. and Alvord, E.C., 1968, Clinicopathological correlations in agenesis of the corpus callosum, *Neurology* 18:745-756.
Ludwin, S.K. and Norman, M.G., 1985, Congenital malformations of the nervous system, *in:* "Textbook of Neuropathology," R.I., Davis and D.M. Robertson, eds., Williams & Wilkins, Baltimore.
Osborn, R.E., Byrd, S.E., Naidich, T.P., Bohan, T.P., and Friedman, H., 1988, MR imaging of neuronal migrational disorders, *AJNR* 9:1101-1106.
Titelbaum, D.S., Hayward, J.C., and Zimmerman, R.A., 1989, Pachygyric like changes: Topographic appearance at MR Imaging and CT and correlation with neurological status, *Radiology* 173:663-667.

COHERENCE PATTERNS OF INFANT SLEEP EEG IN ABSENCE OF THE CORPUS CALLOSUM

Jan B.M. Kuks* and Johan E. Vos

Department of Developmental Neurology
University Hospital
Oostersingel 59, 9713 EZ Groningen
The Netherlands
*Presently at the Department of Neurology

INTRODUCTION

Neurophysiological techniques are generally used as a complement to clinical neurological and neuropsychological investigations. Since the patient's cooperation and the effect of training are of lesser importance, neurophysiological methods can be advantageous for assessing the function of neuronal systems both in clinical and experimental settings. However, the relationship between neurophysiological findings and clinical signs is not always obvious, thus reducing the clinical application of neurophysiology. On the other hand, such methods may yield otherwise concealed data and can therefore contribute to our knowledge of the nervous system.

Electro-encephalography (EEG) can be used to study the function of both cortical and subcortical systems. It can be applied during most functional states of the subject and there are relatively few limitations with regard to environmental conditions. In comparison to several other methods, it is safe and produces little inconvenience to the subject. On the other hand, several difficulties are associated with the evaluation and interpretation of an electro-encephalogram. Conventional reading of an EEG is liable to subjectivity; interobserver variability is high, even between well-experienced EEG readers. Several aspects of the EEG may remain unnoticed if analysis is performed only *de visu*. Furthermore, surface EEGs produce complex signals from remote sources and thus deformations (loss of energy, filtering, interference with other signals) will inevitably influence the finally recorded signal.

In the past, EEG has been overrated in clinical practice, thus leading to skepticism in many clinicians and scientists. There is even a trend nowadays to ignore this method of investigation. Frequently and erroneously, EEG results are compared to advanced radiological techniques whereas these two techniques have a completely different nature and function. In this context, the present chapter will not advocate the use of EEG in diagnosing the absence of the corpus callosum. Radiological techniques - e.g. ultrasound, CT-scanning and MR imaging - produce fine images of the corpus callosum and can precisely detect defects of the cerebral commissures. However, these

Callosal Agenesis, Edited by M. Lassonde and
M.A. Jeeves, Plenum Press, New York, 1994

anatomical methods cannot provide information about the functional effect of total or partial absence of the corpus callosum, which the EEG may do.

The corpus callosum ensures interhemispheric communication allowing both inhibitory and facilitatory interaction between homotopic areas of the cerebral cortex. This structure may thus be expected to keep in balance the electrical activity of both hemispheres. This activity, in turn, would result in the production of similar EEG traces from corresponding locations in the two hemispheres. To verify this assumption, the most obvious method consists of comparing EEG-signals recorded from one hemisphere with corresponding signals from the other. This indeed has been done by several investigators but their results are inconsistent.

PREVIOUS EEG ANALYSIS OF CALLOSAL FUNCTIONS

Animal studies

Synchrony of EEG patterns from homologous regions of the two hemispheres was examined before and after section of the corpus callosum in animals. Results varied; some authors reported no difference in synchrony (Batini et al., 1959; Catier et al., 1971; De Lucchi et al., 1961; Singer and Creutzfeldt, 1969), while others (Berlucchi, 1966; Bremer and Stoupel, 1957; Magni et al., 1960) found a permanent disruption of the normal bilateral synchrony after selective sections of the corpus callosum. These findings only concerned background stationary activity and can therefore not be compared with studies concerned with the propagation of epileptic discharges.

In animals, interhemispheric transmission of epileptic discharges has often been reported to be diminished after callosotomy (Catier et al., 1971 Kopeloff et al., 1950; Kusske and Rush, 1978; Marcus and Watson, 1966; Mares et al., 1975; Musgrave and Gloor, 1980), although subcortical structures also play an important role in the propagation to the contralateral cortex (Kopeloff and Kennard, 1950; Kusske and Rush, 1978). On the other hand, an inhibitory action of the corpus callosum on a contralateral epileptic focus has sometimes been reported (Kopeloff and Kennard, 1950; Kusske and Rush, 1978; McCaughran et al., 1978; Spencer et al., 1985).

Sleep patterns could only be mismatched after section of deeper midline structures such as the rostral midbrain (Berlucchi, 1966; Singer and Creutzfeldt, 1969). Finally, one cat with agenesis of the corpus callosum, without other neurological and behavioral abnormalities, showed a persistent lack of bilateral synchrony of background activity (Berlucchi, 1966).

Human studies

In man, interhemispheric interaction has been studied mostly in patients who underwent callosotomy for treatment of intractable seizures and, to a lesser extent, in patients with agenesis of the corpus callosum. Total section of the corpus callosum can prevent the development of secondary generalized seizures (Gates et al., 1984; Spencer et al., 1988). Partial (anterior) callosotomy has also been reported to be effective in restricting epileptic discharges to one hemisphere (Huck et al., 1980; Purves et al., 1988). It is important to note, however, that diencephalic and mesencephalic structures also play an important role in epileptic propagation (Spencer et al., 1985). In this context, the corpus callosum should perhaps best be considered as only one element of an interhemispheric network involving several cortical and subcortical structures.

Agenesis of the corpus callosum is frequently associated with epilepsy and with other congenital malformations (Barkovich and Norman, 1988; Ettlinger, 1977; Njiokiktjien, 1991; Parrish et al., 1979) This makes it difficult to study exclusively the effect of callosal absence on the synchrony of background activity of both hemispheres. Use of medication in these patients may create an additional problem in the EEG study. Several EEG patterns have been described in callosal agenesis, some reporting normal activity (Lilienthal and Tarlau, 1969; Loeser and Alvord, 1968) and others, non-specific dysrhythmic background activity (Chusid et al., 1949; Fariello et al., 1977; Guizzaro et al., 1985; Loeser and Alvord, 1968), asynchrony in multifocal epileptiform abnormalities (Carpenter and Druckemiller, 1953; Fariello et al., 1977; Fermaglich and O'Doherty, 1972; Green and Russell, 1966; Guizzaro et al., 1985; Johnson and Jones, 1985), asynchronous sleep spindles (Guizzaro et al., 1985; Lynn et al., 1980), asynchrony of non-epileptic activity between both hemispheres (Chusid et al., 1949; Guizzaro et al., 1985), and others yet, independence of alpha activity in both hemispheres (Green and Russell, 1966).

Obviously, these studies are inconsistent and an EEG abnormality specific to callosal agenesis has yet to be described. In fact, many of the reported anomalies were of an intermittent nature. On the other hand, abnormalities such as multiple independent spike foci can often be found in children (Blume, 1978; Noriega-Sanchez and Markand, 1976) and thus may not be specific to callosal absence. Finally, normal EEG patterns were also described. These findings suggest that the corpus callosum is not always indispensable for bilateral EEG symmetry and that other concomitant anomalies may have been responsible for the asymmetries observed in some patients.

A problem common to those studies is the lack of objectivity in the evaluation of synchrony between the EEG signals of the two hemispheres. This is especially true for the analysis of the ongoing signal during the non-stimulated condition. Results are often hard to reproduce and it is difficult to prove that the differences observed through visual analysis only are statistically significant. Finally, minor asynchronies may be overlooked when only visual examination of the EEG traces is carried out.

A solution to this problem may be found in computer-assisted analyses. Essentially, these methods allow the conversion of analogue EEG signals to numerical (digital) data (Vos, 1977). Statistical methods can then be applied to measure the conformity between the two signals. Cross-covariance and cross-correlation functions (Dumermuth, 1971) reflecting similarity between the two signals may be obtained. In spite of being objective measurements, such time-shift functions remain difficult to interpret and treat statistically. This is why, in most of our studies, we have chosen to consider the EEG signals in the frequency domain, i.e. by spectral analysis based upon Fourier transformations.

This procedure yields exact knowledge of the amount of electrical energy within a frequency band with a well-defined width, as expressed by a power spectrum. While time-dependent EEG signals can be mutually compared in terms of a cross-correlation function, the relation between spectral components can be described by a coherence function (Nunez, 1981; Vos, 1975; Vos, 1977). Comparing two signals derived (without additional noise) from the same source - and not necessarily without any phase lag - yields a coherence value of 1; calculating the coherence between two completely independent signals results in a zero value. The high degree of resolution of this procedure prompted us to use a frequency-based analysis of EEG in the present study. Details about the numerous factors that can affect coherence results can be found elsewhere (e.g. Bullock and McClune, 1989).

COHERENCE STUDIES IN ACALLOSAL NEONATES

Introduction

In a previous paper (Kuks et al., 1988), we have described normative data on coherence in normal infants during their first postnatal week. During Quiet Sleep, coherence is relatively high in the low frequency band of 0 to 1 Hz with a decrease to levels not significantly different from zero at frequencies above 4 Hz. This results in a so-called 'descending' pattern of the coherence function during the Quiet Sleep state (as described below). Another pattern was found during Active Sleep (REM-sleep, see below): coherence between 0 and 1 Hz and above 4 Hz was rather low while maximum values were found between 1 and 3 Hz leading to an 'ascending' pattern (Figure 1A). These differences between the two sleep states were quite consistent and accordingly, sleep state cycling was clearly reflected in three-dimensional plots of a long-lasting period (Figure 2) or survey plots of coherence from a single frequency range (Figure 3). Frontocentral coherences were usually higher than centroparietal ones; thus, state cycling was clearer in the anterior interhemispheric coherence. There was, however, a large intersubject variability, even in the median values of long recordings. To overcome this factor, we used a larger sample of normal infants as a control group to the acallosal subjects.

Our frequency-based analysis of the neonates' EEG revealed that coherence between two locations within one hemisphere of a normal neonate was generally lower than interhemispheric coherence between homotopic regions (Kuks et al., 1987). This effect could easily be attributed to the corpus callosum (Batini et al., 1959) although the influence of subcortical pacemakers on interhemispheric synchrony could also have had an influence (Aird and Garoutte, 1958; Enomoto, 1958).

Indeed, abnormal interhemispheric coherence may be found in cases with intact corpora callosa. Prechtl et al. (Prechtl and Vos, 1973; Prechtl et al., 1968, 1974) found relatively low coherence in neurologically abnormal neonates born to mothers who developed toxemia during pregnancy. These children also exhibited a lack of sleep state cycling and this finding emphasizes the role of structures other (lower) than the corpus callosum in the genesis of synchrony between both hemispheres. On the other hand, the incidental observation of an acallosal infant in whom unreasonably low interhemispheric coherence was found (Vos and Kuks, 1991) also tended to confirm the callosal involvement in such process. The present study was carried out to further confirm this finding in cases of callosal agenesis without concomitant pathology that could produce diffuse brain damage and abnormal sleep state cycling.

Subjects

We had the opportunity to study two children who had solely, or nearly-solely, callosal agenesis as confirmed by clinical and radiological examinations. Both infants were admitted to the pediatric department of the University Hospital Groningen. An extensive report of the radiological findings appeared earlier (Baarsma et al., 1987).

Patient A (polygram 721) is a boy born at a gestational age of 30 weeks with a birth weight of 1520 grams. On his 5th day the infant began to develop apneic spells. Physical examination showed no abnormalities. Limb movements were noted to be jerky and abrupt but further neurologic examination revealed no asymmetries. Ultrasonography of the brain revealed total agenesis of the corpus callosum. The apneic spells subsided gradually and the infant's development was characterized by a slight psychomotor retardation without focal neurologic abnormalities. The polygraphy was carried out in our department at the gestational age of 40 weeks.

Patient B (polygram 722) is a boy, born at term, after a normal pregnancy, with a birth weight of 3340 grams. On physical examination he had a large head circumference (above the 90th percentile). No significant abnormalities were observed during general and neurological examination. Ultrasonography revealed a total agenesis of the corpus callosum. His development has been uneventful up to date (age 8 years). The polygraphy was carried out at the gestational age of 41 weeks.

Thirteen neurologically normal (Prechtl, 1977) infants were used as controls for the two acallosals.

Methodology

Polygraphic recordings that lasted five to six hours were performed in a special polygraphy room under standard conditions (temperature 27-30°C, humidity 30-50%, dimmed lighting). In addition to the EEG signal, electrooculograms, electromyograms of the respiratory muscles, heart rate and respirograms were recorded (Prechtl et al.,

1974). These additional measurements were taken to assess the behavioral state independently from EEG criteria. A system composed of 22 quasi-unipolar or bipolar pre-amplifiers/impedance transformers and 16 differential amplifiers was used. Technical details of the equipment have been described previously (Kuks et al., 1987). Low pass filters were set at 16 Hz and a time constant of 1 sec was chosen because neonatal EEG consists mainly of slow activity, and the risk of slow artifact is relatively small in

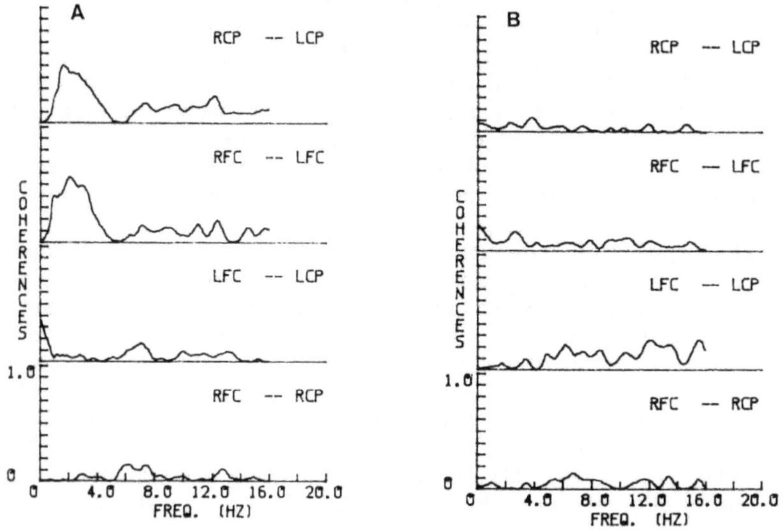

Figure 1. Inter- and intrahemispheric coherence functions (epoch: 2 min.) of the EEG during state II (active sleep). from **A**- a normal infant (Polygram No. 0705) and **B**- an acallosal infant (Polygram No. 0721). Coherence values exceeding 0.04 are significantly different from zero (p<.05). Note the 'ascending' pattern (see text) in the interhemispheric coherence of the infants. Interhemispheric coherences in the acallosal infant are considerably lower than in the normal infant.

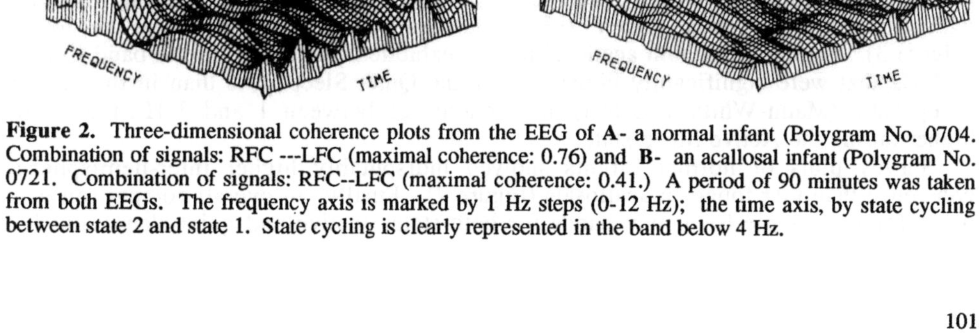

Figure 2. Three-dimensional coherence plots from the EEG of **A**- a normal infant (Polygram No. 0704. Combination of signals: RFC ---LFC (maximal coherence: 0.76) and **B**- an acallosal infant (Polygram No. 0721. Combination of signals: RFC--LFC (maximal coherence: 0.41.) A period of 90 minutes was taken from both EEGs. The frequency axis is marked by 1 Hz steps (0-12 Hz); the time axis, by state cycling between state 2 and state 1. State cycling is clearly represented in the band below 4 Hz.

our procedure. Four bipolar EEG signals were derived, namely F5-C5, C5-P5, F6-C6, and C6-P6 from the modified 10-20 system (Hellström et al., 1963).

Only sleep-recordings were considered both in order to avoid movement artifacts and because the best coherence are found during sleep (Willekens and Dumermuth, 1984). Neonatal sleep can be divided into two states. State 1, "quiet sleep" is characterized by regular respiration, eye closure and absence of movements. State 2, "rapid eye-movement sleep", also called Active Sleep or REM-sleep, is characterized by irregular respiration, eye closure and incidental movements (Prechtl, 1974). State assessment was carried out by visual analysis of the paper recordings, mainly the respiratory signal and electro-oculogram, and on the basis of visual observation of the infant's behavior during polygraphy (Prechtl, 1974).

Data were recorded on magnetic tape and on paper. Analogue to digital conversion was done off-line with a sampling rate of 68.3 Hz so that 15 seconds of signal was stored in a block of 1024 digital data points. Computation of power spectra and coherence functions was done by a Fortran-77 program using the FFT and spectral estimation subroutines in assembler provided by Cranfield Institute of Technology (U.K.). Remaining artifacts and episodic activities in the EEG signals were assessed by visual inspection of the paper recordings and eliminated by a special Fortran-77 subroutine before further processing.

Consecutive epochs of 2 minutes duration were processed and divided in eight sections of 15 seconds. After tapering, complex (two-channel) Fourier transformations were carried out for four pairs of the four bipolar signals mentioned above and triangular smoothing was then performed over 19 adjacent spectral components resulting in spectral estimates of at least 20 degrees of freedom. Thereafter the eight sections from one epoch were averaged, leading to estimates with at least 160 degrees of freedom, and then four coherence functions were computed: two interhemispheric, fronto-central (F5C5-F6C6) and centro-parietal (C5P5-C6P6), and two intrahemispheric (F5C5-C5P5 and F6C6-C6P6) (Vos, 1975). Details of artifact handling and marginal conditions for acceptation or rejection of an epoch have been described previously (Kuks et al., 1987, 1988).

All coherence functions were plotted in single graphs (e.g. Figure 1) and in three-dimensional survey figures for each infant (e.g. Figure 2). Median coherence values for 1 Hz frequency bands were calculated and stored in a database for further data-handling with another Fortran-77 special purpose package.

Results

Visual inspection of the EEGs of the two acallosal infants did not reveal background abnormalities, deviant paroxysms (especially no spike-wave abnormalities or other signs of epileptic activity) or significant hemispheric asymmetries. Consequently, power spectra of the single derivations were essentially the same as those found in normal infants.

When compared to normal infants, no differences in synchronization of both hemispheres were observed at visual examination of the EEG traces. By contrast, single interhemispheric coherence functions from the two acallosal infants were generally low in the low frequency bands (0-4 Hz), where higher coherence values are usually found in normals (Figure 1). Sleep state cycling was hardly apparent when using the normal scaling of the y-axis for coherence values from 0 to 1; using rescaling with a maximum coherence value of 0.3 the same pattern as in normal infants was found (Figure 3). Furthermore, both acallosal infants exhibited coherences in the band from 0 to 1 Hz that were significantly higher during the Quiet Sleep state than in the Active Sleep state (Mann-Whitney U-test) and coherences between 1 and 3 Hz that were higher during the Active Sleep state.

Survey plots of normal infants showed indisputably higher inter-hemispheric coherence than that found in acallosals (Figure 2). Finally, an overall comparison was made for four single frequency bands for both states (active and quiet sleep) and for both pairs of interhemispheric derivations [FC and CP (Figure 4)]. This resulted in 16 series of coherence values for each child. Both acallosals were then compared with all 13 normals using the Mann-Whitney U-test.

Differences between the acallosals and most of the normal infants were statistically significant. In a few series, however, the acallosals were indistinguishable from some of the normal infants. The number of normal infants who showed coherences significantly higher than those of the two acallosals are given at the top of the frames in Figure 4.

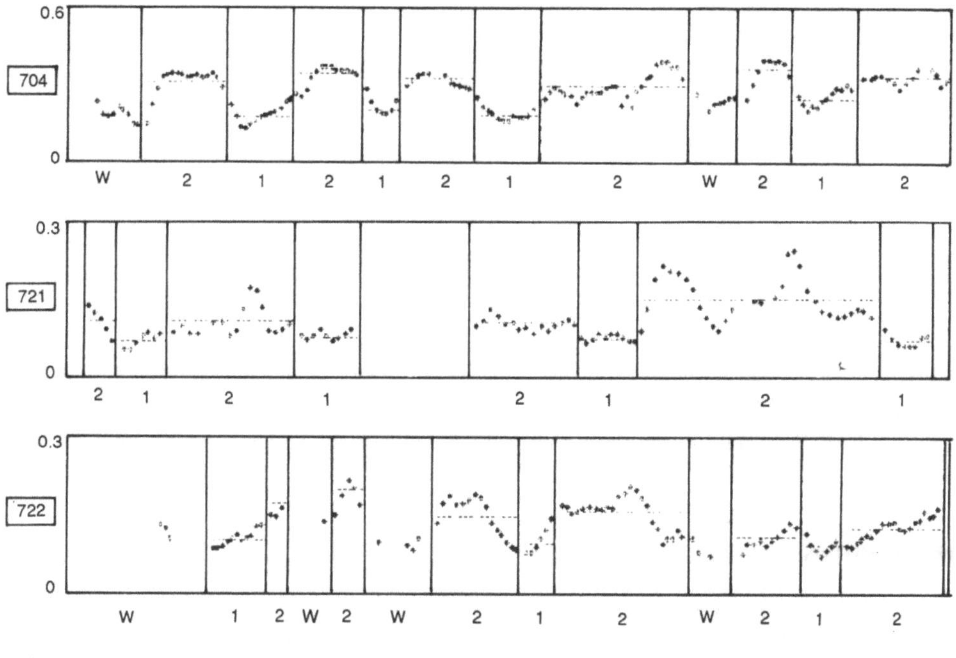

Figure 3. Survey of mean interhemispheric coherences between 1 and 4 Hz from one control (704) and the two acallosals (721 and 722) during state cycling. Data were balanced by triangular smoothing with a factor 3. Horizontal scale: recording time in minutes. W indicates the periods in which the infant was awake. Vertical scale: coherence. Note the difference in scaling between the control (0-0.6) and the acallosals (0-0.3). State cycling can be seen in all three subjects. Mean coherences per state are depicted by dotted lines.

Figure 4 also illustrates that only eight normal children had centroparietal coherences in the 3 to 4 Hz frequency range that were significantly higher during the Active Sleep state. Two normal infants showed rather low coherences overall and thus cannot be distinguished from the acallosals in most of the frequency bands. It is important, however, to realize that the highest coherences are normally found between 0 and 1 Hz during Quiet Sleep and between 1 and 3 Hz during Active Sleep in the frontocentral derivations (Kuks et al., 1988). In these conditions all normal children had unequivocally higher coherences than the acallosals, as can also be seen in Figure 4.

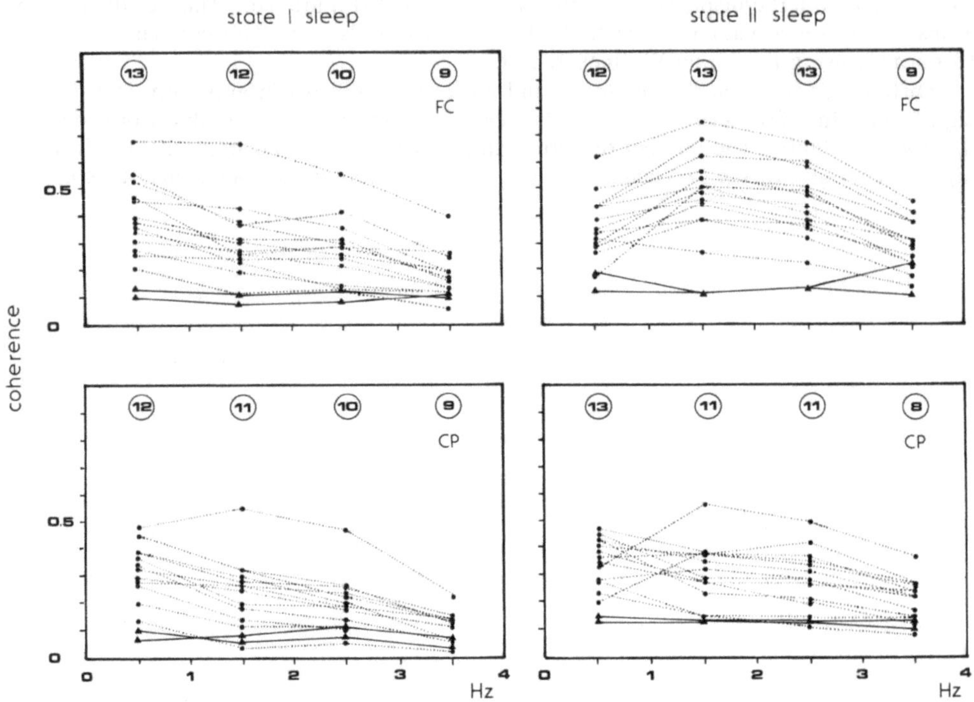

Figure 4. Median coherences in the lowest four frequency ranges for both interhemispheric combinations of channels and for both sleep states. Within each frame, the circled figures indicate the number of normal infants whose coherences are significantly higher (P<.05, Wilcoxon's Rank Sum test) than those of both the acallosal infants. Normal infants are represented by the dashed lines, the acallosals, by full lines.

DISCUSSION

A perfect coherence - equal to 1.0 - between two signals stands for a common generator for both signals when no additional noise is present. Therefore, a coherence of, say, 0.5 reflects only a partial sharing of generators for both signals. For cortical EEG, such a common generator could be a deep cerebral pacemaker. If, however, such subcortical pacemakers were the only or the most important cause of coherence, no lack of coherence should have been found in acallosals. Alternately, coherence can be found between signals derived from two points that are anatomically connected - e.g. two cortical points connected with each other by means of interhemispheric fibers. Such a model could explain the low level of interhemispheric coherence seen in callosal agenesis. In this context, the coherence observed in acallosals - insofar as it is significantly different from zero - may either originate from common subcortical sources or be caused by electrical transfer through other commissural pathways such as the anterior commissure.

It could be argued that interhemispheric coherence was also unacceptably low in some of the normal infants who presumably have a functional corpus callosum. It is, however, not realistic to consider the - still unmyelinated - corpus callosum as a perfect conductor between the two hemispheres in infancy. Furthermore, slight differences in electrode positions and variations in the electrode material, etc. may have influenced absolute coherence values, thus explaining some of the intersubject variability in normals (Bullock and McClune, 1989).

The variance in normal values notwithstanding, we found significantly lower coherences in the two acallosal infants with respect to all normal infants in the conditions where normally high interhemispheric coherence can be expected (frontocentral derivations: 1-3 Hz during Active Sleep, 0-1 Hz during Quiet Sleep). This confirms the hypothetical role of the corpus callosum in establishing coherence between two signals from different hemispheres. Conversely, it may be said that interhemispheric coherence is a measure of the function of the corpus callosum.

The routine used in our laboratory did not include the analysis of occipital and temporal EEG signals. It might be worthwhile to extend the study to these regions especially because coherence studies in human patients with an intact corpus callosum indicate that the corpus callosum may be relatively unimportant for the interhemispheric propagation of epileptic activity from the mesial temporal regions (Gotman, 1987; Lieb et al., 1987).

As already mentioned, the presence of associated neural defects must be considered in evaluating the EEG of patients with agenesis of the corpus callosum. With respect to the present study, one might ask whether the low coherences observed in our acallosal infants could be a result of unrevealed concomitant brain stem abnormalities. However, sleep state cycling was found in the acallosals and this finding militates against the supposition that the low coherences were a result of brain abnormalities other than callosal absence. Studies on neonates are scarce but our results may be compared to data obtained in adult callosotomized patients. The high coherences at and around the alpha frequency range found in normals were found to be subnormal in split-brain patients (Hillyard et al., 1981). Furthermore, in a recent report on two epileptic patients (Montplaisir et al., 1990), measurements of coherence before and after callosotomy indicated changes in interhemispheric coherence, confirming the fact that the electrical connectivity between both hemispheres is mediated by the corpus callosum.

It is remarkable that, in the neonate, interhemispheric electrical transport should be carried via the corpus callosum considering that myelination is only scanty during infancy. On the other hand, the total amount of callosal fibers is greater in the early postnatal period than later in the development (Innocenti, 1981a; Innocenti, 1981b; Innocenti, 1986; Innocenti and Caminiti, 1980; Innocenti and Frost, 1979; Knaap van der and Valk, 1990; McArdle et al., 1987; Ramaekers, 1991; Valk and Njiokiktjien, 1991). The abundance of callosal fibers in early development has led to postulate the existence of transitory 'wrong' connections between functionally non-related or heterotopic cortical areas (Innocenti and Clarke, 1984a, 1984b). The question remains as to whether there is significant interhemispheric coherence between heterotopic areas in infancy. We did not measure coherences between frontocentral and centroparietal channels from opposite hemispheres but since intrahemispheric coherence is usually very low, it is not likely that interhemispheric coherence from these heterotopic derivations should be significant.

As development continues, myelination progresses and a great deal of callosal axons are eliminated. These two processes should lead to an augmentation in interhemispheric coherence with maturation and, in fact, interhemispheric bioelectric synchrony increases with age in pre-term and full-term neonates (Dreyfus-Brisac and Curzi-Dascalova, 1975; Dreyfus-Brisac and Monod, 1975) and continues to increase between 2 and 17 years of age (Farber and Knyazeva, 1991). Besides the technical factors reported above, the dynamic development of interhemispheric connections may account for some of the variance seen in normal coherence values. Our normal group was very consistent with respect to postnatal age (5 days) but gestational age varied from 37 to 41 weeks and indeed, there is a variation in the development of myelination in normal subjects of the same gestational age (Ramaekers, 1991).

The two acallosal children were carefully matched according to gestational age with the 13 controls. It is therefore unlikely that the overall lowering of coherence found in both cases could be explained by a difference in the developmental stage of myelination.

In fact, it should be interesting to study how coherences develop in a follow-up study of controls and acallosals. Is there a compensatory mechanism that would allow coherence to increase with age in acallosals albeit, perhaps, to a lesser degree than in normals ?

CONCLUSION

We feel that quantitative analysis of EEG by means of coherence functions is a valuable tool for the evaluation of interhemispheric bioelectric synchrony, and as such, of the function of the corpus callosum. Furthermore, its use can be recommended for both human and animal studies. One of the major features of our study was the large amount of data used to overcome normally occurring variance. Data taken from a small period of time may, in fact, show differences between individual infants simply attributable to chance. As we have seen in our survey plots, discrimination between acallosal and normal infants may be problematic if it is carried out using small samples or using data coming from a narrow frequency range or from only one behavioral state. Some basic work has been done in this field, and with proper control over methodology many of the unsolved questions will be answered.

REFERENCES

Aird, R.B. and Garoutte, B., 1958, Studies on the 'cerebral pacemaker', *Neurology* 8:581-589.

Baarsma, R., Martijn, A., and Okken, A., 1987, The missing pericallosal artery on sonography: a sign of agenesis of the corpus callosum in the neonatal brain? *Neuroradiology* 29:47-49.

Barkovich, A.J. and Norman, D., 1988, Anomalies of the corpus callosum: correlation with further anomalies of the brain, *AJNR* 9:493-501.

Batini, C., Moruzzi, G., Palstini, M., Rossi, G.G., and Zanchetti, A., 1959, Effect of completed pontine transections on the sleep-wakefulness rhythm: the midpontine-pretrigeminal preparation, *Arch. Ital. Biol.* 97:1-12.

Batini, C., Radulovacki, M., Kado, R.T., and Adey, W.R., 1967, Effect of interhemispheric transection on the EEG patterns in sleep and wakefulness in monkeys, *Electroencephalogr. Clin. Neurophysiol.* 22:101-112.

Berlucchi, G., 1966, Electroencephalographic studies in 'split brain' cats, *Electroencephalogr. Clin. Neurophysiol.* 20:348-356.

Blume, W.T., 1978, Clinical and electroencephalographic correlates of the multiple independent spike foci pattern in children, *Ann. Neurol.* 4:541-547.

Bremer, F. and Stoupel, N., 1957, Recherche d'une participation du corps calleux au mécanisme de la synergie bio-électrique des hémisphères cérébraux, *J. Physiol. (Paris)* 49:66-67.

Bullock, T.H. and McClune, M.C., 1989, Lateral coherence of the electrocorticogram: a new measure of brain synchrony, *Electroencephalogr. Clin. Neurophysiol.* 73:479-498.

Carpenter, M.B. and Druckemiller, W.H., 1953, Agenesis of the corpus callosum diagnosed during life, *Arch. Neurol.* 69:305-322.

Catier. J., Cordeau, J.P., Dimov, S., Riche, D., Eberhard, A., and Naquet, R., 1971, Preliminary studies of the electrical changes following section of the corpus callosum in photosensitive Papio Papio, *Electroencephalogr. Clin. Neurophysiol.* 30:248.

Chusid, J.G., Gutierrez-Mahoney, C.G., and de Chaffee, J.S., 1949, Agenesis of the corpus callosum. An electroencephalographic study, *Arch. Neurol.* 62:840-846.

De Lucchi, M.R., Garoutte, B., and Aird, R.B., 1961, Lack of effect of midsagittal section through corpus callosum and massa intermedia on bilateral synchrony of the EEG of the cat, *Electroencephalogr. Clin. Neurophysiol.* 13:306-307.

Dreyfus-Brisac, C. and Curzi-Dascalova, L., 1975, The EEG during the first year of life, *in:* "Handbook of Electroencephalography and Clinical Neurophysiology," Vol. 6B, A. Rémond, ed., Elsevier, North Holland, Amsterdam.

Dreyfus-Brisac, C. and Monod, N., 1975, The EEG of full-term newborns and premature infants, *in:* "Handbook of Electroencephalography and Clinical Neurophysiology," Vol. 6B, A. Rémond, ed., Elsevier, North Holland, Amsterdam.

Dumermuth, G., 1971, Electronic data processing in pediatric EEG research, *Neuropediatrics* 2:349-374.

Enomoto, T.F., 1958, Unilateral activation of the non-specific thalamic system and bilateral cortical responses, *Electroencephalogr. Clin. Neurophysiol.* 11:219-232.

Ettlinger, G., 1977, Agenesis of the corpus callosum, *in:* "Handbook of Clinical Neurology," Vol. 30, P. Vinken and G. Bruyn, eds., Elsevier, North Holland, Amsterdam.

Farber, D.A. and Knyazeva, M.G., 1991, Electrophysiological correlates of interhemispheric interaction in ontogenesis, *in:* "The Child's Corpus Callosum," G. Ramaekers and C. Njiokiktjien, eds., Suyi, Amsterdam.

Fariello, R.G., Chun, R.W.M., Doro, J.M., Buncic, J.R., and Prichard, J.S., 1977, EEG recognition of Aicardi's syndrome, *Arch. Neurol.* 34:563-566.

Fermaglich, J. and O'Doherty, D.S., Agenesis of the corpus callosum: An electroencephalographic study, *Epilepsia* 13:345.

Gates, J.R., Leppik, I.E., Yap, J., and Gumnit, R.J., 1984, Corpus callosotomy: Clinical and electroencephalographic effects, *Epilepsia* 25:308-316.

Gotman, J,. Interhemispheric interactions in seizures of focal onset: data from human intracranial recordings, *Electroencephalogr. Clin. Neurophysiol.* 67:120-133.

Green, J.B. and Russell, D.J., 1966, Electroencephalographic asymmetry with midline cyst and deficient corpus callosum, *Neurology* 16:541-545.

Guizzaro, A., Vizioli, R., and Paolozzi, C., 1985, Clinical and EEG findings in agenesis of the corpus callosum, Acta *Neurol. (Napoli)* 7:252-269.

Hellström, B., Karlsson, B., and Müssbichler, H., 1963, Electrode placement in EEG of infants and its anatomical relationship studied radiographically, *Electroencephalogr. Clin. Neurophysiol.* 15:115-117.

Hillyard, S., Kutas, M., Nunez, P.L., Kammer, Y., and Kruetz, K., 1981, EEG data from split brain patients compared to normal subjects, Unpublished report, cited *in*: "Nunez, P.: Electric Fields of the Brain. The Neurophysics of EEG," Oxford University Press, New York, Oxford.

Huck, F.R., Radvany, J., Avila, J.O., et al., 1980, Anterior callosotomy in epileptics with multiform seizures and bilateral synchronous spike and wave EEG pattern, *Acta Neurochir.* 30:127-135.

Innocenti, G.M., 1981a, Growth and reshaping of axons in the establishment of visual callosal connections, *Science* 218:824-827.

Innocenti, G.M.,1981b, The development of interhemispheric connections, *Trends Neurosci.* 4:142-144.

Innocenti, G.M., 1986, General organization of callosal connections in the cerebral cortex, *Cortex* 22:291-353.

Innocenti, G.M. and Caminiti, R., 1980, Postnatal shaping of callosal connections from sensory areas, *Exp. Brain Res.* 38:382-394.

Innocenti, G.M. and Clarke, S., 1984a, Bilateral transitory projection to visual areas from auditory cortex in kittens, *Dev. Brain Res.* 14:143-148.

Innocenti, G.M. and Clarke, S., 1984b, The organization of immature callosal connections, *J. Comp. Neurol.* 230:287-309.

Innocenti, G.M. and Frost, D.O., 1979, Effects of visual experience on the maturation of the efferent system to the corpus callosum, *Nature* 280:231-234.

Johnson, M.H. and Jones, S.N., 1985, Status epilepticus, hypothermia and metabolic chaos in a man with agenesis of the corpus callosum, *J. Neurol. Neurosurg. Psychiatry* 48:480-483.

Knaap, M.S. van der and Valk, J., 1990, MR imaging of the various stages of normal myelination during the first year of life, *Neuroradiology* 31:459-470.

Kopeloff, N., Kennard, M.A., Pacellal, B.L., Kopeloff, L.M., and Chusid, J.G., 1950, Section of corpus callosum in experimental epilepsy in the monkey, *Arch. Neurol.* 63:719-727.

Kuks, J.B.M., Vos, J.E., and O'Brien, M.J., 1987, Coherence patterns of the infant sleep EEG in absence of the corpus callosum, *Electroencephalogr. Clin. Neurophysiol.* 66:8-14.

Kuks, J.B.M., Vos, J.E., and O'Brien, M.J., 1988, EEG coherence functions for normal newborns in relation to their sleep state, *Electroencephalogr. Clin. Neurophysiol.* 69:295-302.

Kusske, J.A. and Rush, J.L., 1978, Corpus callosum and propagation of afterdischarge to contralateral cortex and thalamus, *Neurology* 28:905-912.

Lieb, J.P., Hoque, K., Skomer, C.E., and Song, X.W., 1987, Inter-hemispheric propagation of human mesial temporal lobe seizures: a coherence/phase analysis, *Environ. Res.* 67:101-119.

Lilienthal, E. and Tarlau, M., 1969, The EEG in congenital absence of the corpus callosum, *Electroencephalogr. Clin. Neurophysiol.* 26:635.

Loeser, J.D. and Alvord, E.C., 1968, Clinicopathological correlations in agenesis of the corpus callosum, *Neurology* 18:745-756.

Lynn, R.B., Buchanan, D.C., Fenichel, G.M., and Freemon, F.R., 1980, Agenesis of the corpus callosum, *Arch. Neurol.* 37:444-445.

Magni, F., Melzack, R., and Smith, C.J., 1960, A stereotaxic method for sectioning the corpus callosum in cat, *Electroencephalogr. Clin. Neurophysiol.* 12:517-518.

Marcus, E.M. and Watson, C., 1966, Bilateral synchronous spike wave electrographic patterns in the cat, *Arch. Neurol.* 14:601-610.

Marcus, E.M. and Watson, C.W., 1968, Symmetrical epileptogenic foci in monkey cerebral cortex, *Arch. Neurol.* 19:99-116.

Mares, J., Mares, J., and Filip, J., 1975, Role of corpus callosum in the interhemispheric transmission of epileptic electrographic phenomena, *Electroencephalogr. Clin. Neurophysiol.* 39:548.

McArdle, C.B., Richardson, C.J., Nicholas, D.A., et al., 1987, Developmental features of the neonatal brain: MR imaging, part I, gray-white matter differentiation and myelination, *Radiology* 162:223-229.

McCaughran, J.A., Corcoran, M.E., and Wada, J.A., 1978, Role of the forebrain commissures in amygdaloid kindling in rats, *Epilepsia* 19:19-33.
Montplaisir, J., Nielsen, T., Coté, J., Boivin, D., Rouleau, I., and Lapierre, G., 1990, Interhemispheric EEG coherence before and after partial callosotomy, *Clin. Electroencephalogr.* 21:42-47.
Musgrave, J. and Gloor, P., 1980, The role of the corpus callosum in bilateral interhemispheric synchrony of spike and wave discharge in feline generalized Penicillin epilepsy, *Epilepsia* 21:369-378.
Njiokiktjien, C.H., 1991, Absence of the corpus callosum: clinicopathological correlations, *in:* "The Child's Corpus Callosum," G. Ramaekers and C. Njiokiktjien, eds., Suyi, Amsterdam.
Noriega-Sanchez, A. and Markand, O.N., 1976, Clinical and electroencephalographic correlation of independent multifocal spike discharges, *Neurology* 26:667-672.
Nunez, P., 1981, "Electric Fields of the Brain," Oxford University Press, New York, Oxford.
Parrish, M.L., Roessmann, U., and Levinsohn, M.W., 1979, Agenesis of the corpus callosum: a study of the frequency of associated malformations, *Ann. Neurol.* 6:349-354.
Prechtl, H.F.R., 1974, The behavioral states of the newborn infant (a review), *Brain Res.* 76:185-212.
Prechtl, H.F.R., 1977, "The Neurological Examination of the Full-Term Newborn Infant. Clinics in Developmental Medicine," Vol. 63, Heinemann, London.
Prechtl, H.F.R. and Vos, J.E., 1973, Verlaufsmuster der Frequenzspektren und Kohärenzen bei schlafenden normalen und neurologisch abnormalen Neugeborenen, *in:* "Symposium der Arbeitsgemeinschaft für Methodik in der Elektroencephalographie. Jongny sur Vevey," G.K. Schenk, ed., AEG-Telefunken, Konstanz.
Prechtl, H.F.R., Akiyama, Y., Zinkin, and Grant, D.,K., 1968 Polygraphic studies of the full-term newborn. I. Technical aspects and quantitative analysis, *in:* "Studies in Infancy. Clinics in Developmental Medicine," Vol. 27, M.C.O. Bax and R. C. MacKeith, eds., Heinemann, London.
Prechtl, H.F.R., Vos, J.E., Akiyama, Y., and Casaer, P., 1974, Neonatal EEG: age codes and coherence functions in relation to intra-uterine growth and oestrogen levels, *in:* "Ontogenesis of the Brain. II Proceedings of the International Symposium Neuroontogeneticum Secundum," G. Jilek and S.Trojan, eds., Universita Karlova, Prague.
Purves, S.J., Wada, J.A., Woodhurst, W.B., et al., 1988, Results of anterior corpus callosum section in 24 patients with medically intractable seizures, *Neurology* 38:1194-1201.
Ramaekers, G., 1991, Embryology and anatomy of the corpus callosum, *in:* "The Child's Corpus Callosum," G. Ramaekers and C. Njiokiktjien, eds., Suyi, Amsterdam.
Singer, W. and Creutzfeldt, O., 1969, Die Bedeutung der Vorderhirnkommissuren für die Koordination bilateraler EEG-Muster, *Exp. Brain Res.* 7:195-213.
Spencer, S.S., Spencer, D.D., Glaser, G.H., Williamson, P.D., and Mattson, R.H., 1984, More intense focal seizure types after callosal section: the role of inhibition, *Ann. Neurol.* 16:686-693.
Spencer, S.S., Spencer, D.D., Williamson, P.D., and Mattson, R.H., 1985, Effects of corpus callosum section on secondary bilaterally synchronous interictal EEG discharges, *Neurology* 35:1689-1694.
Spencer, S.S., Spencer, D.D., Williamson, P.D., Sass, K., Novelly, R.A., and Mattson, R.H, 1988, Corpus callosotomy for epilepsy. I Seizure effects, *Neurology* 38:19-24.
Valk, J. and Njiokiktjien, C.H., 1991, Abnormal morphogenesis of the corpus callosum I: MR imaging, *in:* "The Child's Corpus Callosum," G. Ramaekers and C. Njiokiktjien, eds., Suyi, Amsterdam.
Vos, J.E., 1975, Representation in the frequency domain of non-stationary EEGs, *in:* "Computerized EEG Analysis," H. Künkel, ed., Gustav Fischer Verlag, Stuttgart.
Vos, J.E., 1977, Between EEG-machine and computer: data storage and data conversion, *in:* "EEG Informatics. A Didactic Review of Methods and Applications of EEG Data Processing," A. Rémond, ed., Elsevier, North Holland, Amsterdam.
Vos, J.E. and Kuks, J.B.M, 1991, Agenesis of the corpus callosum is reflected by very low interhemispheric EEG coherence in early life *in:* "The Child's Corpus Callosum," G. Ramaekers and C. Njiokiktjien, eds., Suyi, Amsterdam.
Vos, J.E., Lammertsma, A.A., and Eykern, L.A. van., 1977, Ordinary and partial coherences of bipolar and quasi-unipolar derivations of infant electroencephalograms, *IEE Conf. Publ.* 159.
Willekens, H. and Dumermuth, G., 1984, EEG spectral power and coherence analysis in healthy full-term neonates, *Neuropediatics* 15:180-190.

SLEEP, DREAMING AND EEG COHERENCE PATTERNS IN AGENESIS OF THE CORPUS CALLOSUM: COMPARISONS WITH CALLOSOTOMY PATIENTS

Tore A. Nielsen,[1] Jacques Montplaisir,[1]
Roger Marcotte,[1] and Maryse Lassonde[2]

[1]Centre d'Etude du Sommeil
Clinique du Sommeil
Hôpital du Sacré Coeur
5400 Boul. Gouin O.
Montréal, Que., H4J 1C5, Canada and
Département de Psychiatrie, Université de Montréal
[2]Groupe de Recherche en Neuropsychologie Expérimentale
Université de Montréal
C.P. 6128, Succ. A
Montréal, Qué., H3C 3J7, Canada

INTRODUCTION

The typical sleep and dreaming patterns of adults with agenesis of the corpus callosum (ACC) have not been well-studied. A recent literature review produced only one study on the sleep EEG of acallosal infants (Kuks et al., 1987) and a short clinical report on the sleep of an acallosal father and his 11-year-old son (Lynn et al., 1980). The medical histories of the latter patients were both complicated by megalencephaly and their sleep patterns were found to be very different from each other. Moreover, there is also little known about the functioning of the corpus callosum during sleep. Berlucchi's study (1965) of callosal activity in the sleep of cats remains unreplicated with either animal or human subjects. The study suggests that tonic activity in the corpus callosum is diminished during sleep, with some phasic bursting activity occurring in conjunction with rapid eye movements. Studies in our laboratory have also produced information about sleep, dreaming and EEG coherence in two patients before and after partial callosotomy (Montplaisir et al., 1990; Montplaisir et al., 1984), but overall there is very little literature on sleep in ACC available.

The study described here was meant to address this lack of information concerning sleep and ACC. Results are reported which summarize analyses of the sleep, EEG coherence, and dreaming of four acallosal subjects and four matched control subjects who were studied in a sleep laboratory. Where possible, data from our previous studies of two callosotomy subjects is also presented for comparison with the acallosal subjects.

Callosal Agenesis, Edited by M. Lassonde and
M.A. Jeeves, Plenum Press, New York, 1994

METHODOLOGY

Subjects

Subjects were two females and two males (M=23.0yrs) with ACC. They were paid participants recruited from the population collaborating with Dr. Lassonde at the Université de Montréal. Three ACC subjects (MG, LG, and SG) are siblings in a French-Canadian family with four children. More extensive descriptions of the psychiatric, neuropsychological, and cognitive status of these siblings has been described elsewhere (Lassonde et al., 1988, 1991; Sauerwein et al., 1981; Sauerwein and Lassonde, 1983). Diagnoses of ACC in all cases were confirmed with computerized axial tomography (CAT) scan and/or nuclear magnetic resonance imaging (MRI). *Case 1:* MG, a 19-year-old left-handed male, is the youngest of the family. He scored a full-scale IQ of 77 (Verbal: 71, Performance: 87) on a French version of the WAIS-R. *Case 2:* LG, a 27-year-old right-handed female, is the second child of the family. She has a full-scale IQ of 78 (Verbal: 81, Performance: 81) on the Ottawa-Wechsler. *Case 3:* SG, a 29-year-old right-handed female, is the second of four children and the sister of MG and LG described above. She has a global IQ of 84 (Verbal: 88, Performance: 82) on the revised WAIS. *Case 4:* SB, a 17-year-old right-handed male, was adopted at 6 months. He scored 68 on the global IQ of the WISC-R (Verbal: 58, Performance: 81). He is dyslexic.

The four control subjects were persons known to members of the research team and were recruited so as to match the age (M=24.5yrs), gender, and hand preference, but not the IQ, of the agenesis subjects. None reported a history of neurological, psychiatric, or sleep problems. Like the acallosal subjects, none abused alcohol or took psychotropic medications.

General design

Subjects spent 2 nights in the sleep laboratory; night one was an adaptation night and was not assessed further. On night 2, EEG, EMG, and EOG were recorded from the standard electrode montage (Rechtschaffen and Kales, 1968) which permits discrimination of stages 2, 3/4, and REM sleep and the waking state. Both inter- and intrahemispheric EEG coherence were calculated for 5-minute epochs in each of these stages. Details of these methods have been published elsewhere (Nielsen et al., 1990). Finally, dreams were collected from both laboratory REM periods and from a 2-week home dream diary. Specific details about methods and the results will be described for each of three major sets of variables: (1) sleep architecture variables, (2) EEG coherence variables, and (3) dream recall and content variables. Group differences were evaluated with standard two-tailed t-tests and normal curve z-tests (Edward, 1972).

Table 1. Dependent variables examined in analysis of sleep architecture.

Time in Bed	Sleep Latency	Time (%) Awake
Total Sleep Time (TST)	Stage 1 Latency	Time (%) Stage 1
# Awakenings	Stage 2 Latency	Time (%) Stage 2
Sleep Efficiency	Stage 3 Latency	Time (%) Stage 3
REM Sleep Efficiency	Stage 4 Latency	Time (%) Stage 4
# REM Sleep Periods	Stage REM Latency	Time (%) Stage REM
REM Cycle Duration	REM Period Duration	

STUDY 1. SLEEP ARCHITECTURE

Methods

Sleep variables scored in the present study are shown in Table 1. Stage REM sleep is characterized by a desynchronized, low-voltage EEG, muscle paralysis and vivid extended dream sequences and is thought to be implicated in the integration of emotionally important experiences with long-term memory. Stages 3 and 4 sleep are characterized by a synchronized EEG with high-voltage slow waves and are thought to play a role in physiological recuperation. Stage 2 sleep is similar in some respects to the previous stages, but is distinguished especially by sleep spindles and K-complexes in the EEG. Its function is unknown, although humans spend about 50% of their sleep time in this stage. Total Sleep Time (TST) was used to calculate percentages of time spent in each of these stages of sleep. Table 1 also shows several additional measures of REM sleep that were calculated, including the number of REM periods per night, the mean duration of individual REM periods, and an estimate of the duration of the ultradian rhythm, which is from 90-100 minutes in length in normal subjects.

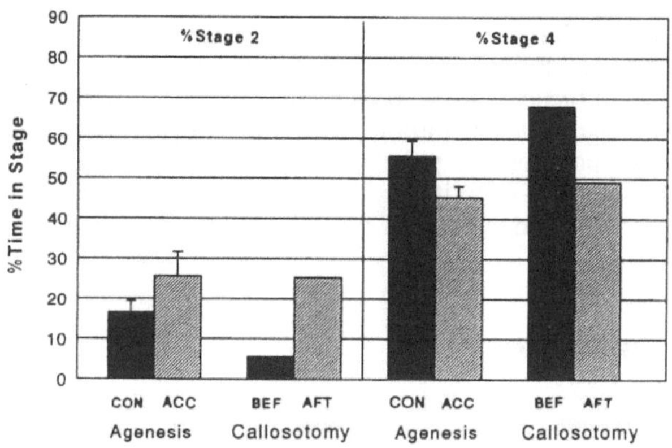

Figure 1. Percent of time spent in stage 2 and stages 3+4 for subjects with agenesis of the corpus callosum (ACC) and controls (CON), and patients before (BEF) and after (AFT) partial callosotomy.

Results

Subjects with ACC spent more minutes in stages 3+4 sleep (M=122.3) than controls (M=84.9; t(6)=2.61, p=.04); this difference was slightly less marked when expressed as a percentage of TST (M=25.6% vs 16.6%; t(6)=2.14, p=.08) (See Figure 1). A complementary difference was observed for stage 2 time (t(6)=-2.36, p=.06) and percent (t(6)=-4.38, p=.005). The magnitude of these differences was about 10% of TST in both cases.

Similar results were obtained in the 2 previously studied partial callosotomy patients. An increase in stage 3+4 sleep and a decrease in stage 2 sleep were observed pre- to post-surgery in these patients. Callosotomy led to a 19.7% increase in stage 3+4 sleep time (from 5.6% to 25.3%) and a corresponding 19.0% decrease in stage 2 sleep time (from 68.0% to 49.0%) (Figure 1).

The acallosal subjects also tended to have more REM sleep periods per night (5.8 v. 4.4; t(6)=2.24, p=.07), even when controlled for TST (t(6)=2.14, p=.08), which suggested a faster REM sleep periodicity. This was corroborated by a significantly shorter duration of REM sleep cycle (t(40)=-2.32, p=.03), about 84 minutes, which was shorter than the 100-min average rhythm shown by the controls. The acallosals did not spend less time in REM sleep than the controls, however.

Discussion

If these results are found to be valid with additional subjects, they would implicate the corpus callosum in at least two functions relevant to the regulation of sleep. One is the propagation of slow waves and may be explained as follows: individual cortical regions receive a synchronizing influence from thalamo-cortical projections, particularly from fibers in non-specific intralaminar thalamic nuclei. Local intracortical mechanisms may further enhance synchronization within limited cortical regions. Contralateral callosal fibers may exert a desynchronizing effect on these cortical regions, thereby reducing slow wave activity and consequently stages 3 and 4 sleep. Nevertheless, the callosal fibers also enhance coherence between homologous interhemispheric regions, which is consistent with a result presented in Study 2 below. In cases of callosotomy or ACC, local cortical regions will be released from the desynchronizing influence of the callosal fibers, producing more slow waves, but these waves will no longer be coherent with regions in the opposite hemisphere.

A second function suggested by the present results is that the corpus callosum may play a role in the timing of ultradian biorhythms. Absence of the corpus callosum may disrupt somewhat the periodicity of sleep-related processes associated with ultradian pacemakers, producing a rhythm with a faster than normal frequency. The present findings that acallosal subjects tend to have more REM sleep periods and a REM cycle shorter than the normal 90-100-minute ultradian cycle are consistent with this possibility. Further support for this possibility comes from a previous study (Nielsen et al., 1989) of the Basic Rest-Activity Cycle (BRAC) in two of the present acallosal subjects: one subject showed a BRAC periodicity shorter than that of the control subject.

STUDY 2. EEG COHERENCE

Methods

Inter- and intrahemispheric coherence was calculated for each sleep stage from a standard, monopolar, parasagittal montage (Fp1, Fp2, F3, F4, C3, C4, P3, P4, O1, O2, T3, T4) referenced to linked ears (A1+A2). Temporal leads were not further analyzed. Interhemispheric coherence scores were means calculated from four frequency bandwidths: delta (0.5 - 4.0Hz), theta (4.0 - 8.0Hz), alpha (8.0 - 13.0Hz), and beta (13.0 - 22.0Hz). These scores were assessed for all homologous pairs of electrodes (e.g. Fp1-Fp2, F3-F4, etc.). Similarly, intrahemispheric coherence was assessed for adjacent pairs of electrodes (Fp1-F3, F3-C3, etc) on the left and right sides. Three types of mean coherence scores were calculated for both inter- and intrahemispheric electrode pairs: (a) coherence scores by individual cortical regions, (b) coherence scores by stage, or an average of all individual coherence scores for a given stage, and (c) global coherence scores, or an average of all individual scores, regardless of stage.

Results

Interhemispheric Coherence. Global interhemispheric coherence was lower in acallosal subjects (M=38.7) than in controls (M=52.0; p=.021). By sleep stage, coherence was lower for acallosal subjects during stage 2 (p=.021), stage 3+4 (p=.021), and stage REM (p=.043) sleep, but not during wakefulness (p=.149) (see Table 2).

By cortical regions, interhemispheric coherence was significantly (p<.05) or marginally (p<.10) lower in acallosal subjects for all regions except 1) the occipital region (O1-O2) during stages 2 and 3+4 sleep, 2) frontal (Fp1-Fp2, F3-F4) and central

Table 2. Mean interhemispheric EEG coherence (in %) by sleep stage.

	Awake		Stage 2		Stages 3+4		Stage REM	
	M	sd	M	sd	M	sd	M	sd
Agenesis	43.9	7.3	40.3*	8.7	37.0*	10.6	47.2*	12.4
Control	56.6	7.2	60.9	3.3	61.0	2.4	64.3	3.4

(C3-C4) regions during stage REM sleep, and 3) the frontal region (F3-F4) during wakefulness. Thus, frontal leads showed the most marked and uniform differences between groups while occipital leads showed the fewest differences. The interhemispheric differences are further reflected in the fact that acallosal subjects scored lower than control subjects on 78 of the 80 coherence measures calculated for the individual channel-bands (5 pairs X 4 stages X 4 bands).

Callosotomy Comparisons. The interhemispheric coherence difference is similar in direction to the decrease in coherence seen in callosotomy patients after surgery. Callosotomy patients showed a decrease of interhemispheric coherence of from 37.7% to 24.5% in stage 2 sleep and of from 36.2% to 26.4% in REM sleep. The magnitude of the difference was also similar in the two studies. In the acallosal subjects, there was a mean difference of 13% relative to controls and in the callosotomy patients, there was a mean reduction of about 11.5% pre-to post-surgery. Further, these differences parallel findings reported by Kuks and colleagues (1987 and this volume) for acallosal newborns.

Intrahemispheric Coherence. Global intrahemispheric coherence did not differentiate between acallosal subjects (M=50.4) and controls (M=50.6; p=.564). By sleep stage, there were also no significant differences between groups. By cortical regions, only two channel-pairs differentiated the groups; mean coherence for C4-P4 during wakefulness and mean coherence for P3-O1 during REM sleep were lower for acallosal subjects (p<.05). However, means for acallosal subjects were lower than those for control subjects on 130 of the 160 measures calculated for individual channel-bands (5 pairs X 2 hemispheres X 4 stages X 4 bands), suggesting a slightly lower overall coherence for the acallosal group.

Discussion

There is some indication in the present results that interhemispheric EEG coherence differentiates the underlying anatomy of acallosal subjects. That occipital channel-pairs were least affected in any stage by ACC may be explained by the fact that the posterior commissure typically remains intact in ACC because of its independent development at the diencephalon-mesencephalon border (Chiarello, 1980). The specificity of these EEG findings is also consistent with those from the callosotomy study, where it was observed that interhemispheric EEG coherence fairly accurately reflected the location of anatomical section produced by surgery. These parallels, together with the findings of Kuks, et al., suggest that in ACC there may be a unique brain organization which is reliably indexed by low interhemispheric coherence and which is not modified by subsequent neuronal changes in the brain.

A post-hoc observation raises the possibility that there might be some interhemispheric compensatory activity reflected in the occipital coherence index. Specifically, 1 of the 80 interhemispheric channel-bands assessed in calculating the mean coherence scores registered a striking 10.6% elevation for the acallosal group. This activity was recorded exclusively over the occipital region during the waking state for the delta band. Although it is very possible that this may be a spurious result attributable to the large number of means calculated, the sheer magnitude of the effect and its association with occipital cortex suggests that it may reflect some specific non-callosal compensatory activity. One might question why the posterior commissure, and not the anterior commissure, compensates for the lack of a corpus callosum. The answer may be that because primary visual cortex (area 17) is devoid of commissural connections (Pandya and Seltzer, 1986), interhemispheric coherence is *normally* subserved by the posterior commissure, and would thus be an available route for compensatory activity.

Finally, the finding that intrahemispheric coherence, in general, was not elevated during any sleep/waking stage in ACC supports the notion (Sauerwein and Lassonde, 1983) that there is no specialized intrahemispheric compensation for the missing corpus callosum in ACC. Rather, apart from a few exceptions, intrahemispheric processes in ACC appear to function at normal levels.

STUDY 3. DREAM RECALL AND DREAM CONTENT

The dreams of callosotomy patients have been of interest because of their relevance to the hypothesis that dreaming is primarily a right hemisphere activity. If this hypothesis is true, then it might be expected that absence of the corpus callosum will be associated with a difficulty in remembering or verbally reporting dream content. Greenwood and colleagues (1977) claimed to have found dreaming in their callosotomy patients, thus failing to support this hypothesis, but others (Kerr and Foulkes, 1981) have argued that biases in verbal reporting may mislead judges to attribute right hemisphere, visuo-spatial determinants to oneiric content that did not, in fact, exist.

To address this issue, we assessed the dream content of acallosal patients with both a verbal measure of dreaming and a checklist procedure which required verbal recognition of categories but not verbal reporting of dream content *per se*. This checklist included two categories which referred specfically to the visuospatial determinants of dream imagery, i.e., colors and movements. Thus visuospatial determinants were assessed in a manner which was relatively independent of verbal reporting abilities.

Methods

Subjects were awakened from the last REM period of each night to report dreams; there were 2 awakenings per subject or a total of 8 awakenings per group. In addition, 3 acallosal subjects and 3 matched controls agreed to keep a home diary after the laboratory study in which they wrote out their spontaneous home dreams over a 2-week period. Measures of recall, verbal elaboration and specific contents were scored in both the laboratory and the home dreams. These measures consisted of number of recalls, number of rêves blancs, number of no recalls, number of words per report and number of content categories per report. Rêves blancs were scored when subjects stated that they had indeed dreamed but that they could not remember the contents of the dream. For each of 10 content categories, subjects were asked to state simply whether or not that category was present in the dream. The categories were presented orally in the laboratory study and in written checklist form on the home diary task. Categories included known persons, unknown persons, animals, movements, emotions, speech, music, sounds other than speech or music, colors, and odors.

Results

In the laboratory, acallosals reported dreams with content on only 50% of trials, rêves blancs on 25%, and no recalls on 25%. One acallosal subject recalled no dreams on either awakening, one recalled a dream on both awakenings, and two recalled one dream and one rêve blanc each. Control subjects, by contrast, recalled dreams on all 8 awakenings (z=-2.30, p=.02).

The frequency of rêves blancs in acallosals paralleled that observed for callosotomy patients. Specifically, acallosals reported rêves blancs on 25% of awakenings, while callosotomy produced an increase in rêves blancs to about 28%. On the other hand, unlike laboratory dream recall for acallosals (57%), post-surgery dream recall among callosotomy patients was only 33%.

On the home diary reports, a different pattern of dream recall was observed. The number of dreams recalled by acallosal subjects at home (24/44 or 57.1%) was not different from the number recalled by controls (24/42 or 54.5%). However, the mean number of rêves blancs reported by acallosal subjects (11/44 or 25.0%) was marginally greater than the number reported by control subjects (5/42 or 11.9%; z=1.56, p=.12) and exactly the same proportion as reported by acallosals in the laboratory (25.0%). The greater frequency of rêves blancs on the home diary task is due mainly to the same 2 acallosal subjects who reported rêves blancs in the laboratory, specifically, MG, who reported 6 more rêves blancs than his matched control on the diary task, and LG, who reported 2 more rêves blancs than her control. By contrast, subject SG, who reported 2 out of 2 no recalls in the laboratory, also reported a high proportion (8/14 or 57.1%) of no recalls on the diary task relative to her matched control.

The dreams that were recalled by acallosal subjects in the laboratory were shorter in length (M=30.0 words) than those of controls (M=107.8; t(10)=-2.30, p=.044). The dreams recalled by acallosal subjects at home were also much shorter in length (M=37.4) than those of control subjects (M=145.8; t(46)=-5.10, p<.0001).

In the laboratory, the mean number of categories on the 10-item checklist scored as 'present' by the acallosal subjects (M=5.8) did not differ from the mean number reported by control subjects (M=4.9). Similarly, at home, the number of categories per dream scored by acallosal subjects (M=3.9) and control subjects (M=4.5) did not differ.

Finally, on the home diary task, acallosals tended to report fewer dreams with unknown characters (29% vs 54%, p= .08), colors (25% vs 46%, p=.11) animals (13% vs 29%, p=.15), and movements (50% vs 68%, p=.12) than did control subjects.

Discussion

The results of these analyses indicate that acallosal subjects have the capacity for experiencing dreaming to much the same extent as do subjects with an intact corpus callosum. This is supported by the fact that all acallosal subjects reported recalling at least one dream on either the laboratory or home diary measures. In fact, the dream recall capacity of most acallosal subjects in the present study was more similar to that of the control subjects than it was to that of the callosotomy patients. The only clear exception to this generalization was SG, who recalled no dreams in the laboratory and very few on the home diary task.

Although acallosal subjects recalled dreams, these dreams were, on average, shorter than those of control subjects. Some factors are consistent with the suggestion that this was very likely due to the lower verbal skills of the acallosal subjects rather than to any differences in their dream processes *per se*. First, control subjects were not systematically matched for IQ with the acallosals and were likely of higher intelligence. Second, the verbal IQ's of the acallosal subjects were lower than their performance IQ's. Third, the relatively non-verbal measure of dream content indicated that the acallosals' dreams contained as many of the basic content categories as the control subjects' dreams. One finding which is not consistent with this interpretation, however, is that SG - the eldest and most verbally gifted of the siblings (verbal IQ=88) - also demonstrated the lowest overall level of dream recall.

Some peculiarities in the dreams of the acallosal subjects were nevertheless observed. First, there occurred a tendency for acallosal subjects to have fewer dreams with colors and movements than control subjects on the home diary task. Second, acallosal subjects displayed higher frequencies of rêves blancs. Their average frequency of 25% on both the laboratory and home diary measures is very similar to the 28% frequency among callosotomy patients after surgery.

Both of these anomalies, though in need of further study, nevertheless provide some possibilities as to how absence of the corpus callosum may affect dreaming. The lower incidence of dreams with colors among acallosals could be seen as support for the notion that the right hemisphere is essential for visuo-spatial dreaming. However, more careful assessment of the visual determinants of dream content in cases of dysfunction of the corpus callosum are required. Second, the higher frequency of rêves blancs in acallosals may be related to the emotional deficit of alexithymia, in which there is a difficulty in expressing emotions verbally. This is suggested by the fact that both alexithymia and rêves blancs have been found to characterize callosotomy patients (Montplaisir et al., 1984; Ten Houten et al., 1988). They have also been found to characterize asthmatic patients (Monday et al., 1987; Feiguine et al., 1982) among whom a 'functional callosotomy' has been postulated as central to the pathophysiology (Hoppe and Bogen, 1977). Finally, a restrictive emotional style has been noted among acallosals (e.g. O'Brien, this volume) although alexithymia per se has not been assessed in this population. It may be speculated that alexithymia and rêves blancs reflect two aspects of the same dysfunctional corpus callosum, specifically, that right hemisphere affective/prosodic processes are only poorly integrated with left hemisphere verbal/narrative processes.

CONCLUSION

To summarize, our analyses provide evidence relevant to how sleep processes may differ in agenesis of the corpus callosum. Because the study sample is small, the differences observed should at present be considered as provisional hypotheses. Nevertheless, these hypotheses implicate the corpus callosum in some functions related to sleep which have not been suggested before in the literature, such as synchronization of cortical slow waves and timing of ultradian rhythms, as well as some other functions which are better-known yet still lacking empirical support, such as interhemispheric integration and the recall and visuo-spatial elaboration of dreams.

Acknowledgments

This research was supported by the 'Fonds de la recherche en Santé du Québec (FRSQ)' and the Medical Research Council of Canada.

REFERENCES

Berlucchi, G., 1965, Callosal activity in unrestrained, unanesthetized cats, *Arch. Ital. Biol.* 103:623-635.

Chiarello, C., 1980, A house divided? Cognitive functioning with callosal agenesis, *Brain Lang.* 11:128-158.

Edwards, A.L. 1972., "Experimental Design in Psychological Research," 4th Edition, Holt, Rinehart and Winston, New York.

Feiguine, R.J., Hulihan, D.M., and Kinsman, R.A., 1982, Alexithymic asthmatics: Age and alexithymia across the life span, *Psychother. Psychosom.* 37:185-188.

Greenwood, P., Wilson, D.H., and Gazzaniga, M.S., 1977, Dream report following commissurotomy, *Cortex* 13:311-316.

Hoppe, K.D. and Bogen, J.E., 1977, Alexithymia in twelve commissurotomized patients, *Psychother. Psychosom.* 28:148-155.

Kerr, N. and Foulkes, D., 1981, Right hemispheric mediation of dream visualization: A case study, *Cortex* 17:603-610.

Kuks, J.B.M., Vos, J.E., and O'Brien, M.J., 1987, Coherence patterns of the infant sleep EEG in absence of the corpus callosum, *Electroencephalogr. Clin. Neurophysiol.* 66:8-14.

Lassonde, M, Sauerwein, H, Chicoine, A.J., and Geoffroy, G., 1991, Absence of disconnexion syndrome in callosal agenesis and early callosotomy: Brain reorganization or lack of structural specificity during ontongeny?, *Neuropsychologia* 29:481-495.

Lassonde, M., Sauerwein, H., McCabe, N., Laurencelle, L., and Geoffroy, G., 1988, Extent and limits of cerebral adjustment to early section or congenital absence of the corpus callosum, *Behav. Brain Res.* 30:165-181.

Lynn, R.B., Buchanan, D.C., Fenichel, G.M., and Freemon, F.R., 1980, Agenesis of the corpus callosum, *Arch. Neurol.* 37:444-445.

Monday, J., Montplaisir, and Malo, J-L., 1987, Dream process in asthmatic subjects with nocturnal attacks, *Am. J. Psychiatry* 144:638-640.

Montplaisir, J., Côté, J., Laverdière, M., et al., 1987, Dream recall before and after partial commissurotomy, *in:* "Sleep '84," W.P. Koella,et al., eds., Gustav Fischer Verlag, Stuttgart, New York.

Montplaisir, J., Nielsen, T., Côté, J., Boivin, D., Rouleau, I., and Lapierre, G., 1990, Interhemispheric EEG coherence before and after partial callosotomy, *Clin. Electroencephalogr.* 21:42-47.

Nielsen, T.A., Abel, A., Lorrain, D., and Montplaisir, J., 1990, Interhemispheric EEG coherence during sleep and wakefulness in left- and right-handed subjects, *Brain Cogn.* 14:113-125.

Nielsen, T.A., Lapierre, G., Montplaisir, J., Boivin, D., Marcotte, R., and Lassonde, M., 1989, Ultradian EEG rhythmicity in two cases of agenesis of the corpus callosum, *Sleep Res.* 18:142.

Pandya, D.N. and Seltzer, B., 1986, The topography of commissural fibers, *in:* "Two Hemispheres-One Brain: Functions of the Corpus Callosum," F. Lepore, M. Ptito, and H.H. Jasper, eds., Alan R. Liss, New York.

Rechtschaffen, A. and Kales, A., 1968, "A Manual of Standardized Terminology Techniques and Scoring System for Sleep Stages of Human Subjects," Nr. 204, United States Government Printing Office (National Institute of Health Publication), Washington.

Sauerwein, H. and Lassonde, M.C., 1983, Intra- and interhemispheric processing of visual information in callosal agenesis, *Neuropsychologia* 21:167-171.

Sauerwein, H.C,. Lassonde, M., Cardu, B., and Geoffroy, G., 1981, Interhemispheric integration of sensory and motor functions in agenesis of the corpus callosum, *Neuropsychologia* 19:445-454.

Ten Houten, W.D., Walter, D.O., Hoppe, K.D., and Bogen, J.E., 1988, Alexithymia and the split brain: VI. EEG correlates of alexithymia, *Psychiatr. Clin. (North Amer.)* 11:317-329.

ASPECTS OF DENDRITIC MATURATION OF CALLOSALLY PROJECTING NEURONS

Giorgio M. Innocenti

Institut d'Anatomie, 9 rue du Bugnon
1005 Lausanne, Switzerland

INTRODUCTION

The focus of the present chapter is on whether the development of cortical neurons is affected by retrograde signals, possibly trophic in nature, originated at the site of termination of their axon. This question is about a fundamental mechanism of brain development (see Purves 1988, for review) and is relevant for understanding-predicting deficits and/or recovery of function in a number of pathological conditions which cause destruction of neocortex or of brain regions associated with it, and/or the interruption of fiber tracts of cortical origin. In the context of this book the question I wish to address is whether maturation and/or maintenance of callosally projecting neurons depends on the integrity of the contralateral areas to which they project and/or on the fate of their axons to these areas. A positive answer to these questions would predict changes in the information-processing characteristics of cortical neurons in the case of callosal agenesis and possibly of callosal transection in man. As it happens, the results are more of a negative kind and therefore, conclusions based on them must be prudent and are worth further testing in the coming years.

CALLOSAL CONNECTIONS AS A MODEL OF CORTICAL DEVELOPMENT

The background for the studies reviewed here lies with the notion that neurons depend for survival as well as for maturation and maintenance of their dendritic arbors on "trophic" substances produced by or near their targets and retrogradely transported to the cell bodies through their axons. The best known substance acting according to this scheme is NGF (Levi-Montalcini and Calissano, 1986) although other substances with presumably similar action are being identified in the nervous system (Barde et al., 1987). These substances may shape the information-processing characteristics of a neural network by acting retrogradely on neuronal connectivity or on the mechanism whereby this is integrated at the single neuron level. Retrograde signals have proven their usefulness as means for shaping artificial "neural" circuits (Hopfield and Tank, 1986), although similarities between "back propagating algorithms" used in models and trophic signals mediated by target released substances remain to be explored and are highly speculative.

Callosal Agenesis, Edited by M. Lassonde and
M.A. Jeeves, Plenum Press, New York, 1994

The second notion underlying these studies is the finding that callosal projections from area 17 develop through a phase of exuberance, during which parts of this area transiently project via the corpus callosum to contralateral areas 17 and 18 (see, for review, Innocenti 1986, 1991). If the morphology of cortical neurons were critically affected by the site where their axons terminate, cortical neurons with transient and with permanent axon to the contralateral hemisphere should have different morphology; furthermore, an early lesion of contralateral areas 17 and 18 should eliminate this difference.

DENDRITES OF CALLOSALLY PROJECTING NEURONS DEVELOP THROUGH BOTH PROGRESSIVE AND REGRESSIVE EVENTS

The dendritic arbor of cortical neurons matures in a characteristic way first described by Cajal (1911). Initially, an apical dendrite forms and establishes a terminal tuft in layer I. Subsequently, basal dendrites grow from the cell body and side branches from the stem of the apical dendrite. The basal dendrites therefore mature after the apical dendrites.

Callosally projecting neurons in areas 17 and 18 of the adult cat conform to this pattern of development, but in addition it was recently found that in some of them, part of the dendritic arbors undergoes regression in development (Assal et al., 1991; Vercelli et al., 1992). These neurons can be classified into two types on the basis of their dendritic pattern: pyramids and stellate cells. The latter are characterized by the absence of an apical dendrite which, in contrast, the pyramidal neurons always possess. They are nearly exclusively of the spinous type and their cell bodies are predominantly located in layer IVa (Innocenti, 1986; Weisskopf and Innocenti, 1991). They represent about 20 % of the total callosally projecting neurons in kittens at P57. Retrograde labeling with the carbocyanine dye DiI applied to the corpus callosum showed that until P6 all callosally projecting neurons in areas 17 and 18 have a pyramidal morphology, that is, they possess an apical dendrite reaching layer I. Between P6 and P10 however, callosally projecting neurons with stellate morphology appear. This late appearance may be due either to delayed arrival of the axon of this type of neuron to the corpus callosum or to the transformation of neurons which initially had an apical dendrite and subsequently lost it. In order to distinguish between the two, callosally projecting neurons were retrogradely labeled by injections of fluorescent beads in contralateral areas 17 and 18, at P1-P3, that is, at an age when callosally projecting neurons with stellate morphology were not present yet, and then relabeled either with DiI applied to the corpus callosum at P10 or with intracellularly injected Lucifer Yellow at P57. The results were clear-cut. Double-labeled callosally projecting neurons with stellate morphology were found. The ratio between stellate cells and pyramids was similar in the double-labeled population and in the whole population of callosally projecting neurons. Therefore it appears that loss of the apical dendrite and not the delayed arrival of their axon to the corpus callosum causes the late appearance of callosally projecting neurons with stellate morphology. This conclusion is strengthened by the finding at P6, that is close to the time when stellate cells appear, of callosally projecting neurons with cell bodies in layer IVa, a morphology similar to that of stellate neurons and an atrophic apical dendrite, probably an apical dendrite in the process of being eliminated.

MATURATION OF CALLOSALLY PROJECTING NEURONS AND THEIR CONTRALATERAL CONNECTIVITY

In order to assess whether maturation and/or maintenance of dendrites of callosally projecting neurons depends on the connections they establish in the contralateral hemisphere, the morphology of area 17 and 18 neurons with a callosal axon to contralateral homotopic areas was analyzed in the following conditions (Weisskopf and Innocenti, 1991). Neurons were labeled by fluorescent beads applied to contralateral areas 17 and 18 at P2-P3 and then relabeled by intracellular injection of Lucifer Yellow at P27 or P32 (Katz et al., 1984). The intracellular injections were performed on perfusion-fixed (4% paraformaldehyde) slices observed through a fluorescent microscope. This method yields a very extensive and, often, probably complete visualization of the dendritic arbors, comparable to the Golgi technique, but is far less erratic in its results.

At the time of fluorescent bead injection, neurons in the medial part of area 17 still had a transient axon to the contralateral hemisphere. By the time of sacrifice they had lost this axon but their cell bodies could still be identified by the presence of the fluorescent beads (Innocenti, 1981; Innocenti et al., 1986). Therefore, the intracellularly labeled neurons could be subdivided into two populations, one near the border of areas 17 and 18 and made up of neurons, most of which probably project permanently to the corpus callosum and one in the medial part of area 17, made up of neurons which all lose the callosally projecting axon by the end of the first postnatal month. Inspection did not show qualitative differences between the two neuronal populations. Therefore we performed a quantitative analysis of the basal dendrites of the pyramidal neurons. The basal dendrites were chosen for two reasons. First, as mentioned above, they mature relatively late and are therefore more likely to be affected by late developmental events and, second, they were often more completely filled by the intracellular application of Lucifer Yellow than the apical dendrites. The following parameters were measured: (i) total length of the basal dendrites, (ii) number of bifurcation points (nodes) in the dendritic arbor (iii) total and median length of the segments peripheral to the most peripheral node (end segments), (iv) total and median length of the segments between two nodes (intermediate segments).

Thirty-five dendrites were analyzed in 9 neurons of area 17 and the same number of neurons and dendrites near the 17/18 border. Total dendritic length was greater for neurons in the permanent projection sector than in the transient projection sector and the difference was due to longer end-segments, while intermediate segments and number of nodes were not different in the two sectors.

In order to determine if the difference between these two groups of neurons was indeed due their maintaining or losing an axon to the contralateral areas 17 and 18, we injected kittens with fluorescent beads as above but in these kittens, we ablated by suction the site of injection at P8-P10. At the time of the lesion, the dendrites of callosally projecting neurons had not yet completed their maturation as shown by the presence of filopodia, swellings and growth cones along their dendrites. The biochemical maturation of cortical neurons was also not finished yet, as indicated by cytoskeletal markers such as the microtubule associated proteins MAP2 and MAP5 which were still undergoing changes until at least the end of the first postnatal month (Riederer et al., 1990; Riederer and Innocenti, in progress).

Three outcomes of this experiment seemed equally probable. Lesions could have (i) modified the dendritic arbors of both neuronal populations, suggesting that the contralateral targets were necessary for growth and maintenance of both cell types; (ii) modified only dendrites of neurons in the permanent projection sector, suggesting that the contralateral targets were necessary for growth and/or maintenance of neurons which maintain the projection to the contralateral areas 17 and 18; (iii) affected neither population, suggesting that the contralateral targets are not necessary for either growth and/or maintenance of the dendritic arbor of callosally projecting neurons.

Three animals and a total of 44 dendrites (11 neurons) near the 17/18 border, 37 dendrites (9 neurons) from area 17 were studied. Compared to normal animals, no changes were found in the number of nodes or the length of intermediate segments in any of the lesioned animals, in either permanently or transiently projecting sectors. Total dendritic length and length of end-segment were also not affected although the end segments were shorter near the 17/18 border in one animal.

CONCLUSION

These results suggest that trophic signals *specifically* originating from the target site in the contralateral hemisphere are probably not critical for growth and/or maintenance of dendritic arbors of callosally projecting neurons in areas 17 and 18 although they may affect other aspects of their phenotype. The most probable interpretation of these experiments is that these neurons receive trophic support through the initial collaterals of their axons in the ipsilateral hemisphere. Both permanent (Innocenti, 1986) and transient (Innocenti et al., 1986) callosally projecting neurons possess such collaterals. This would however suggest that the trophic signals are not crucial for the decisions of whether the neurons will maintain or lose the contralateral projection but rather act as general signal not directly involved in the specification of connections. Unfortunately, these experiments were performed before we became aware of the loss of the apical dendrites of some callosally projecting neurons. Therefore, they do not provide a direct answer to whether interactions with the contralateral hemisphere affect the fate of the apical dendrite of some callosal neurons, mainly because at the age (P8-P10) at which the lesion of the contralateral target was performed, most of the apical dendrites are already in the process of disappearing.

Within the context of this book, our findings indicate that lesions of one hemisphere which cause interruption of callosal axons, even if performed in early life, do not necessarily affect survival and/or maturation of cortical neurons in the other hemisphere. These conclusions might be generalized to other corticocortical projection neurons and if so, they narrow the interpretations of the causes of deficits and recovery after cortical lesions. From our findings, however, it does not necessarily follow that callosal agenesis has no effects on the morphology of callosal neurons. First, callosal agenesis deprives callosally projecting neurons of their normal contralateral target earlier than our experimental lesions. Second, lesions of areas 17 and 18 could have been followed, in our experiments, by rerouting of callosal axons to other contralateral areas, e.g. area 19 or the visual areas in the suprasylvian sulcus. An answer to these questions requires, in the cat, an extension of the present experiments to the prenatal development.

Acknowledgments

The studies whose results are reviewed here were performed in collaboration with Frederic Assal, Alessandro Vercelli and Marc Weisskopf and with the precious technical help of Marceline Robert-Tissot. Supported by Swiss National Science Foundation Grant n 3129948.90. I wish to thank Directors and members of the Institute for Neuroscience and the Department of Histology at the University of Alicante (Spain) for their hospitality during the preparation of this text.

REFERENCES

Assal, F., Vercelli, A., and Innocenti, G.M., 1991, The morphology of immature callosal neurons in area 17 and 18 of the cat, *Eur. J. Neurosci. Abstr. (Suppl.)* 4:283.

Barde, Y.A., Davies, A.M., Johnson, J.E., Lindsay, R.M., and Thoenen, H., 1987, Brain derived neurotrophic factor, *Progr. Brain. Res.* 71:185-189.

Cajal, S. and Ramon, Y., 1955, "Histologie du Système Nerveux de l'Homme et des Vertébrés," Vol. II, Consejo Superior de Investigaciones Cientificas, Madrid.

Hopfield, J.J and Tank, D.W., 1986, Computing with neural circuits: a model, *Science* 233:625-633.

Innocenti, G.M., 1981, Growth and reshaping of axons in the establishment of visual callosal connections, *Science* 212:824-827.

Innocenti, G.M., 1986, General organization of callosal connections in the cerebral cortex, *in:* "Cerebral Cortex," Vol 5, E.G. Jonesand and A. Peters, eds., Plenum Publishing Corporation, New York.

Innocenti, G.M., 1991, The development of projections from cerebral cortex, *Prog. Sens. Physiol.* 12:65-114.

Innocenti, G.M., Clarke, S., and Kraftsik, R.,1986, Interchange of callosal and association projections in the developing visual cortex, *J. Neurosci.* 6:1384-1409.

Katz, L.C., Burkhalter, A., and Dreyer, W.J., 1984, Fluorescent latex microspheres as a retrograde neuronal marker for in vivo and in vitro studies of visual cortex, *Nature* 310:498-500.
Levi-Montalcini, R. and Calissano, P., 1986, Nerve growth factor as a paradigm for other polypeptide growth factors, *Trends Neurosci.* 9:473-477.
Purves, D., 1988, "Body and Brain," Harvard University Press, Cambridge, Mass., London, England.
Riederer, B.M., Guadano-Ferraz, A., and Innocenti, G.M., 1990, Difference in distribution of microtubule-associated proteins 5a and 5b during the development of cerebral cortex and corpus callosum in cats: dependence on phosphorylation, *Dev. Brain Res.* 56:235-243.
Vercelli, A., Assal, F., and Innocenti, G.M., 1992, Emergence of callosally projecting neurons with stellate morpholoy in the visual cortex of the kitten, *Exp. Brain Res.* 90:346-358.
Weisskopf, M. and Innocenti, G.M. , 1991, Neurons with callosal projections in visual areas of newborn kittens: an analysis of their dendritic phenotype with respect to the fate of the callosal axon and of its target, *Exp. Brain Res.* 86:151-158.

DEFECTS OF THE FETAL FOREBRAIN IN ACALLOSAL MICE

Douglas Wahlsten[1] and Hiroki S. Ozaki[1,2]

[1]Department of Psychology, University of Alberta,
Edmonton, AB, T6G 2E9, Canada
[2]Department of Anatomy, Kagawa Medical School Kagawa, Japan

INTRODUCTION

The corpus callosum (CC) normally unites the cerebral hemispheres and is obvious in the mouse as a large bundle of heavily myelinated axons (Figure 1A). In several strains, however, the corpus callosum is entirely absent and a deep chasm is apparent between the hemispheres (Figure 1B). The putative callosal axons do not all perish; instead, they travel in a whorl ipsilaterally and eventually form the longitudinal Probst bundle. The adult defect is not simply an all-or-none affair. Most often there is a fragment of the corpus callosum located dorsal to the hippocampal commissure. In a large sample of BALB/c mice, there is a continuum of CC sizes (Wahlsten, 1989) and the distinction between normal and abnormal is somewhat arbitrary.

GENETIC VARIABILITY IN ACALLOSAL MICE

Four major inbred strains and many of their substrains are known to suffer from agenesis of the corpus callosum (Table 1). Three were originally discovered by Richard E. Wimer and the ddN was detected by Hiroki S. Ozaki. They differ greatly in the degree of penetrance. I/LnJ never has even a small bundle of callosal axons joining the hemispheres (Lipp and Waanders, 1990; Livy and Wahlsten, 1991), whereas an immense sample of ddN mice is needed to detect a few with complete agenesis (Ozaki et al., 1984). Within the BALB/c strain, several substrains differ in penetrance (Wahlsten, 1989), most notably the BALB/cWahl which has a bimodal distribution of CC size, as does the 129/J strain (Wahlsten, 1982a).

These dramatic individual differences within a highly inbred strain do not result from any residual hereditary variation; there is no correlation between the presence of a defect in the two BALB/c parents and their offspring (Wahlsten, 1989). This fact offers strong evidence against the notion that the genes encode a rigid blueprint for brain structure. The defect is strain-specific and thus hereditary, yet heredity is only one aspect of forebrain ontogeny.

Incomplete penetrance of the BALB/c heredity also provides an excellent experiment of nature for assessing consequences of callosal agenesis. Genetically identical siblings conceived at the same time and reared in the same maternal environment, one with and the other lacking a corpus callosum, can be compared on a variety of interesting

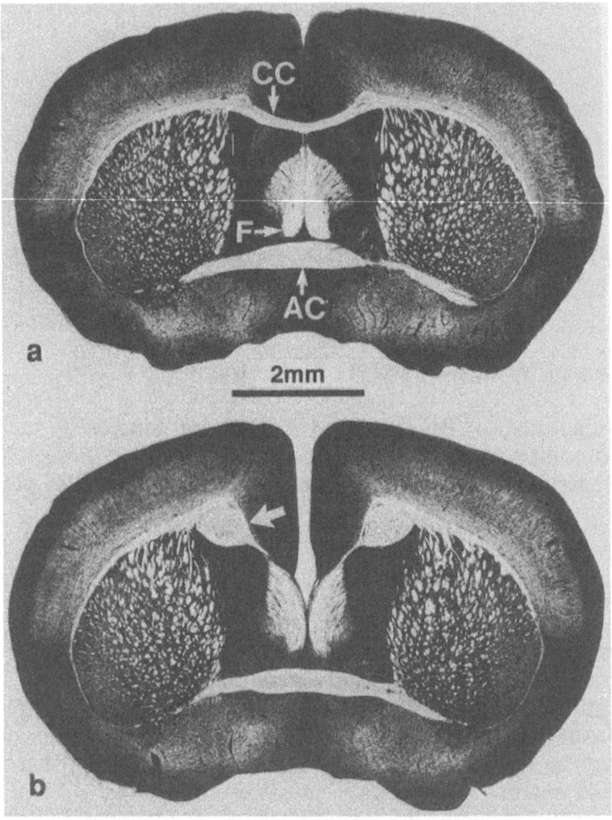

Figure 1. Coronal sections at the level of the anterior commissure (AC) and columns of the fornix (F) stained with gold chloride for myelin (Schmued, 1990) and printed directly on photographic paper from the stained slide to give a negative image. **A.** Normal adult B6D2F_1/J female with brain weight 0.420 g at 103 days of age. The corpus callosum (CC) spans the two cerebral hemispheres. **B.** I/LnJ male brain weighing 0.405 g at 178 days of age. There is no CC and the callosal axons have grown longitudinally to form the Probst bundle (arrow).

measures. For example, in 99 matched pairs of BALB/c litter mates where only one has an absent or abnormally small corpus callosum, there is definitely no expansion of the sizes of the hippocampal or anterior commissures (see Table 2) and the sizes of the commissures do not correlate significantly with size of the abnormal corpus callosum (Wahlsten and Jones, 1983).

Table 1. Penetrance (%) of corpus callosum (CC) defects in four inbred strains.*

Strain	Abnormally Small or Absent CC	Totally Absent CC
I/LnJ	100	100
129/J	70	17
BALB/cWahl	55	20
ddN	8	4

* Lipp and Wahlsten (1992), plus unpublished data from our recent research.

Table 2. Mean cross-sectional areas (sq-mm) of commissures at the midsagittal plane for 99 matched litter mate pairs of BALB/c mice and t tests of significance.*

Measure	Sib with Normal CC	Sib with Abnormal CC	t	P(2-tail)
Area of CC	1.180	0.478	22.91	–
Area of ACA	0.079	0.079	0.70	.48
Area of ACP	0.057	0.056	1.54	.13
Area of HC	0.267	0.270	-0.99	.32

* Data from Wahlsten and Jones (1983). HC=Hippocampal commissure; ACA, ACP=Anterior and posterior part of anterior commissure..

Contrary to earlier claims that absence of the mouse corpus callosum is a single locus characteristic (Keeler, 1933), crosses of BALB/c, 129 or I/Ln mice with a normal strain such as C57BL/6J reveal that inheritance is autosomal, recessive and polygenic (Lipp and Waanders, 1990; Wahlsten, 1982b). In a backcross to the abnormal parent strain, offspring with obvious defects of the corpus callosum are much less common than is expected for a single mendelian gene. Major strains with callosal agenesis have genetic similarities of about 60% across a large number of loci (Table 3; see Roderick, 1980), indicating they are not closely related genetically, but when they are crossed to form F_1 hybrids (Table 4), only the BALB/c x 129 groups yields entirely normal offspring (Livy and Wahlsten, 1991). This demonstrates that mice of different strains can suffer from an anatomically similar defect for genetically different reasons, and it confirms that inheritance involves more than one locus.

Table 3. Genetic similarity (% of loci with the same allele) of five inbred mouse strains.*

Strain	129	BALB/c	ddN	C57BL/6
I/Ln	72	68	60	56
129	–	71	64	58
BALB/c	–	–	64	58
ddN	–	–	–	27

* Data for 129/J, BALBcJ, I/LnJ and C57BL/6J are based on Roderick (1980). At least 29 loci are involved for each comparison. Data for ddN are from Wahlsten et al. (1992) and involve 12 loci.

Table 4. Defective corpus callosum (% of mice) in F_1 hybrid crosses of inbred strains.*

Strain	I/Ln	129	BALB/c	ddN	C57BL/6
I/Ln	100	73	69	44	0
129	–	70	0	8	0
BALB/c	–	–	55	17	0
ddN	–	–	–	8	0

* Data for crosses of I/Ln, 129 and BALBc are from Livy and Wahlsten (1991), whereas those involving ddN are from Wahlsten et al. (1992). Data for inbred matings are from Table 1.

EARLY CALLOSAL CONNECTIONS

In order to better understand the nature of the hereditary defect, we have found it helpful to study the corpus callosum at the earliest stages of its formation. The corpus callosum forms prenatally in mice and humans alike (Glas, 1975; Rakic and Yakovlev, 1968). Until recently it was very difficult to state with any confidence precisely when the callosal axons first reach and cross the midsagittal plane or which regions of the cerebral cortex contribute to the earliest corpus callosum. Now it has been discovered that carbocyanine dyes applied to paraformaldehyde fixed tissue slowly diffuse along the lipid bilayer of the cell membrane and eventually extend along the entire axon as well as dendritic processes (Godement et al., 1987; Honig and Hume, 1989). This technique can be used to trace early callosal connections in the embryo and fetus.

In a series of 153 normal B6D2F_2/J fetuses varying in developmental age or body weight, a small crystal of the dye is placed in either frontal, parietal, temporal or occipital cortex, and after three to four weeks it has diffused to the tips of the callosal axons where growth cones can be seen (Ozaki and Wahlsten, 1992a). We have found that the earliest callosal axons begin to grow out from their cortical cells of origin (Figure 2A) at about 15.5 days after conception for all four cortical regions. After a few hours the axons have grown towards the midsagittal crossing point (Figure 2B) and before long they reach midplane (Figure 2C) just ventral to the longitudinal cerebral fissure. Usually there are a few pioneer axons advancing beyond the main bundle of axons, most notably when the axons have crossed to the opposite hemisphere (Figure 2D). Axons from frontal cortex grow more rapidly than those from the other regions of cortex and reach midplane at about 16.2 days from conception. This sample of genetically heterogeneous embryos provides a standard for normal callosal development.

We have found that the emergence and growth of callosal axons are evidently normal in acallosal mice until the axons reach midplane (Ozaki and Wahlsten, 1992a). In an I/Ln fetus at El7.0 or 0.7 g body weight (Figure 3A), a large bundle of axons coming from the left frontal cortex has reached midplane and made contact with the meninges but cannot cross; meanwhile, axons from the right parietal cortex are approaching midplane on schedule and show abundant growth cones. Soon after reaching midplane and finding no bridge to the other side, in I/Ln the axons often loop back and form a large whorl that is evident as the Probst bundle (Figure 3B). Apparently, the problem with callosal agenesis in these mice occurs in the substrates of axon guidance at the midsagittal plane, not in the cells of origin or the axons themselves (Silver et al., 1982; Wahlsten, 1987). This conclusion is supported by the finding that in many BALB/c and ddN mice the callosal axons eventually *are* able to cross to the other side and form normal connections (Olavarria et al., 1988; Ozaki et al., 1987).

In the I/Ln embryo at 0.5 g there is a distinct cleft between the hemispheres and no hippocampal commissure has yet formed (Wahlsten and Bulman-Fleming, 1990). In BALB/c embryos there is a similar gap, and in more mature BALB/c fetuses at 0.8 g, it is sometimes apparent that the narrow cleft has become a large chasm and the Probst bundles have started to form. In many BALB/c mice the callosal axons are able to cross to the opposite hemisphere despite the chasm. This occurs when the callosal axons cross over the dorsal surface of the hippocampal commissure (Wahlsten, 1987). The corpus callosum at midplane then grows by fasciculation of incoming axons along those axons which have already crossed.

This intimate association between callosal and hippocampal axons at midplane early in development makes it almost impossible to discriminate between the two structures until they are more mature. The two commissures can be treated as one large structure for the purpose of measuring cross-sectional area for morphometry and genetic analysis. This is quite appropriate, because in BALB/c the formation of the hippocampal commissure is markedly retarded (Wahlsten, 1987), even though the structure attains normal size after birth. For statistical analysis, it is convenient to relate the areas of the combined corpus callosum and hippocampal commissure to fetal body weight. In a series of normal mice the relation is obviously quadratic and variance around the line of

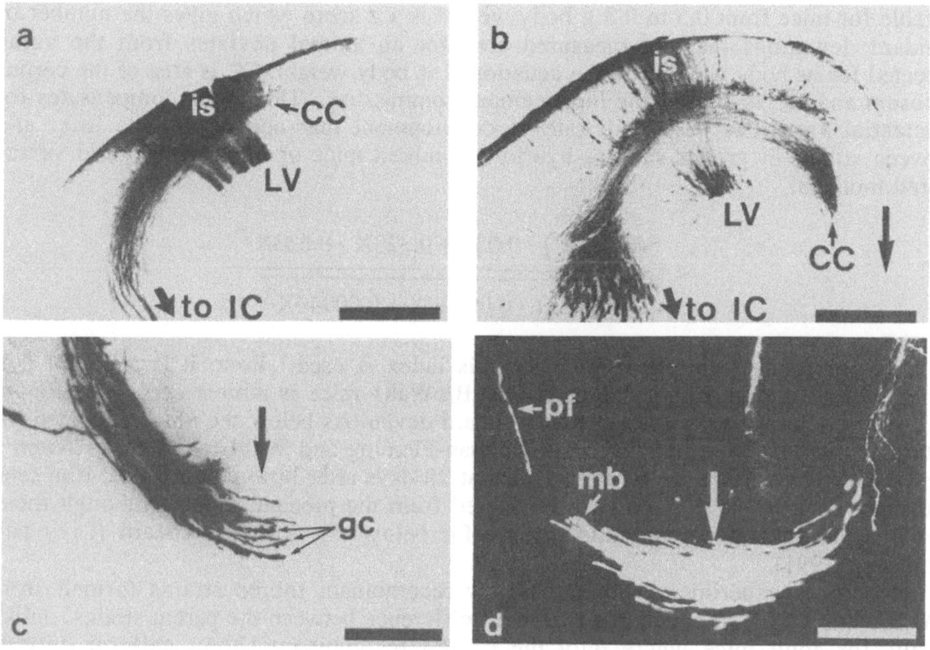

Figure 2. Fetal brains from normal hybrid $B6D2F_2$/J or $B6D2F_1$/J x BALB/cWahl mice labeled with carbocyanine dyes DiA (a,b,c) and DiI (d). **A.** Embryo weighing 0.42 g at E15.5 days after conception. The dye crystal was placed in occipital cortex and has diffused along axons going to the thalamus via the internal capsule (IC). The axons of the corpus callosum (CC) have just emerged from cells near the injection site (is). Cell processes extending to the surface of the lateral ventricle (LV) have also acquired the dye. Bar = 0.5 mm. **B.** Embryo weighing 0,59 g at E17 with dye in frontal cortex. CC axons have grown a considerable distance towards the midsagittal plan (large arrow). Bar = 0.5 mm. **C.** Embryo weighing 0.67 g at E17 with dye in frontal cortex. The CC axons have just crossed midplane (large arrow) and growth cones (gc) are evident. Bar = 0.2 mm. **D.** Embryo weighing 0.70 g at E16.2 with dye in frontal cortex. The main bundle (mb) of CC axons has crossed midplane (large arrow) and a pioneering fiber (pf) has advanced far ahead of the main bundle. Bar = 0.2 mm.

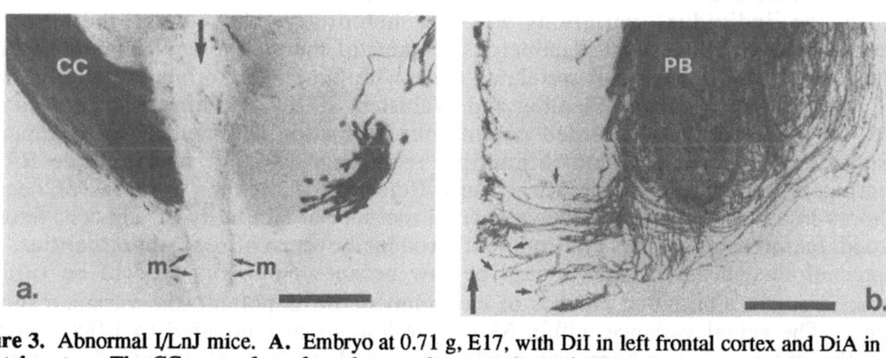

Figure 3. Abnormal I/LnJ mice. **A.** Embryo at 0.71 g, E17, with DiI in left frontal cortex and DiA in right parietal cortex. The CC axons from frontal cortex have reached midplane (large arrow) and made contact with the menings (m), whereas those from parietal cortex are approaching midplane. **B.** Fetus at 0.98 g, E17, with DiI in parietal cortex. CC axons (small arrows) approach midplane (large arrow) and then turn abruptly, returning to a large whorl of axons in the Probst bundle (PB). Bars = 0.2 mm.

best fit increases with body size (Wahlsten and Smith, 1989). An index of abnormality suitable for mice from 0.5 to 0.8 g body weight is a z score which gives the number of standard deviations that the measured area for an animal deviates from the value expected for its body weight. In the equation X is body weight, CC is area of the corpus callosum and HC is area of the hippocampal commissure. This index compensates for substantial variations in overall rate of development that occur within a litter and between strains or groups such as hybrid and inbred mice or mice with hybrid versus inbred mothers.

$$z = \frac{(CC+HC) - 0.050 + 0.323X - 0.633X^2}{\sqrt{0.00231 - 0.00920X + 0.00940X^2}}$$

Two surprising results emerge when this index is used. First, it is apparent that penetrance of the hereditary defect in BALB/cWah1 mice is almost complete (Figure 4A). Most fetuses are more than four standard deviations below the value expected for hybrid fetuses of the same body size (Bulman-Fleming and Wahlsten, 1991). When a similar index is devised for weanling mice at 28 days after birth (Figure 4B), it is seen that most BALB/cWah1 mice have recovered from the prenatal defect, although those which never form a corpus callosum are far below the normal standard (Livy and Wahlsten, 1991).

Second, an experiment with the Bailey recombinant inbred strains formed from BALB/c and C57BL/6 points to a two locus difference between the parent strains, unlike results for adult mice where only the CXBG recombinant shows callosal deficits (Wahlsten and Smith, 1989). The mode of inheritance appears to be simpler during early formation of the corpus callosum.

Combining all the data currently available, it appears that there is a threshold for the formation of a normal corpus callosum. Specifically, if in the early fetus weighing less than 0.8 g the index of abnormality is 5.5 standard deviations or more below normal, then that individual's corpus callosum will never grow to normal adult size; a few axons may cross to the other side on the 19th day after conception, but most axons will divert away from the midsagittal plane and form a Probst bundle. For several abnormal strains, there is a high correlation between the severity of the defect seen in fetuses and the frequency of adult callosal defects.

DETERMINANTS OF CALLOSAL AGENESIS

Given that there is a threshold, what determines which fetuses in a litter will suffer callosal agenesis and which will be spared? Dogma in behavioral genetics requires that, because the individual variations are not hereditary, they must therefore be environmental. We investigated numerous features of the uterine environment of large samples of BALB/c fetuses and correlated these with the z index of abnormality of each fetus within a litter (Bulman-Fleming and Wahlsten, 1991). Nothing was consistently related to the severity of retarded commissure formation. When all environmental measures were incorporated into a multiple regression equation, the multiple R^2 for predicting z score was a paltry 5%. The difference (d) between z scores of nearest neighbors in a uterine horn provides a general measure of similarity of adjacent fetuses. If broad features of uterine environment produce severe callosal abnormalities, the variance of this measure (s_d^2) should be low because neighbors tend to be similar, whereas if small, alternating patches of environment are important, the variance should be high. The actual variance within 63 uterine horns was compared to 1,000 samples where the order within a horn was randomized, and it was well within the limits of random orders (Bulman-Fleming and Wahlsten, 1991). Apparently, the variation in severity of corpus callosum deficits arises from sources within the embryo and is not environmental (see also Gärtner, 1990; Kurnit et al., 1987).

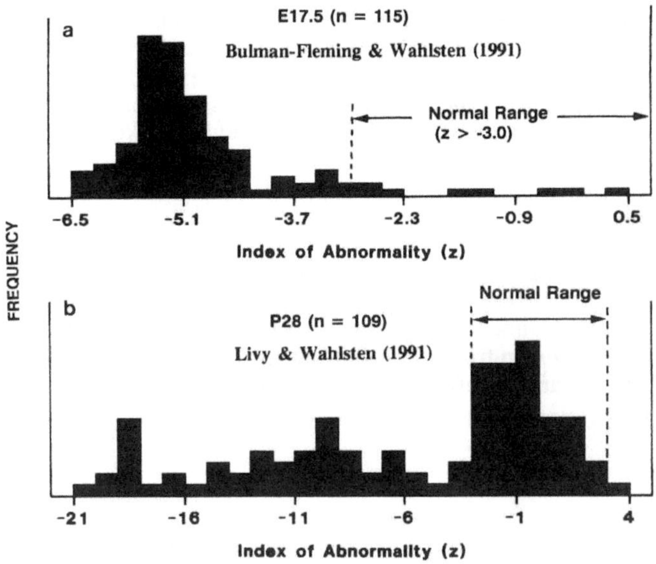

Figure 4. Index of abnormality for BALB/cWah1 mice, indicating the number of standard deviations each animal is from the value expected for its body weight, using a reference group of hybrid mice for comparison. Animals with a z score less than -3 are considered abnormal. **A.** The embryos at E17.5 are almost all less than z = -4. **B.** Most of the weanling mice at P28 have recovered and have a CC of normal size. Based on the data of Bulman-Fleming and Wahlsten (1991) and Livy and Wahlsten (1991).

In summary, inheritance of absent corpus callosum in mice is multifactorial; it depends on differences at several genetic loci as well as inherent variability in embryonic development. Whether a specific animal fails to form an adequate corpus callosum depends on the interaction between at least two processes in the embryo. First, there is a deep cleft between the cerebral hemispheres in abnormal strains at the time when callosal axons arrive at the midsagittal plane, and this cleft widens to form a large chasm two days later. Second, the hippocampal commissure is crossing between the hemispheres during this same period, and if this commissure becomes large enough, the callosal axons can traverse the chasm by growing across the dorsal surface of this commissure. However, formation of the hippocampal commissure is also retarded in strains which sometimes lack the corpus callosum. If the hippocampal commissure has not achieved a sufficient size by the time that the chasm becomes too large, the callosal axons will never be able to cross. Instead, they form a whorl near the longitudinal cerebral fissure for a day or two and then grow longitudinally to form ipsilateral connections. Once the callosal axons have diverted into a longitudinal growth pattern, there is no turning back. The I/Ln strain never has a corpus callosum because it suffers severe impairment of the hippocampal commissure, so severe that this commissure usually does not reach its normal adult size (Livy and Wahlsten, 1991). Putative callosal axons which fail to cross to the opposite hemisphere generally do not reroute through the anterior commissure (Olavarria et al., 1988; Ozaki et al., 1987; Wahlsten and Jones, 1983).

CONCLUSION

There is no arrested development in callosal agenesis. Instead, there are two dynamic and interacting systems whose joint functioning determines the fate of callosal axons. The timing of development of each of these systems varies quantitatively, and sufficiently extreme variations in both processes can exceed a threshold for recovery from an early defect, the midplane cleft. If the substrates for normal axon growth are absent at midplane, the callosal axons follow other cues and form a novel structure, the Probst bundle. Plasticity of prenatal development holds the key to understanding the absent corpus callosum.

Acknowledgments

This research was supported in part by grants EG13953 from the Alberta Heritage Foundation for Medical Research and OGP45825 from the Natural Sciences and Engineering Research Council of Canada to D.W., and an International Research Exchange Program fellowship from the Japan Ministry of Education, Science and Culture to H.S.O. The authors are grateful to Andy Gee, Roy Hillhouse and Violet Sparks for technical assistance.

REFERENCES

Bulman-Fleming, B. and Wahlsten, D., 1991, The effects of intrauterine position on the degree of corpus callosum deficiency in two substrains of BALB/c mice, *Dev. Psychobiol.* 24:395-412.

Gärtner, K., 1990, A third component causing random variability beside environment and genotype. A reason for the limited success of a 30 year long effort to standardize laboratory animals?, *Lab. Anim.* 24: 71-77.

Glas, P., 1975, "Onderzoek naar de Vroege Ontwikkeling van de Commissuren in het Mediane Gebied van het Telencephalon bij de Witte Muis," Drukkerij van Denderen, Groningen.

Godement, P., Vaselow, J., Thanos, S., and Bonhoeffer, F., 1987, A study in developing visual systems with a new method of staining neurones and their processes in fixed tissue, *Development* 101: 697-713.

Honig, M.G. and Hume, R.I., 1989, DiI and DiO: versatile fluorescent dyes for neuronal labelling and pathway tracing, *Trends Neurosci.* 12: 222-341.

Keeler, C.E., 1933, Absence of the corpus callosum as a Mendelizing character in the house mouse, *Proc. Natl. Acad. Sci. U.S.A.* 19: 609-611.

Kurnit, D.M., Layton, W.M., and Matthysse, S., 1987, Genetics, chance and morphogenesis, *Am. J. Hum. Genet.* 41: 33-40.

Lipp, H.P. and Waanders, R., 1990, The acallosal mouse strain I/Ln: Behavioral comparisons and effects of cross-breeding *Behav. Genet.* 20: 728-729.

Lipp, H.P. and Wahlsten, D., 1991, Absence of the corpus callosum, *in*: "Genetically-Defined Animal Models of Neurobehavioral Dysfunction," P. Driscoll, ed., Birkhäuser, Boston.

Livy, D.J. and Wahlsten, D., 1991, Tests of genetic allelism between four inbred mouse strains with absent corpus callosum, *J, Hered.* 82:4599-464.

Olavarria, J., Serra-Oller, M.M., Yee, K.T., and Van Sluyters, R.C., 1988, Topography of interhemispheric connections in neocortex of mice with congenital deficiencies of the callosal commissure, *J. Comp. Neurol.* 270: 575-590.

Ozaki, H.S. and Wahlsten, D., 1992a, Prenatal formation of the normal mouse corpus callosum: a quantitative study with carbocyanine dyes, *J. Comp. Neurol.* 323: 81-90.

Ozaki, H.S. and Wahlsten, D.,1992b, Prenatal development of callosal axons in acallosal mice: a quantitative study with carbocyanine dyes, *Soc. Neurosci. Abstr.* 18: 221.

Ozaki, H.S., Murakami, T.H., Toyoshima, T., and Shimada, M., 1984, Agenesis of the corpus callosum in ddN strain mouse associated with unusual facial appearance (flat face), *Neurosci. Res.* 1: 81-87.

Ozaki, H.S., Murakami, T.H., Toyoshima, T., and Shimada. M., 1987, The fibers which leave the Probst's longitudinal bundle seen in the brain of an acallosal mouse: a study with the horseradish peroxidase technique, *Brain Res.* 400: 239-246.

Rakic, P. and Yakovlev, P.I., 1968, Development of the corpus callosum and cavum septi in man, *J. Comp. Neurol.* 132: 45-72.

Roderick, T.H., 1980, Strain distributions of genetic polymorphisms in the mouse, *in*: "Handbook of Genetically Standardized JAX Mice," H.J. Heiniger and J.J. Dorey, eds., The Jackson Laboratory, Bar Harbor.

Schmued, L.C., 1990, A rapid, sensitive histochemical stain for myelin in frozen brain sections, *J. Histochem. Cytochem.* 38: 717-720.

Silver, J., Lorenz, S.E., Wahlsten, D., and Coughlin, J., 1982, Axonal guidance during development of the great cerebral commissures: Descriptive and experimental studies, in vivo, on the role of preformed glial pathways, *J. Comp. Neurol.* 210: 10-29.

Wahlsten, D., 1982a, Deficiency of corpus callosum varies with strain and supplier of the mice, *Brain Res.* 239: 329-347.
Wahlsten. D., 1982b, Mode of inheritance of deficient corpus callosum in mice, *J. Hered.* 73:281-285.
Wahlsten, D., 1987, Defects of the fetal forebrain in mice with hereditary agenesis of the corpus callosum, *J. Comp. Neurol.* 262: 227-241.
Wahlsten, D., 1989, Deficiency of the corpus callosum: incomplete penetrance and substrain differentiation in BALB/c mice, *J. Neurogenet.* 5: 61-76.
Wahlsten, D. and Bulman-Fleming, B., 1990, Commissure formation in six mouse strains with absent corpus callosum, *Soc. Neurosci. Abstr.* 16: 925.
Wahlsten, D. and Jones, G.B., 1983, Structural changes in brains of mice with agenesis of the corpus callosum, *Soc. Neurosci. Abstr.* 9: 494.
Wahlsten, D. and Smith, G., 1989, Inheritance of retarded forebrain commissure development in fetal mice: Results from classical crosses and recombinant inbred strains, *J. Hered.* 80: 11-16.
Wahlsten, D., Ozaki, H.S., and Livy, D., 1992, Deficient corpus callosum in hybrids between ddN and three other abnormal mouse strains, *Neurosci. Lett.* 136:99-101.

PATTERN OF INTERHEMISPHERIC CONNECTIONS IN MICE WITH CONGENITAL DEFICIENCIES OF THE CORPUS CALLOSUM

Jaime Olavarria,[1] Marcos M. Serra-Oller,[2] Kathleen T. Yee,[3] and Richard C. Van Sluyters[4]

[1]Department of Psychology, NI-25, University of Washington
Seattle, WA 98195, U.S.A.
[2]Remei: 10-12, Esc. B Atica 2B, 08014 Barcelona, Spain
[3]Department of Anatomy, University of Cambridge
Downing St., Cambridge CB2 3DY, U.K.
[4]School of Optometry, University of California
Berkeley, CA 94720, U.S.A.

INTRODUCTION

In the past decades, the main forebrain commissure, the corpus callosum, has received considerable attention from scientists with varied research interests. Much effort has been devoted to understanding the nature of the information shared between the hemispheres via the callosal pathway. In addition, its intrinsic organization and accessibility to experimental manipulations have made this commissural pathway a specially suitable system for investigating the development of highly organized patterns of cortico-cortical connections. This book focuses on analysis of agenesis or malformation of the corpus callosum from clinicopathological and experimental viewpoints as a way of increasing our knowledge about the functional deficits and developmental factors associated with these anomalies, and about the compensatory mechanisms that can become apparent in acallosal subjects.

Factors that disrupt the development of the callosal pathway do not always completely obliterate the formation of the corpus callosum. Often a variable portion of this commissure remains throughout the life of the patient and it can be detected with NMR images or at postmortem examinations. In these cases, it is of interest to determine whether a small callosal remnant can offer an adequate substrate for the retention of function or for the development of compensatory mechanisms. However, our lack of knowledge about the organization of anomalous callosal pathways makes it difficult at present to assess the functional role of callosal remnants.

Clues about the organization of anomalous callosal pathways come from our knowledge of this commissural system in normal individuals. Studies in several mammalian species have revealed that fibers projecting through the callosal commissure are ordered according to the cortical regions from which they originate, so that visual

Callosal Agenesis, Edited by M. Lassonde and
M.A. Jeeves, Plenum Press, New York, 1994

fibers from the occipital cortex cross through the caudal portion of the corpus callosum, while fibers from more rostral cortical regions pass through successively more rostral portions of this commissure (e.g. Pandya et al., 1971; Innocenti, 1980; Olavarria and Van Sluyters, 1986). Furthermore, in adult animals callosal cells and terminations are not distributed uniformly across the expanse of the cortex, but instead they are aggregated in elaborate patterns that are closely related to the underlying topographic representations of the sensory periphery (e.g. Yorke and Caviness, 1975; Wise and Jones, 1976; Innocenti, 1980; Cusick and Lund, 1981; Olavarria and Montero, 1984; Olavarria and Van Sluyters, 1985). Based on these data from normal animals, at least three patterns of interhemispheric connections, each having different functional implications, can be postulated to exist in subjects with a small corpus callosum.

One possible pattern is that in which a small corpus callosum subserves only a limited sector of the neocortex, whose location corresponds to the location of the callosal remnant. Thus, an anterior callosal remnant would connect only portions of the frontal neocortex, leaving the rest of the cortex devoid of callosal connections, while a posterior remnant would connect only portions of the occipital cortex. In these cases, functional deficits would originate preferentially from areas without callosal connections. Alternatively, it is possible that a small corpus callosum connects all areas of the neocortex according to the normal plan, but with a reduced density of innervation. In this case the functional deficits are likely to represent all cortical modalities, with the severity of the deficits being related to the size of the callosal remnant. Finally, connections through an anomalous corpus callosum can bear no resemblance to the normal organization, with the consequence that their presence does not help, but rather interferes with, interhemispheric coordination.

In a recent study (Olavarria et al., 1988), we investigated the organization of anomalous callosal pathways using as a model mice of the BALB/c strain, in which about 30% of individuals show congenital deficiencies of the callosal commissure ranging from a total absence of the corpus callosum to a moderate reduction in its size (Wahlsten, 1982). In these mutant mice, due to prenatal factors that cause a delay in the fusion of the hemispheres and in the formation of a glial guidance mechanism (Silver et al., 1982; Katz et al., 1983; Wahlsten, 1984, 1987), a variable contingent of the callosal fibers that approach the midplane fail to cross over to the other hemisphere. Instead, these fibers are diverted ipsilaterally along a pair of anomalous elongated neuromas, called longitudinal bundles or bundles of Probst (Probst, 1901; King, 1936), that develop adjacent to the midsagittal fissure. The purpose of our study was to determine the pattern of interhemispheric connections that is present in mice with a reduced corpus callosum.

Multiple intracortical injections of horseradish peroxidase (HRP) were used to reveal the overall distribution of callosal connections in BALB/c mice with varying degrees of deficiency of the callosal commissure. In addition, fine details of the topography of the callosal pathway in these mice were studied by making single, restricted injections of HRP conjugated with wheat germ agglutinin (WGA-HRP) into different regions of the cortex. The details of this work, including complete descriptions of the methods used and the number of animals studied in each experimental condition, are presented elsewhere (Olavarria et al., 1988).

MULTIPLE INJECTIONS OF HRP

We first investigated the overall distribution of callosal connections in mice with severe anomalies of the corpus callosum. We placed multiple injections of HRP into the occipital cortex of a number of normal and callosally deficient mice and then examined the resulting patterns of callosal connections, either in tangential sections through the flattened cortex (Figure 1), or in serial reconstructions from coronal sections (Figure 2). With this approach we expected to be able to determine whether callosal connections originated from all areas of the neocortex, as in normal animals, or whether a small corpus callosum carried projections from only a restricted sector of the neocortex. In addition, we wanted to reveal the fine features of the overall pattern of callosal cells in animals with a reduced corpus callosum, both in the direction tangential to the cortical surface as well as with respect to the cortical layers.

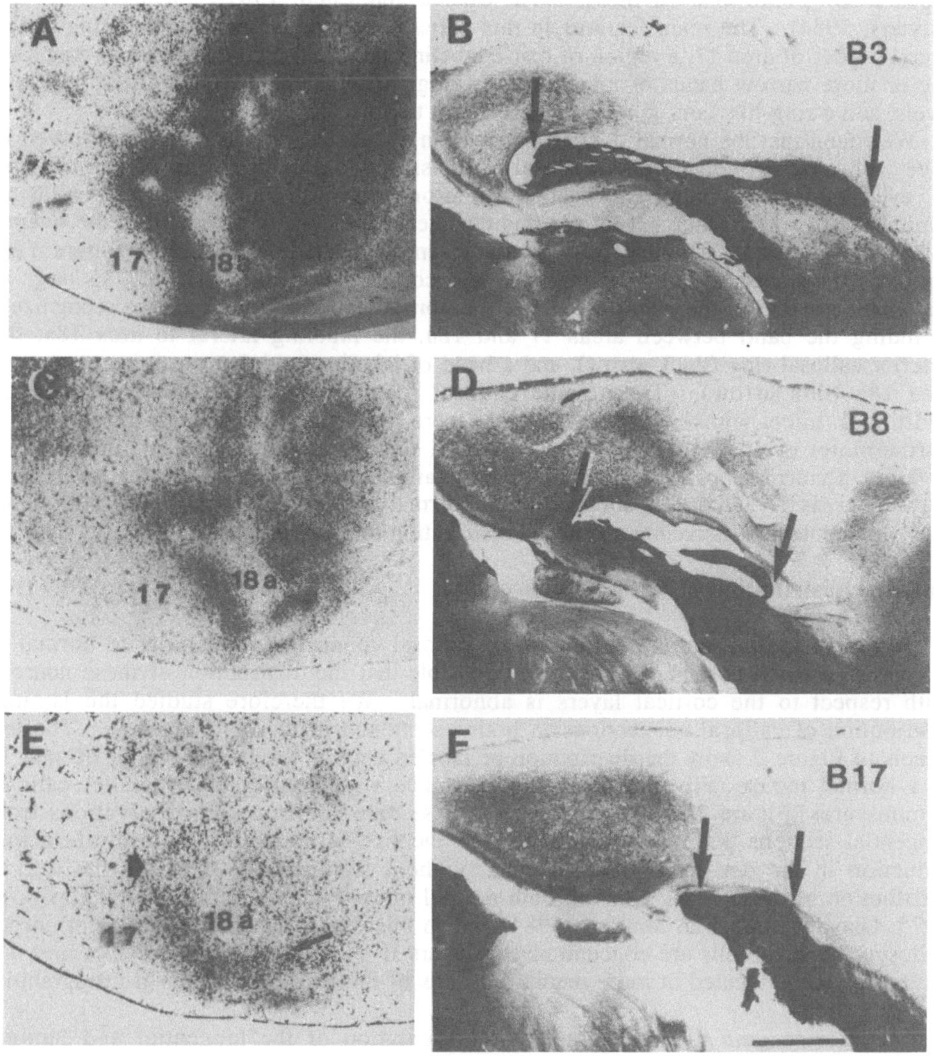

Figure 1. Overall pattern of callosal connections in mice with callosal commissures of different size. Left panels show brightfield photographs of unstained, HRP-tested, tangential sections through supragranular layers of the right hemisphere after multiple injections of HRP into the opposite hemisphere. These sections show the distribution of HRP-labeled cells (dark areas) in posterior neocortex and portions of parietal cortex. Anterior is up, medial to the left. Right panels show brightfield photographs of myelin-stained midsagittal sections showing the size of the corpus callosum (between arrows) in each mouse. Posterior is to the left. Photographs in **A,B** are from a mouse whose corpus callosum is of normal size, while those in **C,D** are from a mouse with moderate deficiency of the corpus callosum, and those in **E,F** are from a mouse with marked reduction of the callosal commissure. Thick arrow in E indicates the anterior callosal ring, and the thin arrow indicates a band of labeling bridging area 18a. These features are more easily distinguished in A and C because of the higher density of the labeling. Scale bar = 1.0 mm. (Olavarria et al., 1988. Reprinted with permission).[1]

Tangential distribution of callosal connections

The left panels in Figure 1 show the distribution of callosal connections over large portions of occipital and parietal cortex in three mice whose callosal commissures (right panels) differed markedly in size. In mouse B3 (Figure 1A), whose corpus callosum (Figure 1B) was normal in length, the accumulation of retrogradely labeled callosal cells and anterogradely labeled terminations forms an elaborate pattern that is normal in appearance (Yorke and Caviness, 1975; Cusick and Lund, 1981; Olavarria and Van Sluyters, 1984). The main features in this pattern are a dense band of labeling at the lateral border of area 17, a region of dense labeling lateral to cytoarchitectonic area 18a, one or more narrow bands of labeling that bridge area 18a at different anteroposterior levels, and a ring-like configuration anterolateral to area 17.

We found that the normal overall distribution of labeled cells is clearly preserved in mice with progressively smaller callosal commissures (e.g. mice B8 and B17 in Figure 1), even though the number of labeled neurons decreases as the corpus callosum diminishes in size. Thus, except for this change in the density of labeling, the overall pattern in mouse B8 (Figure 1C) closely resembles that in mouse B3 (Figure 1A). Similarly, in mouse B17 (Figure 1E, F) whose corpus callosum is only 20% as large as that in mouse B3, all the main features of the normal callosal pattern can be recognized, including the band between areas 17 and 18a, the labeling lateral to area 18a, the anterior callosal ring (thick arrow), and a band of labeling bridging posterior aspects of area 18a (thin arrow). These same callosal features were also clearly visible in additional mice whose corpora callosa were as small as that in mouse B17. Furthermore, even fine details of the callosal pattern are preserved in mice with a severely abnormal corpus callosum. For example, the width of the callosal band between areas 17 and 18a is similar in both normal and mutant mice, and in both cases this band contains a fairly homogeneous distribution of labeled cells.

Laminar distribution of callosal connections

Although the tangential distribution of callosal connections is similar to normal in mice with a small corpus callosum, it is possible that the distribution of these neurons with respect to the cortical layers is abnormal. We therefore studied the laminar distribution of callosal connections in brains sectioned in the coronal plane. The left panels in Figure 2 show the distribution of labeled cells at the lateral border of area 17 in a normal mouse (Figure 2A) and in two mice with progressively smaller callosal commissures (Figure 2C, E). Analysis of these data confirms our observations from tangential sections that marked reductions in the size of the corpus callosum lead to a reduction in the density of labeled cells, without significantly affecting the overall distribution of these cells. Thus in both normal mice (Figure 2A; Yorke and Caviness, 1975; Olavarria and Van Sluyters, 1984) and in mice with a markedly deficient corpus callosum, labeled cells are concentrated in layers II-VI at the lateral border of area 17, with a few cells located in more medial portions of this area, especially in infragranular layers.

In order to gain a more complete representation of the tangential and laminar distributions of callosal cells in mice with a very reduced corpus callosum, the brain of mouse B7 (Figure 2E, F) was serially reconstructed by using a camera lucida. Labeled cells were found over an area of neocortex that is roughly equal in size to the area infiltrated with HRP in the opposite hemisphere. Moreover, both the tangential and laminar distributions of cells are very similar to those observed following multiple injections of HRP in normal mice. For example, as in normal cases (Yorke and Caviness, 1975; Olavarria and Van Sluyters, 1984; Olavarria et al., 1984), labeled cells accumulate at the lateral border of area 17 and around primary somatosensory cortex (SI), with fewer cells within these areas.

It is interesting to note that in mice with a small corpus callosum, this commissure is often located anteriorly (e. g. mouse B17, in Figure 1F, and mouse B7, in Figure 2F),

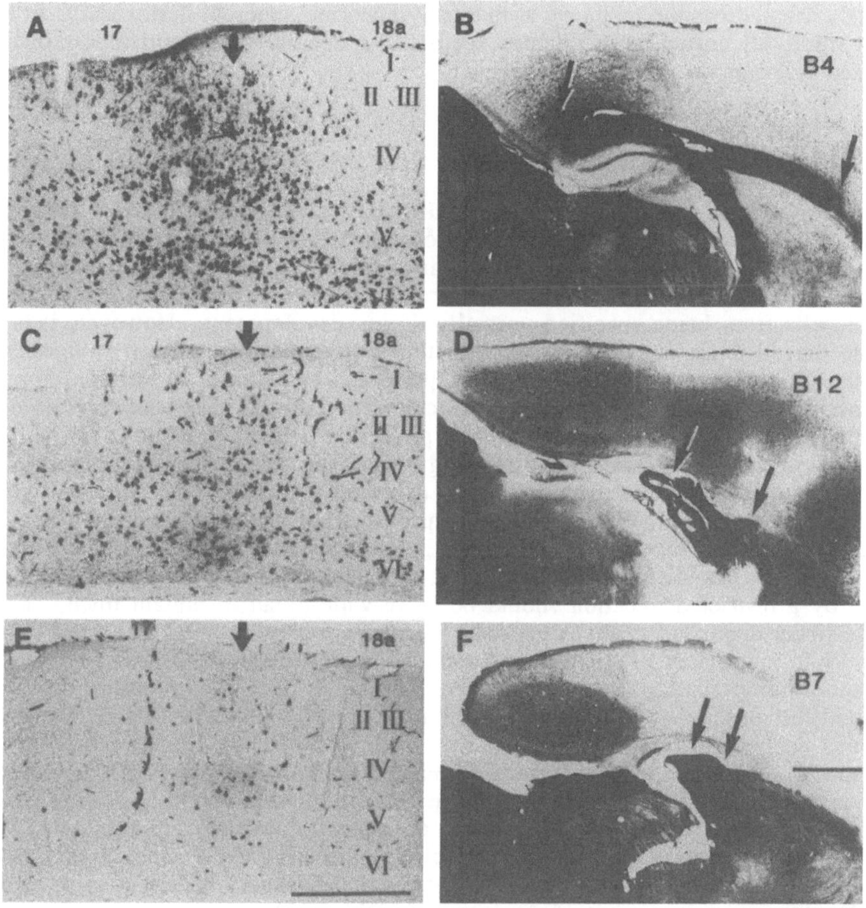

Figure 2. Laminar pattern of callosal connections in mice with callosal commissures of different sizes. Left panels (**A**,**C**,**E**) are brightfield photographs of coronal, HRP-tested sections, showing the laminar distribution of retrogradely labeled cells at the 17/18a border region after multiple injections of HRP into the opposite hemisphere. Medial is to the left. Arrows indicate the border between areas 17 and 18a as determined from adjacent Nissl- or myelin-stained sections. Roman numerals indicate cortical layers. Scale bar in E = 0.5 mm. Right panels (**B**,**D**,**F**) show the size of the corpus callosum (between arrows) in myelin-stained midsagittal sections from each mouse. Panels **A**,**B** are from a mouse whose corpus callosum is of normal size, while panels **C**,**D** are from a mouse with a moderate reduction (greater than 50%) in the size of this commissure, and panels **E**,**F** are from a mouse with a marked reduction (greater than 80%) in the size of the corpus callosum. Scale bar in F = 1.0 mm. (Olavarria et al., 1988. Reprinted with permission).[1]

indicating that visual callosal fibers, which normally course through posterior levels of the interhemispheric fissure, must follow a more anterior route in these mice.

In summary, our observations of the overall distribution of callosal connections following multiple injections of the tracer HRP indicate that the small callosal remnant in mutant mice does not contain fibers from only a restricted sector of the cortex. Instead, we found that all major subdivisions of the neocortex, including frontal, parietal, temporal, and occipital cortex, send projections of comparable strength through markedly deficient callosal remnants. In addition, the cells of origin of these projections aggregate into tangential and laminar patters that closely resemble the corresponding patterns in normal mice. Thus, despite the fact that fibers from some cortical regions, most notably occipital cortex, cross the midplane at markedly abnormal anteroposterior levels in adult mutant mice, all cortical subdivisions seem to be equally successful at establishing organized callosal connections in these mice.

SINGLE INJECTIONS OF WGA-HRP

Our results from multiple injections of HRP show that although they are reduced in number, labeled cells in mice with a small corpus callosum are distributed in regions known to contain dense callosal connections in normal mice. These results indicate that gross abnormalities of the callosal commissure do not prevent the overall distribution of callosal cells from being specified normally during development. However, it is still possible that in these mutant mice, cells within a given cortical region have aberrant contralateral projections. For instance, the presumed availability of terminal space in the neocortex of mice with reduced callosal connections could cause the commissurally connected areas of the cortex to expand their projection fields (e.g. Lent, 1983).

As an initial step toward investigating this issue, we carefully compared the distributions of commissural cells in normal and mutant mice that had received single, restricted injections of WGA-HRP in approximately the same cortical site. If availability of extra terminal space causes a generalized expansion of contralateral projection fields into distant cortical regions, then the distribution of contralateral cells labeled by a restricted injection should be more widespread in mutant mice. This is because tracer deposited at any restricted cortical locus should be taken up not only by cells within contralateral regions that normally project to this locus but also by cells from more remote areas that now project aberrantly into the injection site. We examined this possibility in mice in which single injections of WGA-HRP were made into either occipital or frontal cortex.

Figure 3 compares the distribution of labeled cortical cells resulting from a frontal injection of WGA-HRP in a normal mouse (B2), with that resulting from a similar injection in a mouse with a severely reduced corpus callosum (B1). Inspection of HRP-tested coronal sections through the injection sites (Figure 4) indicates that the injections were of roughly the same size and restricted to gray matter in each of these mice. Furthermore, assessment of the location of the HRP deposits with respect to the cytoarchitectonic map of Caviness (1975) indicates that they are restricted to portions of medial area 6, area 8, and lateral area 24 (see Figure 4). As expected (e.g. Porter and White, 1983), in the normal mouse the WGA-HRP injection produced a densely labeled field in the corresponding contralateral site (see sections a and b from mouse B2 in Figure 3B, and Figure 4A). In addition, less densely labeled fields were observed in more caudal regions of the contralateral cortex, including portions of parietal, occipital, and entorhinal cortex (see sections d-g from mouse B2 in Figure 3B). Because of the small size of the corpus callosum in mouse B1, the number of labeled cells in this animal was greatly reduced in comparison to that in mouse B2. However, the WGA-HRP injection in mouse B1 produced a main field of labeled connections (see sections a and b from mouse B1 in Figure 3A, and field between arrows in Figure 4B) that is similar in size and location to the corresponding field in the normal mouse, as well as an additional number of labeled cells scattered in parietal, occipital, and entorhinal cortex (cf. mice B1 and B2 in Figure 3A and B).

A similar comparison of the distribution of labeled cells in a normal mouse with that in a mouse with a very reduced corpus callosum was carried out after single injections of WGA-HRP into occipital cortex. In both the normal and the mutant mouse, the injections of WGA-HRP fell at the border between architectonic areas 17 and 18a, which in normal animals has dense callosal connections (e.g. Yorke and Caviness, 1975; Olavarria and Van Sluyters, 1984), with some spread into the underlying white matter. In the normal mouse the labeling is largely confined to the posterior neocortex and is densest at the border between areas 17 and 18a and lateral to area 18a. In addition, a

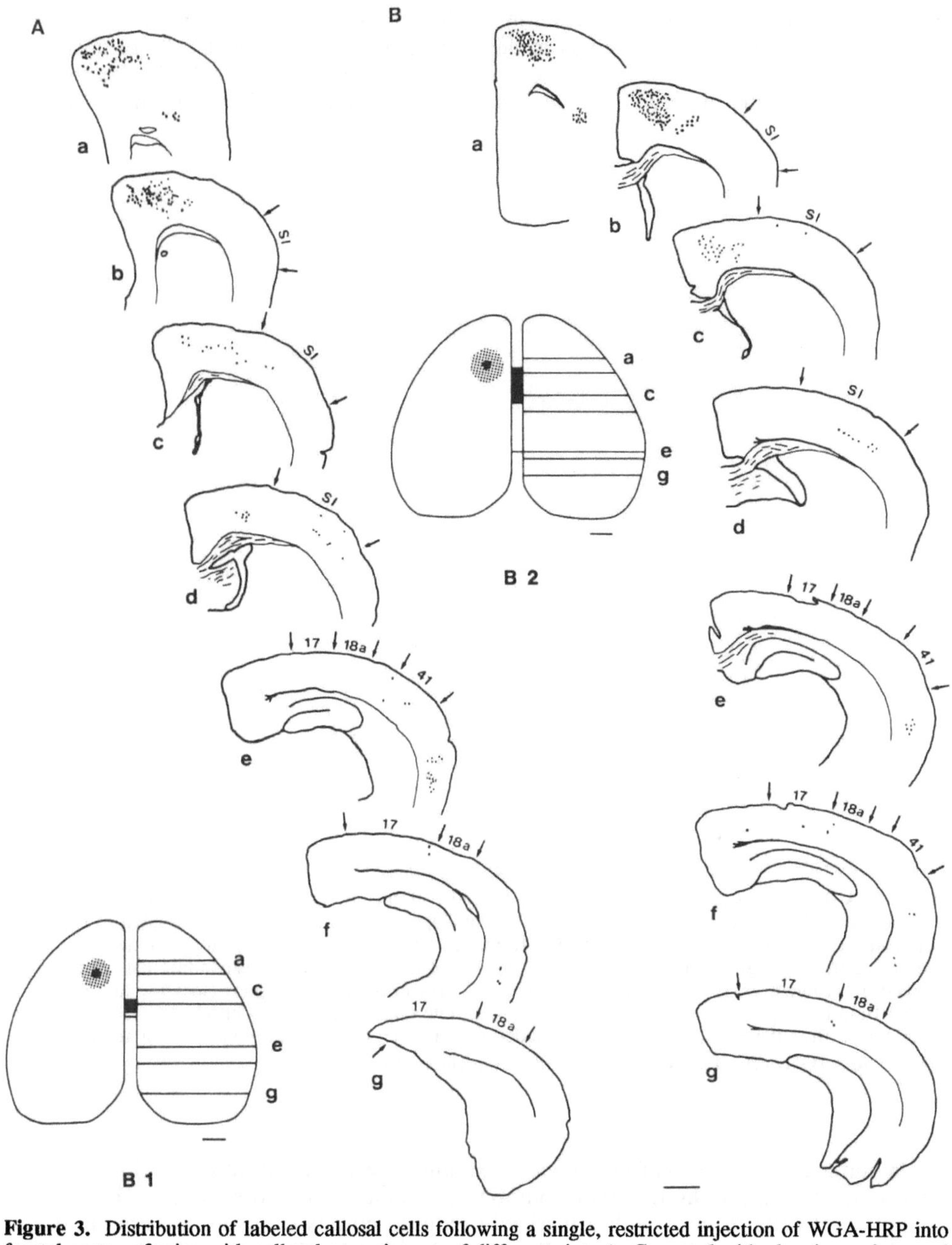

Figure 3. Distribution of labeled callosal cells following a single, restricted injection of WGA-HRP into frontal cortex of mice with callosal commissures of different size. **A.** Camera lucida drawings of coronal sections from mouse B1, which has a markedly deficient corpus callosum. **B.** Drawings from mouse B2, whose corpus callosum was of normal size. The large black dots and shaded areas shown in the inset diagrams indicate the injection sites and regions of enzyme diffusion (see Figure 4). While in the series of drawings from mouse B1 each dot represents one labeled cell, in the series from mouse B2 the density of dots underestimates the density of labeled cells. Portions of the corpus callosum containing HRP-labeled fibers are indicated as black regions in the reconstructions of the corpus callosum shown in the inset. Arrows indicate locations of the borders between cortical subdivisions. SI, primary somatosensory cortex. Scale bars = 1.0 mm. (Olavarria et al., 1988. Reprinted with permission).[1]

few labeled cells are found in frontal cortex, especially within area 8 (Miller and Vogt, 1984). Except for the marked reduction in labeling density in the mutant mouse, labeled cells in this animal are distributed as in the normal mouse, with most cells in posterior neocortex and only a few in frontal cortex. Furthermore, labeled cells in the mutant mouse are found only in cortical regions that contain callosal cells in the normal mouse.

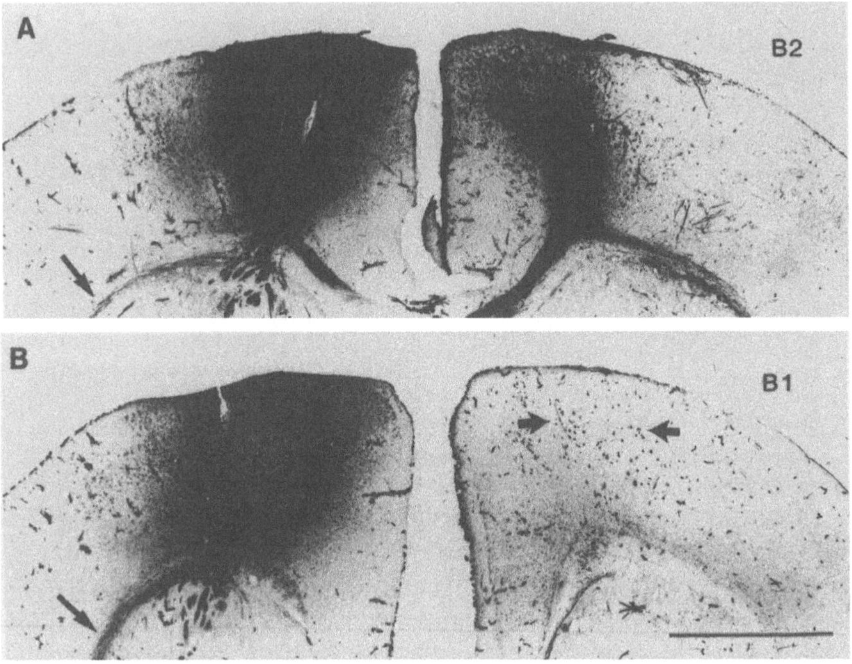

Figure 4. A,B. HRP-tested coronal sections from the brains of mice B2 and B1. These sections were taken at the level of the injections of WGA-HRP, shown in the diagrams of these two mice presented in Figure 3. Note that injections of WGA-HRP into the left frontal cortex of both mice are of approximately the same size and restricted to gray matter. However, the density of labeling in the corresponding part of the right hemisphere is much greater in the normal mouse (**A**) than in the mutant mouse (location and extent of labeled field in **B** is indicated by short arrows). Also note that in both mice densely labeled bundles of fibers extend from the injection sites into the underlying striatum or into white matter (long arrows). Scale bar = 1.0 mm. (Olavarria et al., 1988. Reprinted with permission).[1]

In summary, our results from single injections of WGA-HRP into frontal or occipital cortex show that, although their number is reduced, the distribution of labeled commissural cells in mice with a markedly reduced corpus callosum is very similar to that in normal mice. More specifically, labeled callosal cells in mice with a deficient corpus callosum were found only in regions that also contained labeled cells in similarly-injected normal animals. Thus, labeled cells in mutant mice are not distributed in a more widespread fashion than in normal mice, which argues against the existence of distant aberrant interhemispheric projections in the mutants. This observation suggests instead that gross congenital deficiencies of the callosal commissure do not markedly alter the topography of interhemispheric projections. However, our data do not rule out the existence of small local expansions of projection fields, since these would be more difficult to detect with our experimental paradigm.

Fiber topography within the corpus callosum

The mice that received single injections of WGA-HRP in different cortical sites also provided us with the opportunity to examine the topographic pattern of the fibers *within* the callosal commissure. In these mice, the portion of the corpus callosum containing HRP-labeled fibers was charted onto serial reconstructions of the rostrocaudal extent of this commissure. The fiber topography within the callosum of normal mice is illustrated by the reconstruction for mouse B2 (Figure 3B), which shows that labeled fibers (blackened portions of the corpus callosum) from frontal cortex are located rostrally within the commissure. A similar reconstruction for the normal mouse injected in posterior cortex indicates that the labeled fibers are located posteriorly within the commissure. In contrast, our data from mutant mice reveal that the distribution of fibers within the corpus callosum of severely affected mice is noticeably less orderly than that in normal mice. For instance, while in normal mouse B2 (Figure 3B) fibers labeled by the frontal injection are restricted to the anterior third of the commissure, fibers labeled by a similar injection in mutant mouse B1 (Figure 3A) occupy approximately the rostral two-thirds of the anomalous callosum. A more dramatic alteration in the fiber topography was observed in another mutant mouse, in which a posterior injection of WGA-HRP labeled fibers that course rostrally through the commissure, rather than caudally as occurs in normal mice.

Anterior commissure

Having failed to cross through an anomalous corpus callosum, it is possible that growing callosal fibers are able to find alternative routes to reach the opposite hemisphere. Determining whether migrating callosal axons have this capability is important not only for studies of pathfinding mechanisms in mammalian central nervous system, but also from the perspective of possible compensatory mechanisms for interhemispheric transfer in acallosal subjects. One possible alternative route is the anterior commissure, and Lent (1983), in his study of interhemispheric connections in hamsters with surgically induced deficiencies of the corpus callosum, reported that misrouted callosal axons are able to cross through this commissure.

We investigated whether misrouted callosal axons cross through the anterior commissure in mice with congenital deficiencies of the corpus callosum by carefully examining sections taken through this commissure in mutant mice that had received injections of HRP into one hemisphere. In agreement with previous studies of mutant mice (Jones, 1979; Ozaki et al., 1987), we did not observe HRP-labeled fibers within the anterior commissure in any mouse, irrespective of the degree of callosal deficiency. Since the anterior commissure is the main route for interhemispheric connections in marsupials (see, e.g. Katz et al., 1983), it is unclear why this commissure is not a viable interhemispheric route for neocortical axons in mutant mice. However, the observation that the developing anterior commissure crosses the midplane almost 2 days before the corpus callosum raises the possibility that deviated callosal axons do not use the anterior commissure because they may be unable to grow through a more mature environment (Wahlsten, 1981). In addition to the anterior commissure, there are a number of other pathways that can potentially mediate some form of interhemispheric communications in the absence of the corpus callosum. Perhaps the enhancement of one of these other alternative pathways, and not the anterior commissure, can explain the apparent lack of impairment in subjects with early loss of the callosal commissure (see, e.g. Lassonde et al., 1988).

ORIGIN OF CALLOSAL DEFICIENCIES IN MUTANT MICE

What causes the deficiencies in the corpus callosum that are observed in BALB/c mice? Recent studies of the development of the corpus callosum in normal and mutant animals (Silver et al., 1982; Katz et al., 1983; Wahlsten, 1984, 1987) indicate that the deficiencies of the callosal commissure in mutant mice are mainly attributable to prenatal factors that cause a delay both in the fusion of the hemispheres and in the formation of midline glial guidance mechanisms. According to this view, greater degrees of callosal deficiencies would result simply from longer delays in midline fusion and formation of the glial mechanism (Wahlsten, 1984, 1987). As a result of these delays, early arriving fibers would be unable to cross over to the other hemisphere. Instead, they would be diverted anteroposteriorly to form the longitudinal bundles, while fibers arriving later, after the hemisphere had fused and the glial mechanism had formed, would be able to cross the midplane successfully. These later-arriving fibers that successfully grow across the midplane would have to circumvent local disturbances caused in part by the formation of the anomalous longitudinal bundles. It appears as if deviated fibers grew longitudinally within Probst's bundles, searching for a place to cross into the contralateral hemisphere, until they find the small callosal bridge that is formed at the level of hemispheric fusion. This would explain, in part, our observation that callosal fibers from certain cortical regions cross the midplane at markedly abnormal anteroposterior levels in mice with deficient commissures. Also, these anomalies in fiber trajectory, coupled with the tortuous arrangement of the fibers within the longitudinal bundles, would create rather chaotic conditions for fibers attempting to pass through the small corpus callosum in mutant mice, and this could explain the anomalies in fiber topography that we observe within the corpus callosum of mutant mice.

CONCLUSION

Taken together, our data from multiple and restricted injections of HRP indicate that, except for a decrease in the number of callosal fibers, the topographic distribution of callosal cells and terminations throughout the neocortex is largely preserved in animals with gross developmental abnormalities of the corpus callosum. Our findings suggest that callosal fibers are able to establish appropriate contralateral connections in spite of alterations in their growth route. They also suggest that fiber topography within the corpus callosum does not play an important role in guiding migrating axons to their appropriate contralateral targets. All these considerations would seem to indicate that correct specification of callosal connectivity does not require the complete integrity of midline guidance mechanisms. In this respect, the callosal pathway may be similar to other neuronal pathways in which establishment of connectivity patterns is largely independent of fiber trajectory (e.g. Grant and Ma, 1986). To the extent that similar mechanisms may be responsible for deficits of the callosal commissure in humans, it is likely that, as in the mice we studied, the overall topography of connections through small callosal remnants is normal in affected patients. This possibility may suggest ways for evaluating the functional deficits of these patients and for establishing meaningful correlations between these deficits and various degrees of callosal hypogenesis.

Although our study revealed that fibers crossing through an anomalous callosal commissure can establish appropriate connections, it does not identify the mechanisms that act to specify normal callosal connectivity in mutant mice. It is conceivable that callosal axons can orient their growth by responding to the presence of specific substances. For instance, cells located at corresponding locations in both hemispheres may share a common chemical marker and, once they arrive at the midline, axons from each group of cells may recognize their contralateral counterparts and use them as substrate along which to grow into the corresponding region. Alternatively, the attracting substances may not be bound to specific substrates, but instead they may simply diffuse freely from target regions (e.g. Gundersen and Barrett, 1980; Berg, 1984). Finally, orderly callosal connectivity may also derive from an initially disorderly distribution of axon collaterals by the selective retention of those collaterals that establish contact with appropriate targets and the subsequent elimination of the others (e.g. O'Leary et al., 1981). Whatever mechanism guides callosal axons to their targets, once they are in the target's vicinity, additional local mechanisms may be necessary to fine-tune the final pattern of connections. For instance, there is ample evidence indicating that during development the pattern of incoming thalamic input plays an important role in guiding the invasion of callosal axons into specific cortical regions (e.g. see Olavarria et al., 1987, for review).

Footnote

[1]This figure was originally presented by J. Olavarria, M.S. Serra-Oller, K. T. Yee, and R.C. Van Sluyters in the article entitled "Topography of interhemispheric connections in neocortex of mice with congenital deficiencies of the callosal commissure," published in the Journal of Comparative Neurology, © 1988 Alan R. Liss, Inc.

Acknowledgments

We thank J. Fiorillo and A Haperin for their skillful assistance. The original research described in this chapter was supported by research grants from the National Eye Institute (EY02193) and the National Science Foundation (BNS8418738) to R.C.V.S., a CORE Facility Support Grant (EY03176) from the National Eye Institute and an Individual National Research Service Award to J. Olavarria from the National Eye Institute. Preparation of the manuscript was supported by a grant (EY09343) from the National Eye Institute to J. Olavarria.

REFERENCES

Berg, D.K., 1984, New neuronal growth factors, *Ann. Rev. Neurosci.* 7:149-170.

Caviness, V.S. jr., 1975, Architectonic map of neocortex of the normal mouse, *J. Comp. Neurol.* 164:247-264.

Cusick, C.G. and R.D. Lund , 1981, The distribution of the callosal projections to the occipital visual cortex in rats and mice, *Brain Res.* 214:239-259.

Grant, P. and Ma, P.A., 1986, Optic fibers follow aberrant pathways from rotated eyes in *Xenopus laevis*, *J. Comp. Neurol.* 250:364-376.

Gundersen, R.W. and Barrett, J.N., 1980, Characterization of the turning response of dorsal root neurites toward nerve growth factor, *J. Cell Biol.* 87:546-554.

Innocenti, G.M., 1980, The primary visual pathway through the corpus callosum: Morphological and functional aspects in the cat, *Arch. Ital. Biol.* 118:124-188.

Jones, G.B., 1979, Plasticity of neural projections From cerebral cortex in mice with hereditary agenesis of corpus callosum: A lesion-degeneration study, Unpublished Ph.D. thesis, University of Waterloo, Waterloo, Canada.

Katz, M.J., Lasek, R.J., and Silver, J., 1983, Ontophyletics of the nervous system: Development of the corpus callosum and evolution of axon tracts, *Proc. Natl. Acad. Sci. USA* 80:5936-5940.

King, L.S., 1936, Hereditary defects of the corpus callosum in the mouse, *Mus musculus*, *J. Comp. Neurol.* 64:337-363.

Lassonde, M., Sauerwein, H., McCabe, N., Laurencelle, L., and Geoffroy, G., 1988, Extent and limits of cerebral adjustment to early section of the corpus callosum, *Beh. Brain Res.* 30, 165-181.

Lent, R., 1983, Cortico-cortical connections reorganize in hamsters after neonatal transection of the callosal bridge, *Dev. Brain Res.* 11:137-142.

Miller, M.W. and Vogt, B.A., 1984, Direct connections of rat visual cortex with sensory, motor, and association cortices, *J. Comp. Neurol.* 226:184-202.

Olavarria, J. and Montero, V.M., 1984, Relation of callosal and striate-extrastriate cortical connections in the rat: Morphological definition of extrastriate visual areas, *Exp. Brain Res.* 54:240-252.

Olavarria, J. and Van Sluyters, R.C., 1984, Callosal connections of the posterior cortex in normal-eyed, congenitally anophthalmic and neonatally enucleated mice, *J. Comp. Neurol.* 230:249-268.

Olavarria, J. and Van Sluyters, R.C., 1985, Organization and postnatal development of callosal connections in the visual cortex of the rat, *J. Comp. Neurol.* 239:1-26.

Olavarria, J. and Van Sluyters, R.C., 1986, Axons from restricted regions of the cortex pass through restricted portions of the corpus callosum in adult and neonatal rats, *Dev. Brain Res.* 25:309-313.

Olavarria, J., Malach, R., and Van Sluyters, R.C., 1987, Development of visual callosal connections in neonatally enucleated rats, *J. Comp. Neurol.* 260:321-348.

Olavarria, J., Serra-Oller, M.M., Yee, K.T., and Van Sluyters, R.C., 1988, Topography of interhemispheric connections in neocortex of mice with congenital deficiencies of the callosal commissure, *J. Comp. Neurol.* 270:575-590.

Olavarria, J., Van Sluyters, R.C., and Killackey, H.P., 1984, Evidence for the complimentary organization of callosal and thalamic connections within rat somatosensory cortex, *Brain Res.* 291:364-368.

O'Leary, D.D.M., Stanfield, H.P., and Cowan, W.M., 1981, Evidence that the early postnatal restriction of the cells of origin of the callosal projection is due to the elimination of axonal collaterals rather than to the death of neurons, *Dev. Brain Res.* 1:607-617.

Ozaki. H.S., Murakami, T.H., Toyoshima, T., and Shimada, M., 1987, The fibers which leave the Probst's longitudinal bundle seen in the brain of an acallosal mouse: A study with the horseradish peroxidase technique, *Brain Res.* 400:239-246.

Pandya, D.N., Karol, E.A., and Heilbronn, D., 1971, The topographical distribution of interhemispheric projections in the corpus callosum of the rhesus monkey, *Brain Res.* 32:31-43.

Porter, L.L. and White, E.L., 1983, Afferent and efferent pathways of the vibrissal region of primary motor cortex in the mouse, *J. Comp. Neurol.* 214: 279-289.

Probst, M., 1901, Über den Bau des balkenlosen Grosshirns, sowie uber Mikrogyrie und Heterotopie der grauen Substanz, *Arch. Psychiat. Nervenkr.* 34:709-786.

Silver, J., Lorenz, S.E., Wahlsten, D., and Coughlin, J., 1982, Axonal guidance during development of the great cerebral commissures: Descriptive and experimental studies, in vivo, of the role of preformed glial pathways, *J. Comp. Neurol.* 210:10-29.

Wahlsten, D., 1981, Prenatal schedule of appearance of mouse brain commissures, *Dev. Brain Res.* 1:461-473.

Wahlsten, D., 1982, Deficiency of corpus callosum varies with strain and supplier of the mice, *Brain Res.* 239:329-347.

Wahlsten, D., 1984, Growth of the mouse corpus callosum, *Dev. Brain Res.* 15:59-67.

Wahlsten, D., 1987, Defects of the fetal forebrain in mice with hereditary agenesis of the corpus callosum, *J. Comp. Neurol.* 262:227-241.

Wise, S.P. and Jones E.G., 1976, The organization and postnatal development of the commissural projection of the rat somatic sensory cortex, *J. Comp. Neurol.* 168:313-344.

Yorke, C.H. and Caviness, V.S., 1975, Interhemispheric neocortical connections of the corpus callosum in the normal mouse: A study based on anterograde and retrograde methods, *J. Comp. Neurol.* 164:233-246.

THREE DIFFERENT ANIMAL MODELS OF EARLY CALLOSAL DEFECTS: MORPHOLOGICAL AND BEHAVIORAL STUDIES

Sergio L. Schmidt

Instituto de Biofísica Carlos Chagas Filho
Universidade Federal do Rio de Janeiro and
Departamento de Ciências Fisiológicas
Universidade do Estado de Rio de Janeiro
Cidade Universitaria-21949
Rio de Janeiro, Brazil*
*Presently at the Department of Psychology
University of Alberta
Edmonton, Al, T6G 2E9
Canada

INTRODUCTION

A number of hypotheses have been made about the possible mechanisms used by acallosal patients to overcome the deficits generally related to the disconnection syndrome. The following are the most frequently reported (e.g. Jeeves 1984):

1. Bilateral representation of functions that are normally lateralized.
2. Elaboration of ipsilateral pathways.
3. Increased use of extra-callosal commissures such as the anterior commissure.
4. Behavioral strategies (cross-cueing).

Animal models may be used to test some of these hypotheses, especially those pertaining to the investigation of the biological foundations of asymmetry. In particular, such models allow the possibility of studying the relationship between early callosal defects and laterality.

This chapter is divided into three parts. First we shall describe three different animal models of early callosal defects that have been extensively studied in our laboratory. Then we shall focus on the description of morphological and behavioral asymmetries in animals with early callosal defects, especially the BALB/cCF mice and animals submitted to prenatal gamma irradiation. Finally, we shall consider a hypothesis derived from our data on morphological and behavioral asymmetries and we shall briefly discuss the predictions regarding the laterality of acallosal patients that may arise from this hypothesis.

Callosal Agenesis, Edited by M. Lassonde and
M.A. Jeeves, Plenum Press, New York, 1994

ANIMAL MODELS OF EARLY CALLOSAL DEFECTS

The first model that we have been investigating consists of a special strain of mice, the BALB/cCF. As previously described by Wahlsten (1974; 1981; 1989), about 30% of these mice, hereafter referred to as abnormal, show a significant reduction of the callosal area in adult life. In fact, some of these animals present a total absence of the corpus callosum. In the abnormal group, an aberrant fiber bundle is present medially, underneath the white matter of both hemispheres (Figure. 1D). This anomalous fiber agglomerate, the Probst bundle, is very prominent rostrally at the level shown in Figure. 1D but it tapers smoothly caudalward to disappear at more posterior levels. Apart from the presence of this unusual fiber tract, histological techniques indicate that the cortical cytoarchitecture of the callosally-deficient mice is comparable to that of the animals presenting a normal callosal area (Lent and Schmidt, 1987; see also Olavarria et al., this volume). Because the anatomical defect seems mostly restricted to the corpus callosum, this model is particularly useful in a study aimed at elucidating the relationship between callosal size and specific behaviors.

The second model is based upon the study of early midline transections in newborn Swiss mice (Lent and Schmidt, 1986; 1987). Results derive from mice submitted during their first 24 hours of life to a surgery that consists of a small longitudinal slit made in a region immediately caudal to the transverse venous sinuses continuing from there caudorostrally. At adulthood the brains of these early callosotomized animals are analyzed using several histological techniques. As in the BALB/cCF mice, there are no obvious cytoarchitectonic abnormalities in the cerebral cortex of these early transected animals. Similarly, the myeloarchitecture of most cortical areas of the adult mice operated on the first postnatal day is comparable with that of non operated adults. Even the Probst bundle is absent in the surgically acallosal animals. In fact, only surgeries performed in utero produce the appearance of a Probst bundle (Silver et al , 1982). Although this model may prove interesting in the study of brain plasticity and cerebral reorganization after early postnatal lesions, its contribution may perhaps be less pertinent to the investigation of callosal agenesis, a malformation already present in the first gestational months in humans (see Vergani and Strobelt, this volume).

The third animal model consists of mice born with callosal defects induced by prenatal gamma irradiation (Schmidt and Lent, 1987). Pregnant females are exposed to a Co^{60} gamma source at E15, E16, E17, and E19 (E1 = first gestational day), receiving a total dose of 2 or 3 Gy (1 Gy= 100 rads). At adulthood the brains of the irradiated animals are analyzed using several histological techniques. In contrast with the two other models, the brains of the irradiated animals may present not only callosal defects but also architectural abnormalities. The magnitude of these effects depends upon both the total dose received and the embryonic age at irradiation. Thus, all mice irradiated at E16 with 2 and 3 Gy are acallosal but only those exposed to 2 Gy display the Probst bundle (Figure 1A, 1C and 1E). A small callosal remnant may even be seen in some mice irradiated at E16 with 2 Gy (Figure. 1E). In general, the corpus callosum of the animals irradiated at E17 with 3 Gy is markedly hypotrophic but milder effects are observed in animals irradiated at the same age with 2 Gy. Finally, the size of the corpus callosum of the animals irradiated at E19 with 2 Gy does not differ significantly from that of the normal control group. In summary, these data indicate that prenatal ionizing irradiation provides an effective means of modeling brains with callosal defects.

The three animal models all have in common the presence of a commissural deficit with a various degree of extra-callosal damage (Schmidt and Caparelli-Dáquer, 1989). We have not seen any evidence for a compensatory enlargement of the other commissures in any of these models although the techniques we have used may not have been sensitive enough to quantitatively assess the presence of additional fibers in other commissures. With this caveat in mind, it would nonetheless appear that these acallosal mice differ from the opossum wherein cortical axons cross to the other hemisphere via an enlarged anterior commissure.

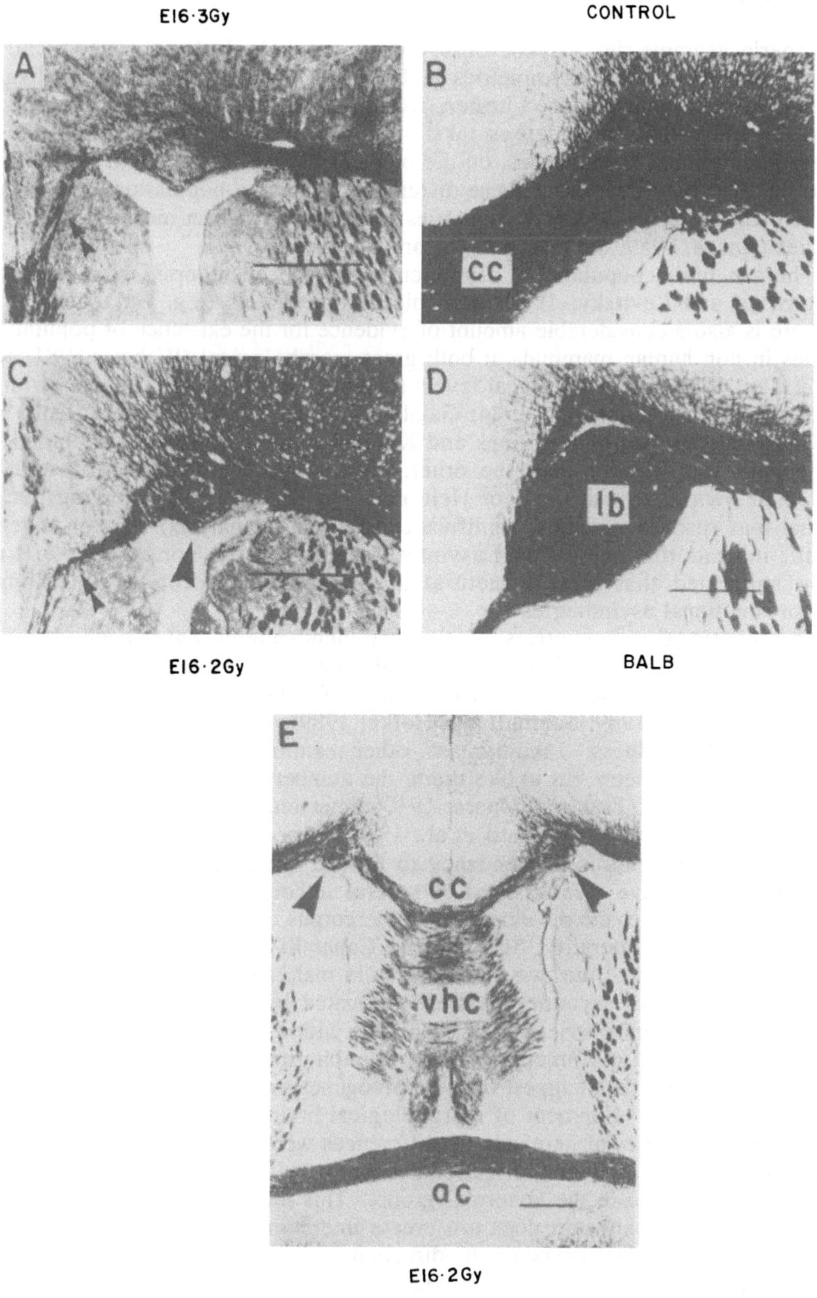

Figure 1- A. The lack of an aberrant longitudinal bundle at the medial sector of the cortical white matter in an animal irradiated at E16 with 3 Gy. The small arrows indicate a few myelinated fibers crossing the midline over the septum. **B.** control brain. **C.** A tiny longitudinal bundle in an animal irradiated at E16 with 2G (arrowhead). **D.** The typical longitudinal bundle (lb) of an acallosal BALB/cCF mouse. **E.** A small callosal remnant and the longitudinal bundle (arrowheads) in a animal irradiated at E16 with 2 Gy. ac = anterior commissure; vhc = ventral hippocampal commissure; Gallyas technique; calibration bar = 300μm.

MORPHOLOGICAL AND BEHAVIORAL ASYMMETRIES IN NORMAL ANIMALS AND IN ANIMALS WITH EARLY CALLOSAL DEFECTS

Hemispheric asymmetries may be observed at morphological, neurochemical and functional levels. Functional asymmetries include not only hemispheric specialization but also lateralized behaviors. Further, asymmetries may be individual and/or populational. Individual asymmetries take only into account the magnitude of the differences; populational asymmetries, on the other hand take into account not only the magnitude but also the direction of these differences. Thus, a populational asymmetry requires the presence of individual asymmetries favoring one side in most animals of the same species (Rosen, 1989; Schmidt and Caparelli-Dáquer, 1989).

In the human brain, populational asymmetries have been reported at macroscopic (e.g. Geschwind and Levitsky, 1968) and microscopic levels (e.g. Galaburda et al., 1978). There is also a considerable amount of evidence for the existence of populational asymmetries in non human mammals at both gross morphological (Webster,1981; Kolb et al.,1982; Tan, 1987) and histological levels (Diamond et al., 1981, 1982; Dowling et al.,1982; Sherman and Galaburda, 1984; Galaburda et al., 1986). In mice, Kolb et al (1982) reanalyzed the data of Heinrichs and Kuhlenkampff (1957) and concluded that one hemisphere was heavier than the other. Similarly, Ward and Collins (1985) combined their own data with those of Heinrichs and Kulenkampff, showing that the populational bias reaches statistical significance (two-tailed analysis). Taken together, these results indicate that populational asymmetries also exist in normal mice. It may further be suggested that these structural asymmetries constitute the anatomical substrate for functional asymmetries .

In man, functional asymmetries of the two hemispheres have been described extensively (see, for instance, Sperry,1982). Among these behavioral asymmetries, hand preference is a remarkable characteristic of adult humans (Oldfield,1971; Levy,1981; Brito et al., 1989; Schmidt and Hofke, 1989), right-handedness being more frequent than left-handedness. Like humans, other mammalian species usually present a clear individual asymmetry but unlike them, the number of left- and right-handers is approximately the same (Tsai and Maurer,1930; Peterson,1934; Cole, 1955; Ettlinger and Moffet, 1964; Collins, 1985; Ward et al., 1987). Another lateralized behavior that has been studied in rodents is the tendency to turn to one side, and again, consistent individual preferences have been reported by several authors (e.g. Glick and Cox,1978).

In order to test whether the development of the corpus callosum exerts any influence on the establishment of laterality, Schmidt and Caparelli-Dáquer (1989) have studied macroscopic asymmetries of the two hemispheres in male adult mice of the BALB/cCF strain (Figure. 2). These asymmetries were evaluated by measuring both the dorsal hemispheric size and hemispheric weight. Animals with a normal callosal area showed a significant populational asymmetry whereas the abnormal group did not display any such asymmetry. These data suggest that the ontogenesis of the corpus callosum plays a role in directing the development of morphological brain asymmetries. On the other hand, when absolute values of hemispheric differences were analyzed regardless of their direction, the presence of a non directional asymmetry was found not only in the "normal" animals but also in the abnormal group. This latter finding indicates that the presence of the corpus callosum does not create interhemispheric asymmetries per se but only has a modulating influence on the direction of these asymmetries.

Functional asymmetries were also investigated in adult male mice of the BALB/cCF strain using paw preference and free swim rotatory behavior (Schmidt and Lent 1991; Schmidt et al., 1991). Both "normal" and abnormal BALB/cCF mice showed individual asymmetries. By contrast, a clear populational tendency for both paw preference (78%) and rotational behavior (75%) favoring the left side were found only in the abnormal group (Figure. 2). Indeed, the "normal" BALB/cCF mice did not present a significant bias to the left or to the right.

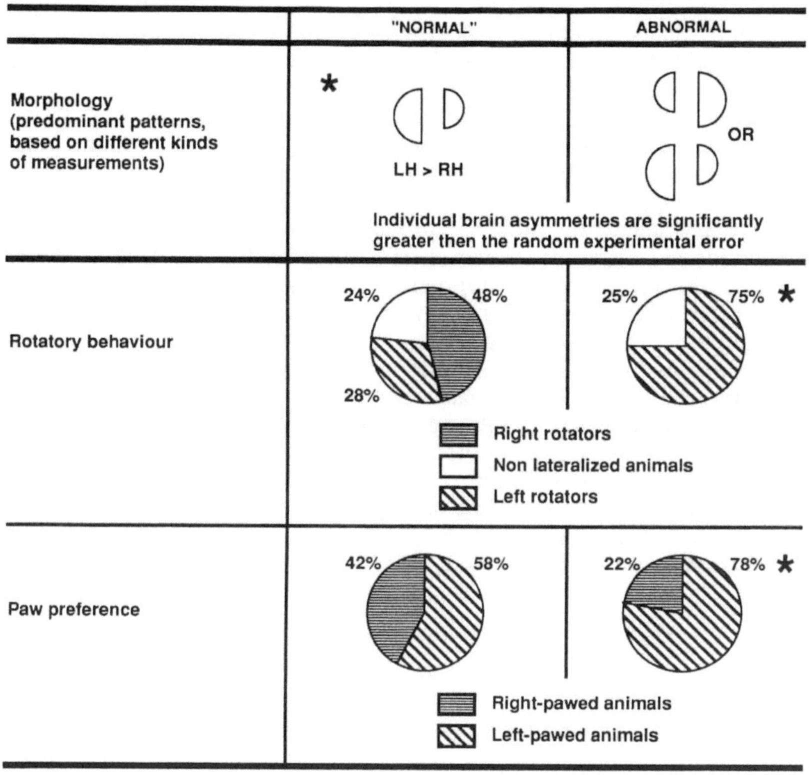

Figure 2. Population asymmetries in the BALB/cCF mouse Note that early callosal defects disrupt normal patterns of populational asymmetries leaving intact individual asymmetries. * indicates a significant populational asymmetry.

MECHANISMS IMPLICATED IN THE ESTABLISHMENT OF POPULATIONAL ASYMMETRIES

The above mentioned results lead to two conclusions. First, absence of cerebral lateralization cannot be postulated as a compensatory mechanism for the absence of the corpus callosum because individual asymmetries were observed even when the corpus callosum was totally absent. Second, the distribution of populational asymmetries may be disturbed as a consequence of early callosal defects. As there is no evidence of a compensatory enlargement of the other commissures, this effect may be directly attributed to the commissural deficit.

On the other hand, it is also possible that the very same factor that prevents the development of the corpus callosum in BALB/cCF mice also disrupts populational asymmetries. Only the study of brain asymmetries in other models of early callosal defects can indicate which of these two explanations accounts for the observed findings. In fact, preliminary results in mice with callosal defects induced by prenatal irradiation (Caparelli-Dáquer and Schmidt, 1991; Manhães et al., 1991) have confirmed the data derived from the BALB/cCF mouse. Thus it would seem that the absence of the corpus callosum per se can account for the disruption of populational asymmetries.

Although this conclusion awaits further confirmation, it nevertheless allows us to speculate about the role of the corpus callosum in the establishment of laterality in humans. Assuming that the early absence of the corpus callosum disrupts the normal pattern of populational asymmetries, leaving intact individual asymmetries, a high rate of left-handed individuals should be expected in cases of callosal agenesis. In fact, some authors (Friefeld, 1988; Schmidt et al., 1990) have reported that the incidence of left-handedness is extremely high (38%- 45%) in cases of callosal agenesis.

If structural asymmetries constitute the anatomical substrate of functional asymmetries, what are the possible mechanisms involved in the establishment of morphological asymmetries? Within the context of cell death occurring because of axonal competition (Oppenheim, 1981), Schmidt and Caparelli-Dáquer (1989) have proposed that the asymmetry of some neocortical areas arises from asymmetrical cell death. The Schmidt and Caparelli-Dáquer (1989)' hypothesis is based upon two sets of findings. First, during early developmental stages, the brain as a whole is asymmetrical, alternating its hemispheric asymmetry (for a review see Bianki, 1988) and second, the number of callosal neurons from each hemisphere follows the general asymmetrical pattern of the brain. Thus, during development, callosal neurons in the larger hemisphere would compete for targets with neurons in the smaller one. The neurons in the smaller side would lose the competition and die, resulting in a permanent smaller hemisphere. Therefore, the corpus callosum would stabilize a preexistent asymmetry by directly influencing the direction of asymmetry toward the normal adult pattern. As most individuals of the same species present similar schedules for neural development, callosal ontogenesis would stabilize a bias toward the same side for most individuals of the same species (populational asymmetry).

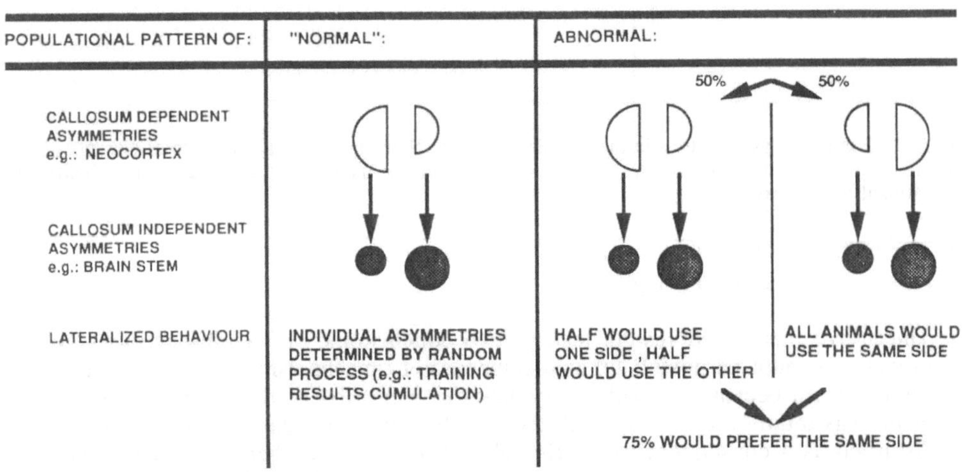

Figure 3. Morphological and behavioral asymmetries in the BALB/cCF mouse. See text for details.

Callosal influence on the establishment of a lateralized behavior may be incomplete in non-human mammals inasmuch as the behavior is under the influence of subcortical as well as cortical mechanisms (Figure. 3). As a consequence, a lateralized behavior that involves these two structural levels may not present a populational asymmetry. Indeed, the callosally-dependent, neocortical asymmetry related to a behavior may act in a direction opposite to the callosally-independent brainstem asymmetry, also involved in the production of the same behavior. By contrast, humans would depend almost totally on neocortical, and thus callosal, influence for most lateralized behaviors, and would consequently show strong populational asymmetries.

CONCLUSION

In conclusion, our data on morphological and behavioral asymmetries in mice with callosal defects indicate that individual asymmetries remain even in the absence of the corpus callosum. However, populational bias favoring one side may be influenced by callosal development and, as a consequence, early callosal damage may disrupt the normal patterns of populational asymmetries.

Acknowledgments

I wish to thank Dr. D. Wahlsten for providing the BALB/cCF breeding pairs. I am also thankful for the comments on the manuscript by my colleagues Alex Manhães and Egas Caparelli-Dáquer. Support for the author's work has been provided by CNNPq-Brazil and FAPERJ.

REFERENCES

Bianki, V.L., 1988, "The Right and Left Hemispheres of the Animal Brain," Gordon and Breach Science Publishers, New York.

Brito, G.N.O. and Brito, L.S.O., Paumgartten, F.J.R. and Lins, M.F.C., 1989, Lateral preferences in Brazilian adults: an analysis with the Edinburgh Inventory, *Cortex* 25:403-415.

Caparelli-Dáquer, E.M. and Schmidt, S.L., 1991, Absence of cerebral asymmetries in mice with callosal defects induced by prenatal gamma irradiation, *Braz. J. Med. Res.* 24:1233-1238.

Cole, J., 1955, Paw preference in cats related to hand preference in animals and men, *J. Comp. Physiol. Psychol.* 48:137-140.

Collins, R.L., 1985, On the inheritance of the direction and degree of asymmetry, *in:* "Cerebral Lateralization in Nonhuman Species," S.D. Glick, ed., Academic Press, New York.

Diamond, M.C., 1985, Rat forebrain morphology: right-left; male-female; young-old; enriched-impoverished, *in:* "Cerebral Lateralization in Nonhuman Species," S.D. Glick, ed., Academic Press, New York.

Diamond, M.C., Dowling, G.A., and Johnson, R.A., 1981, Morphologic cerebral cortical asymmetry in male and female rats, *Exp. Neurol.* 71:261-268.

Dowling, D.A., Diamond, M.C., Murphy, G.M., and Johnson, R.E., 1982, A morphological study of male rat cerebral asymmetry, *Exp. Neurol.* 75:51-68.

Ettlinger, G. and Moffett, A., 1964, Lateral preferences in the monkey, *Nature* 204:4958.

Friefeld S.J., 1988, Somatosensory and motor functions in children with partial and complete agenesis of the corpus callosum, Unpublished Master's thesis, University of Toronto, Toronto, Canada.

Galaburda, A.M., Aboitz, F., Rosen, G.D., and Sherman, G.F., 1986, Histological asymmetry in the primary visual cortex of the rat: Implications for mechanisms of cerebral asymmetry, *Cortex* 22:151-160.

Galaburda, A.M., Sanides, F., and Geschwind, N., 1978, Human brain. Cytoarchitectonic left-right asymmetries in the temporal speech region, *Arch. Neurol.* 35:812-817.

Geschwind, N. and Levitsky, W., 1968, Human Brain: left-right asymmetries in temporal speech region, *Science* 16:186-189.

Glick, S.D. and Cox, R.D., 1978, Nocturnal rotation in normal rats: correlation with amphetamine-induced rotation and effects of nigrostriatal lesions, *Brain Res.* 150:149-161.

Heinrichs, L. and Kuhlenkampff, H., 1957, Die Anzahle der Betzscher Zellen. Über rechts-links Unterschiede im Nervensystem der weissen Maus, *Z. Anat. Entwickl. Gesch.* 120:145-149.

Jeeves, M.A., 1984, Functional and neuronal plasticity: The evidence from callosal agenesis, *in:* "Early Brain Damage," R. Almli, and S. Finger, eds., Academic Press, New York.

Kolb, B., Sutherland, P.J., Nonneman, A.J., and Whishaw, I.Q., 1982, Asymmetry in the cerebral hemispheres of the rat, mouse, rabbit and cat: the right hemisphere is larger, *Exp. Neurol.* 78:348-359.

Lent, R. and Schmidt, S.L., 1986, Dose-dependent occurrence of the aberrant longitudinal bundle in the brains of mice born acallosal after prenatal gamma irradiation, *Dev. Brain Res.* 25:127-132.

Lent, R. and Schmidt, S.L., 1987, Early disconnected brains: developmental plasticity of the cerebral cortex in the absence of the corpus callosum, *in:* "Developmental Neurobiology of Mammals," C. Chagas and R. Linden, eds., Pontifical Academy of Sciences, Città del Vaticano.

Levy, J., 1981, Lateralization and its implication for variation in development, *in:* "Developmental Plasticity," E.S. Goulin, ed., Academic Press, New York.

Manhães , A.C., Schmidt, S.L., Lent, R., Caparelli-Dáquer, E.M., and De Moraes, V.Z., 1991, Lateralized behaviors in mice with callosal defects induced by prenatal ionizing radiation, *IBRO Satellite Symposium on Callosal Agenesis*, Quebec City, Canada (Abstract).

Oldfield, R.C., 1971, The assessment and analysis of handedness: The Edinburgh Inventory, *Neuropsychologia* 9:97-113.

Oppenhein, R.W., 1981, Neuronal death and some related regressive phenomena during neurogenesis: A selective historical review and progress report, *in:* "Developmental Neurobiology - Essays in Honor of Victor Hamburger," W.M. Cowan, ed., Oxford University Press, Oxford, New York.

Peterson, G.M., 1934, Mechanisms of handedness in the rat, *Comp. Psychol. Monogr.* 9:1-67.

Rosen, G.D., Sherman, G.F., Mehler, C., Emsbo, K., and Galaburda, A.M., 1989, The effect of developmental neurophatology on neocortical asymmetry in the New Zealand black mice, *Intern. J. Neurosci.* 45:247-254.

Scherman, G.E. and Galaburda, A.M., 1984, Neocortical asymmetries and open-field behaviour in the rat, *Exp. Neurol.* 86:473-482.

Schmidt, S.L. and Caparelli-Dáquer, E.M., 1989, The effects of total and partial callosal agenesis on the development of morphological brain asymmetries in the BALB/cCF mouse, *Exp. Neurol.* 104:172-180.

Schmidt, S.L. and Hofke, A.A., 1989, Laterality of Brazilian adults and its performance in learning a second language, *Soc. Neurosci. Abstr.* 15:1061.

Schmidt, S.L. and Lent, R., 1987, Effects of prenatal irradiation on the development of cerebral cortex and corpus callosum of the mouse, *J. Comp. Neurol.* 264:193-204.

Schmidt, S.L. and Lent, R., 1991, The effects of partial and total callosal agenesis on the rotatory behavior of the BALB/cCF mice, *Braz. J. Med Res.* 24:417-420.

Schmidt, S.L., Manhães , A.C., and De Moraes, V.Z., 1991, The effects of partial and total callosal agenesis on the paw preference performance in the BALB/cCF mouse, *Brain Res.* 545:123-130.

Schmidt, S.L., Oliveira-Souza, R., Zirreta, J.C., and Caparelli-Dáquer E.M., 1990, Desenvolvimento do corpo caloso e estabelecimento da lateralidade: Análise em pacientes portadores de agenesia do corpo caloso, *Arch. Neuro. Psiquiat.* 48:397.

Silver, J., Lorenz, S.S., Wahlsten, D., and Coughlin, J., 1982, Axonal guidance during development of the great cerebral commissures: Descriptive and experimental studies in vivo or the role of preformed glial pathways, *J. Comp. Neurol.* 210:10-29.

Sperry, R.W., 1982, Some effects of disconnecting the cerebral hemispheres, *Science* 217:1223-1226.

Tan, Ú. and Çaliskan, S., 1987, Allometry and asymmetry in the dog brain: the right hemisphere is heavier regardless of paw preference, *Int. J. Neurosci.* 235:189-194.

Tsai, L.S. and Maurier, S., 1930, 'Right-handedness' in white rats, *Science* 72:436-438.

Wahlsten, D., 1974, Heritable aspects of anomalous myelinated fiber tracts in the forebrain of the laboratory mouse, *Brain. Res.* 68:1-18.

Wahlsten, D., 1981, Prenatal schedule of appearance of mouse brain commissures, *Dev. Brain Res.* 1:461-474.

Wahlsten, D., 1989, Genetic and developmental defects of the mouse corpus callosum, *Experientia* 45:828-837.

Ward, R. and Collins, R.L., 1985, Brain size and shape in strongly and weakly lateralized mice, *Brain Res.* 328:243-249.

Ward, R., Tremblay, L., and Lassonde, M., 1987, The relationship between callosal variation and lateralization in mice is genotype-dependent, *Brain Res.* 424:84-88.

Webster, W., 1981, Morphological asymmetries of the cat brain, *Brain Behav. Evol.* 18:72-78.

MIDLINE SENSORY INTEGRATION IN CALLOSAL AGENESIS

Franco Lepore, Maryse Lassonde, Pierre Poirier,
Alessandra Schiavetto, and Nathalie Veillette

Groupe de Recherche en Neuropsychologie Expérimentale
Département de Psychologie
Université de Montréal
C.P. 6128, Succ. A
Montréal, Qué., H3C 3J7, Canada

INTRODUCTION

Since the pioneering work of Sperry in the '50s and early '60s, a large body of knowledge has accumulated confirming callosal involvement in interhemispheric transfer of lateralized information. The necessity of having a structure that ensures communication between the two half-brains derives from the fact that sensory and motor functions are essentially lateralized, and also from the fact that some higher-order processes, such as language, appear to be primarily controlled by only one hemisphere. The interruption of this interhemispheric link, through surgical intervention to alleviate uncontrollable epilepsy or because of genetic or developmental problems, as in callosal agenesis, typically results in the subjects showing some form of disconnection syndrome.

Data derived from animal studies have ascribed yet another important function to the corpus callosum. Essentially, these findings indicate that the corpus callosum is involved in a number of midline functions, including the process of fusion of the two hemispaces or hemibodies. Just as for mediation in interhemispheric communication, the need for the corpus callosum in midline fusion can be related to the fact that higher-order sensory analysis is essentially lateralized. Thus, somatosensory experience from each half-body projects to the contralateral hemisphere, as does vision for each half-field in superior mammals. In the auditory system, the pattern of separation in the projection of information coming from the two ears is more intermixed. However, the contralateral input to one hemisphere dominates the ipsilateral one by a proportion of approximately 2:1.

When a stimulus is presented to one side of the body or in one half of the visual (and perhaps, auditory) field, and then crosses the midline, it also excites neurons first in one hemisphere and then in the other. Midline fusion ensures that this transition between the hemispheres proceeds smoothly in order to provide continuity of sensation. Similarly, when a stimulus is presented around the midline, some portions project to one hemisphere and some, to the other. To ensure accurate discrimination, it is essential

Callosal Agenesis, Edited by M. Lassonde and
M.A. Jeeves, Plenum Press, New York, 1994

that the lateralized portions be rapidly compared with one another. The following studies examine some of these midline functions for the visual, auditory and somatosensory modalities in callosal agenesis.

VISION

Anatomical experiments carried out on a number of animal species show that callosal efferent neurons and callosal terminal fields are all situated in that portion of visual cortex which contains the vertical meridian (VM) representation of the visual field (Innocenti, 1986b). Electrophysiologically, callosal efferent neurons or callosal recipient cells in lower-order areas, such as the striate cortex, have receptive fields which abut or straddle the vertical meridian (see Figure 1 and Berlucchi, 1972, 1981;

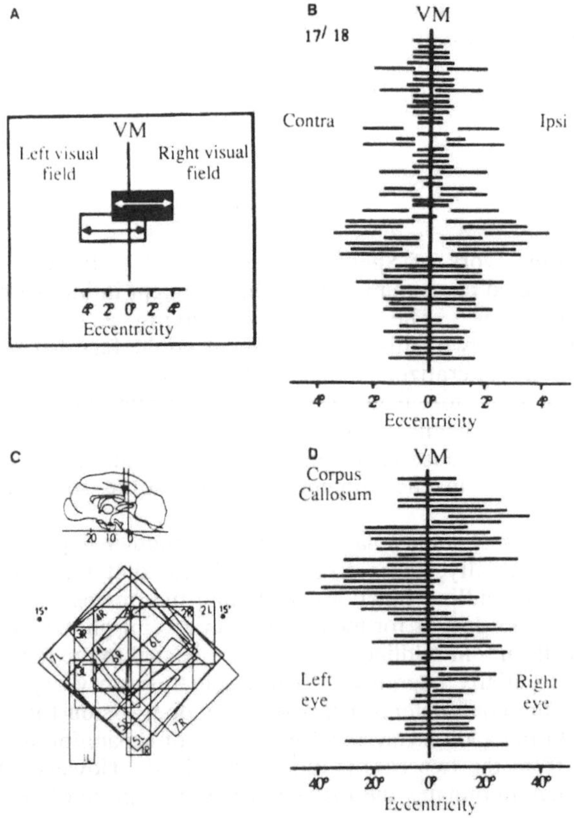

Figure 1. Illustration of the relationship between the vertical meridian and the visual receptive fields (RFs) recorded from the callosal recipient cells of areas 17-18 and from the fibers of the corpus callosum in the cat. **A-** Graphic representation of the treatment to which the RFs were submitted: hypothetical binocular cell RFs are shown for each eye. A horizontal line was drawn across each RF at the point where the medial borders were closest to the vertical meridian (VM). These lines were then aligned with the VM for callosal recipient cells of areas 17-18 (**B**) and for the fibers of the corpus callosum (**D**). **C** shows examples of actual RFs recorded from the corpus callosum. Note, in parts **B** and **D**, that most lines cross the VM (when the two RFs overlapped, only line is shown) or are quite close to it. This is especially remarkable in **D**, where RFs are quite large (compare abscissae in **B** and **D**) and yet overlap or abut the VM.

Lepore and Guillemot, 1982). This finding also applies to some higher-order areas, such as the lateral suprasylvian cortex (Antonini et al., 1983). These anatomical and electrophysiological results have generally been interpreted within the framework of the callosal implication in midline fusion.

The importance of this mechanism for bidimensional stimuli is quite straightforward: it ensures continuity across the midline. For tridimensional objects, however, it may have the additional function of allowing the comparison of the relative depth of stimuli straddling the vertical meridian.

The evidence regarding the involvement of the corpus callosum in the processing of depth information is still a matter of debate. On the one hand, single cell studies carried out in many laboratories, including ours (Guillemot et al., 1993; Lepore et al., 1992), have shown that binocular neurons driven by the ipsilateral thalamo-cortical and the callosal inputs are sensitive to retinal disparity. Disparity is produced experimentally by presenting the visual stimuli to each eye at slightly non-corresponding spatial coordinates. This procedure mimics the situation which exists when a three-dimensional object is viewed by the two spatially separated eyes. The callosum, therefore, presumably is involved in depth discrimination. On the other hand, behavioral experiments attempting to confirm this function show that sectioning the callosum does not always affect depth discrimination. In cats, for example, stereoperception assessed using either random-dot patterns (Lepore et al., 1986b; Ptito et al., 1986) or gratings positioned at different distances from the animal (Timney et al., 1985) is unaffected by callosal transection. Similar results were obtained in the monkey, where a commissurotomy had little effect on a random-dot stereogram discrimination (Hamilton and Vermeire, 1986), and where a callosotomy left intact the estimation of the distance separating two vertical needles situated on either side of VM (Cowey and Wilkinson, 1991).

Contrary to the electrophysiological data, all these behavioral results suggest that the callosum is either not involved in depth discrimination or that it shares this function in a redundant fashion with other neural structures. Both explanations seem to find some support in behavioral experiments carried out in patients. Mitchell and Blakemore (1970), for example, showed that a callosotomized subject failed to discriminate depth for disparate stimuli presented on either side of the midline but had no problem if both stimuli were presented eccentrically on one side. Using random dot stereograms, similar results were obtained by Hamilton and Vermeire (1986) and Hamilton et al. (1987) with commissurotomized patients. However, Ledoux et al. (1977) found no deficit in a callosotomized subject whose anterior commissure was intact, leading to their suggestion that this pathway might suffice to mediate the function (although patient L.B. of Hamilton and Vermeire (1986), with total commissurotomy, thus including the section of the anterior commissure, could discriminate stereopatterns at the midline). Jeeves (1991) compared acallosals and posterior callosotomized subjects to appropriate controls in their ability to stereoscopically evaluate the distance between line stimuli. The two experimental groups were found to have midline deficits for stimuli having uncrossed disparities. When crossed disparities were used or when the stimuli were presented eccentrically, no such problems were observed.

The experiments reported above assessed depth perception based essentially on binocular spatial disparity. There are, however, a number of other cues to depth, some of which may operate even under monocular viewing conditions. One such cue is motion parallax. It is difficult to predict, for stimuli situated on the two sides of midline, how this cue may be integrated interhemispherically in subjects with callosal pathology since perfect performance would require that they compare relative motion between the hemispheres. The only relevant experiment attempting to assess the importance of the callosum for interhemispheric motion processing was carried out by Ramachandran et al. (1986), who showed that the direction of apparent motion produced by alternately flashing spots of light (similar to the phi phenomenon) presented on either side of VM could be discriminated by callosotomized subjects.

Some of us (Rivest et al., in press) examined whether callosal agenesis subjects could use motion parallax to judge the depth of two plates situated on each side of the midline. Four callosal agenesis subjects, two males and two females, with total absence of the corpus callosum were compared to four IQ-matched subjects in this distance estimation task. The plates measured approximately 3° when viewed from a distance of 95 cm. They were equiluminous and were covered with equal numbers of small, randomly dispersed, black and white dots of various sizes. The plates were contained in a box and only their medial edges were visible (interplate separation: 0.6°). A thin white horizontal line drawn on the back of the apparatus served as a reference for scanning in order to produce the motion parallax. The head of the subject was held in a restraining apparatus, which only allowed for small horizontal or vertical movements. During a trial, the two plates were positioned at various distances from each other and the subject had to adjust their relative depth so that they appeared to be on the same plane. The subjects viewed the stimuli monocularly or binocularly, keeping their head still or moving it either up and down or from side to side. Viewing time was not limited.

When compared to controls, acallosals performed normally under binocular viewing conditions, whether head movements were present or not. This suggests that when binocular spatial disparity is available, acallosal subjects show no deficit in estimating the relative depth of two stimuli, each presented on one side of the VM. In the fixed-head condition, the two groups performed poorly, indicating that the task could not be resolved in the absence of both binocular and monocular depth cues. When horizontal and vertical movements were permitted, performance improved dramatically for the control subjects, but not for the agenesis subjects. This failure to improve discrimination with movement cannot be attributed to some nonspecific motor problem present in acallosals since their performance improved to the level of the control subjects if the two plates were presented eccentrically, thus falling within the same hemisphere. Therefore, it appears that motion parallax as a cue for depth perception cannot be integrated across the hemispheres in the absence of the corpus callosum.

These results are instructive in a number of ways. On the one hand, the normal performance manifested when spatial disparity cues were present agree with those of Lassonde et al (1988), who showed that these subjects can discriminate the relative depth of stimuli placed on opposite sides of VM, provided that exposition time is sufficiently long. The results are also in agreement with the animal studies reported above, which show no deficits in disparity discrimination in cats or monkeys following callosal section. The fact that information derived from the relative motion parallax generated by each of the two stimuli cannot be compared across the hemispheres suggests that this task is probably more complex than the one used by Ramachandran et al. (1986) in a commissurotomized subject. The apparent motion produced in the latter case may perhaps be perceived via subcortical commissures, whereas depth perception based on motion parallax might require higher-order cortical processing, thus rendering the comparison impossible without the callosal commissure.

SOMATOSENSORY SYSTEM

As is the case with the visual modality, the somatosensory callosal system has been shown to be closely associated with midline function (Berlucchi, 1972; Innocenti, 1986; Jacobson and Trojanowski, 1974; Pandya and Seltzer, 1986). At the electrophysiological level, most callosal neurons in the anterior CC are functionally related to the midline (Manzoni et al., 1980; McKenna et al., 1981). For instance, over 65 % of all fibers recorded in the callosum represent the trunk and the proximal part of the limbs (Guillemot et al., 1988), a finding which concurs with anatomical work on the distribution of their cells of origin within the callosal zones (Caminiti et al., 1979, Jones and Powell, 1968). Furthermore, the medial borders of most of the unilateral RFs from the trunk and many from the head extend to the midline. Comparable results have been reported for the monkey (Guillemot et al., 1987; Lepore et al., 1986a). In fact, in this animal, nearly 4/5 of all axons recorded in the callosum had RFs corresponding to the trunk and head area. Most RFs representing the trunk were found to be large, covering a significant surface of the skin or deep tissue. The unilateral RF, moreover, either touched or straddled the midline, signifying that the recipient cells would have bilateral RFs which would cover a wide section of the trunk. A number of callosal axons were also found which actually extended symmetrically across the whole trunk on both sides of the midline.

Given these anatomo-physiological data, it would be expected that any pathology of the callosal system which perturbs somatosensory function should mainly affect processing related to the midline. However, almost no attempts have been made to validate this hypothesis in either animals or humans. The experiments about to be described were thus carried out to examine how somatosensory analysis may be affected by callosal pathology.

Somatosensory Acuity Thresholds: Two-Point Discrimination

Two-point discrimination is generally considered as reflecting the capacity of the somatosensory system to detect the fine spatial details inherent in a tactile stimulus, that is, the system's acuity. This measure is influenced by a number of factors, the most important being the region of the body to which the stimulus is applied. The psychometric functions describing these areal differences are best exemplified by the studies of Weinstein carried out some three decades ago (Weinstein, 1968). They have shown that sensitivity is directly proportional to receptor density (and hence to the amount of cortical representation) such that the fingers and face regions have the lowest thresholds and the axial structures, the highest. Irrespective of these organizational constraints, because of the privileged relationship between the callosum and the body midline, it is expected that if acuity changes follow callosal pathology, these would mainly manifest themselves at midline and less at more distal regions.

To test this hypothesis, Corkin's (1964) two-point discrimination procedure was used to measure thresholds on distal, proximal, cranio-axial and axial body regions. Two groups of subjects with callosal pathology were studied. These were matched to two control groups on the basis of age, IQ and handedness. The two experimental groups examined in this study were callosal agenesis subjects (the same subjects as above) and patients who underwent callosal section for the relief of intractable epilepsy (Sauerwein et al., 1981). Two of the patients had total callosotomy while in the other two, the splenium was preserved. In the latter subjects, however, interhemispheric transfer of tactile information was absent since the somatosensory areas were disconnected by the section of the anterior two-thirds of the callosum. One of the subjects with complete callosotomy, moreover, underwent the operation when he was still quite young. Some of our studies have shown that both acallosals and patients whose callosotomy was carried out before puberty do not always show a disconnection syndrome (Lassonde, 1986; Lassonde et al., 1991). This young patient with total callosal section did not present the disconnection syndrome. Midline deficits in this subject might, therefore, not be as serious as in the three other callosotomized subjects and might, in fact, resemble those of the agenesis subjects.

A detailed description of this procedure has been reported elsewhere (Schiavetto et al., in press). Briefly, two point thresholds were determined by means of a Weinstein Two-Point Aesthesiometer, consisting of a ruler calibrated in millimeters, having at one of the extremities a single probe on one side (0.8 mm tip diameter) and two probes on the other (combined tip surface: 1.6 mm, maximum aperture: 120 mm). A specific skin area was touched with either the single or the double probe, in random order, and the subject had to report whether he (she) felt one or two points. Using a modified staircase procedure, the two-point threshold was defined as the smallest separation at which the subject was able to discriminate the presence of two points in 70% of the trials. All subjects were blindfolded during testing. Different regions of the body were stimulated: distal areas (index finger and palm of both hands), proximal area (the forearm), cranio-axial area (six forehead measurements were taken: on the midline and on each side, using both vertical and horizontal stimulations) and axial region (six measurements were taken in the region of the spine: at midline, and on each side, both vertically and horizontally).

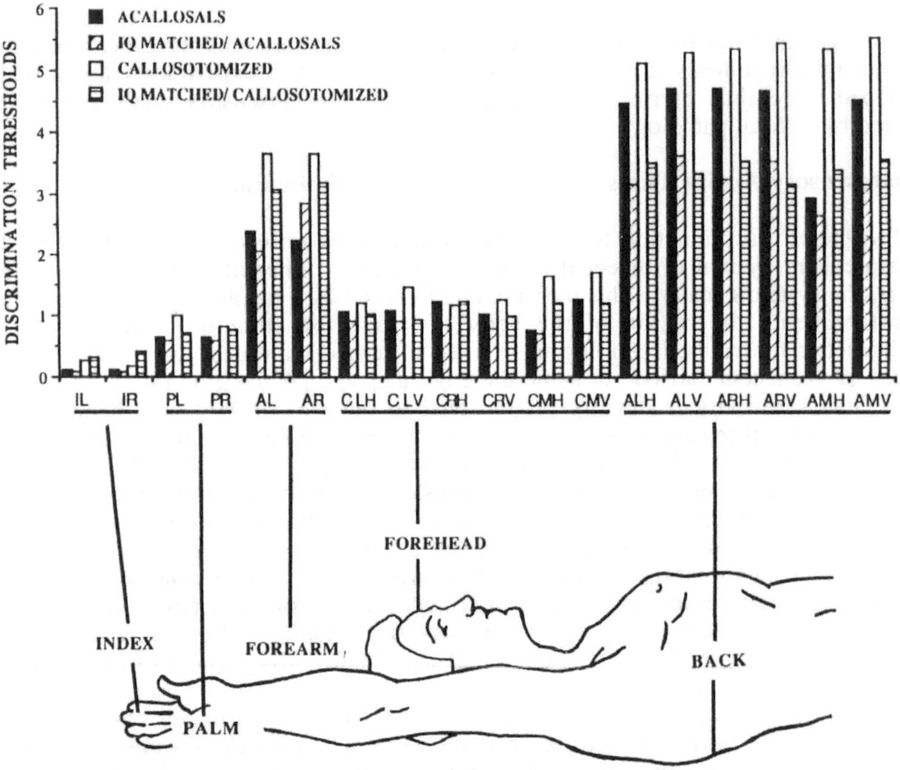

Figure 2. Two-point discrimination thresholds for acallosals, callosotomized and IQ-matched controls for each group. IL=index left; IR= index right; PL=palm left; PR=palm right; AL=arm left; AR=arm right; CLH=cranio-axial left horizontal; CLV=cranio-axial left vertical; CRH=cranio-axial right horizontal; CRV=cranio-axial right vertical; CMH=cranio-axial midline horizontal; CMV=cranio-axial midline vertical; ALH=axial left horizontal; ALV=axial left vertical; ARH= axial right horizontal; ARV=axial right vertical; AMH=axial midline horizontal; AMV=axial midline vertical.

The results obtained by the experimental groups and their matched controls are displayed in Figure 2. To determine whether the observed differences were statistically significant, comparisons were made between these matched groups using ANOVAs. Results obtained from the distal (index and palm) regions show, that sensitivity was extremely high in all groups. There were no significant differences in thresholds between the acallosals and their IQ-matched controls nor between callosotomized subjects and their controls. As for the proximal body region, the mean thresholds also did not differ between the two experimental groups and their respective IQ-matched control groups. Similarly, no significant differences were found between the experimental groups and their respective controls as far as the cranio-axial body region was concerned.

The results were quite different for the axial regions of the body. Mean thresholds for the two groups with callosal pathology were significantly higher in the dorsal trunk area when compared to those of their IQ-matched controls. Differences, moreover, were found at all locations of the trunk tested, including the lateral positions. The early callosotomized patient had thresholds that were closer to those found in the acallosals than those of the adult callosotomized patients. This finding, coupled with evidence from other studies (Lassonde et al., 1991) showing that patients having undergone callosotomy in childhood performed as well as acallosals in interhemispheric integration of tactile tasks, suggests that some compensation might have occurred (Lassonde, 1986; 1988; 1991).

One of the main questions that arise from these results is: why should callosal pathology raise thresholds for this measure of acuity and why only in the trunk region? There may be two, possibly complementary, answers to this first question. On the one hand, a transection of callosal axons (or their genetically determined absence) may result in the loss (or lack of development) of a substantial number of cells in their cortical zones of origin. Indeed, pyramidal cells in layers III and IV, from which callosal axons originate, are less numerous in primary visual areas of callosotomized or acallosal cats and humans (Glickstein and Whitteridge, 1976; Shoumura et al., 1975). This might modify the intrinsic neuronal architecture within these zones so that precise function is perturbed. If similar alterations also occur in somatosensory areas, then higher two-point discrimination thresholds may ensue. Callosal pathology may, on the other hand, affect the excitability of the recipient cells. In the visual system, following a callosotomy, cells in area 17 show decreased responsivity to ipsilateral eye activation, despite the fact that they are still presumably receiving binocular inputs (Payne, 1986). In the somatosensory system, this surgery also reduces the responsivity of bilaterally activated cells in SII, particularly to stimulation of the ipsilateral side of the body (Petit et al., 1990). Higher thresholds to the two-point stimulus may result from this altered excitability of the callosal recipient cells.

Another question to be addressed is why callosal pathology does not modify sensitivity in the other regions of the body. The lack of effect for the distal and proximal areas may simply reflect their under-representation in the callosum. However, the extensive representation of the head area in the callosum, as demonstrated in animals (Guillemot et al., 1987; 1988; 1991), would suggest that this region should also be affected and yet, the results do not bear out this prediction. This may be due to the fact that the head region, in comparison to the trunk, receives more bilateral activation through structures other than the callosum (Kelly, 1985). This redundancy of inputs from the two sides might attenuate the effects of callosal pathology. In order to verify this assumption, the next experiment was designed to evaluate the callosal implication in a somatosensory system known to have a more important bilateral representation, namely the spino-thalamic pathway.

Somatosensory Thermal Thresholds

The anatomical structures mediating tactile discrimination in the study outlined above probably involve the dorsal column-medial lemniscus system. This system is highly lateralized up to its terminal fields in the cortex. The disconnection syndrome observed following callosal transection for a number of tasks, such as stereognosis and texture discrimination, testifies to the importance of this commissure for ensuring bilateral comparisons. The somatosensory system, however, is also made up of a second major system, involving the anterolateral spino-thalamic tract, which may not be as highly lateralized. Thus, sensory information is not only carried to central structures by the controlateral pathway but also to a limited extent via the ipsilateral pathway. In fact, several authors have postulated that the absence of disconnection signs in the tactile modality in callosal agnensis may be due to the increased use of the bilateral representation afforded by this parallel system (for a review, see Jeeves, this volume). Given this lack of strict lateralization and the fact that the proficiency of the ipsilateral pathway may be increased in acallosal subjects, it is possible that the effects of this pathology on sensory function might not be as severe for the body regions normally represented within the callosum.

Thermal information is transmitted to central structures through the spino-thalamic tract. Assessing thresholds with thermal stimulation would therefore constitute an ideal method for examining this parallel somatosensory system. This is precisely what we have done in an experiment in which thermal stimuli were presented to essentially the same regions of the body for which we tested two-point discrimination, namely, distal areas (the index finger), proximal areas (the arm distal to the elbow), the cranial area (the face) and axial regions (the back).

The stimuli were delivered via two thermostatically controlled Peltier-type thermoprobes measuring approximately 2x3 cm. One probe served as reference and was maintained at the relatively neutral temperature of 30° C. The temperature of the second probe was set at various levels above and below that of the reference probe. The two probes were applied, as close as possible to one another, on one of the different body regions for approximately 2 sec. Since the two probes were too large to be placed on a single finger, they were placed on adjacent fingers, namely, the index and the middle fingers. Intertrial interval was at least 10 sec. The psychophysical method of constant stimuli was used to determine the differential thresholds. Generally, six comparison temperatures were used, three above and three below 30°C. A total of eight trials were given for each of the six paired comparisons. The upper and lower limits were set such that the subject detected the difference on almost all trials. The intermediate temperatures of the test probe were chosen at regular intervals between these two limits. Threshold was defined as the difference in the temperature of the two probes which was detected by the subject 70% of the time.

The experiment was carried out on the four callosal agenesis subjects and their matched controls. The results indicated that, although the absolute value of the differential thresholds varied from region to region, the acallosals and their controls had comparable thresholds for all regions tested, including the axial areas (see Figure 3).

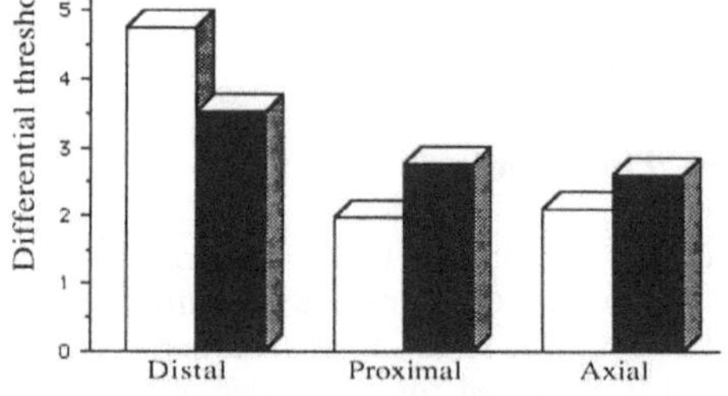

Figure 3. Differential temperature thresholds in agenic and IQ-matched subjects during unilateral presentation of the stimuli.

These results suggest that, in spite of the absence of the callosum, no midline deficits are observed in a system that has a relatively large bilateral representation. In this sense, they corroborate the finding of adequate midline threshold observed in those regions of the lemniscal system which have strong bilateral projections (namely, the head area).

This study therefore indicates that in the absence of the corpus callosum compensation for midline integration can be observed in a sensory system which has bilateral representations. The auditory system is also characterized by such bilateral projections. The next experiments were thus carried out to verify whether midline fusion could equally be achieved in the auditory modality and whether similar compensatory mechanisms are present within this system.

AUDITION

Few studies have examined callosal functions in audition. Electrophysiological and behavioral data obtained in animals point to two complementary contributions of this structure to auditory perception, namely, midline fusion and sound localization. As was the case for the visual and somatosensory modalities, the first function would ensure continuity of sensation across the midline for a frontally moving sound. In the cat, Imig et al. (1986) described bands of cells in primary auditory cortex which are connected with the corpus callosum and which correspond functionally to the omnidirectional cells of Middlebrooks and Pettigrew (1981), involved in the analysis of sound at the midline. This function has also been examined more directly by recording in the callosal pathway while presenting, dichotically and with variable interaural delays, equal-intensity stimuli. The results indicate that a large number of fibers show the greatest excitatory or inhibitory interactions at null interaural time delay, a condition which supports the notion of a callosal contribution to auditory midline-fusion (Poirier et al., 1989).

Auditory neurons that either project their axons through the callosum or receive callosal afferents from the contralateral cortex also seem to be involved in sound localization. Besides the omnidirectional neurons described above (Middlebrooks and Pettigrew, 1981), a number of callosal fibers were found (Poirier et al., 1989) which, when stimulated dichotically, were excited maximally at specific non-zero interaural time delays, suggesting that they preferred sounds situated at various spatial locations. On the other hand, neurobehavioral experiments with cats and dogs (Girden, 1939; Nauman, 1958, cited by Moore et al., 1974; Moore et al., 1974; Casseday and Neff, 1975; Brunk et al., 1979) do not confirm this expectation since a callosal section produces little or no effect on the localization behavior of the operated animals. This lack of effect might, however, result from the use of inappropriate experimental paradigms. One of the difficulties in some of these studies resides in the use of a two-choice test, where the sound source is lateralized to the left or to the right of the animal's midline. This is considered to be a poor test of higher auditory processing (Jenkins and Masterton, 1982; Jenkins and Merzenich, 1984). Other factors which might also have masked the presence of deficits include small sample size (Girden, 1939; Naumann, 1958), the type of stimulus (pure tones, which may be mainly treated subcortically) and long stimulus duration, which may have allowed scanning of the head (Moore et al., 1974; Casseday and Neff, 1975; Brunk et al., 1979).

No experiment with the objective of assessing the importance of the callosum for midline fusion or sound localization has been carried out in humans. The purpose of the experiments presented below, therefore, was to shed some light on the possible callosal involvement in these auditory functions. Two experiments were carried out on the four acallosal subjects described above and their matched controls. In the first, the subjects had to point to a fixed auditory target presented on a sound perimeter. It was expected, on one hand, that all subjects would localize more accurately at peri-central than at lateral positions and, on the other, that acallosals would show a poorer performance at peri-central positions than control subjects. The second experiment used simulated moving stimuli. Basically, the subjects had to point to the beginning and the end of the perceived sound trajectory. Since this experiment required the localization of some auditory stimuli moving across the midline, and thus a shift in the relative contribution of each hemisphere to the discrimination, it was expected to reveal even larger deficits of midline integration than the fixed-sound conditions. The procedures are described in detail elsewhere (Poirier et al., 1993) but will be summarized below.

The apparatus used to test sound localization consisted of sixteen loudspeakers (Tweeters) mounted on a graduated semi-circular wood perimeter having a radius of 50 cm inserted in a double-wall quasi-anechoic chamber. The loudspeakers were separated by approximately 10° and were positioned at ±5°, ±16°, ±26°, ±37°, ±47°, ±58°, ±68° and ±78° to the left and to the right of the veridical midline. The stimuli were broadband noise (BBN) bursts maintained at 52 db SPL (re 20μPa). In the fixed-sound source experiment, the stationary stimuli consisted of a double-gated white noise burst, 90 ms in total duration, delivered randomly in each trial through one of the 16 loudspeakers. In this experiment, the peri-central field was arbitrarily defined as the space contained between loudspeakers located at ±5° and ±16° (size of the field: 32°) whereas the loudspeakers situated at ±26°, ±37°, ±47°, ±58°, ±68° and ±78° were considered as making up the lateral fields (size of the left and right fields: 52°). In the second experiment, apparent movement was produced by activating successive loudspeakers. This apparent motion traveled in either direction (left to right or right to left) over a distance of 10°, 32° or 52°, depending on the number of loudspeakers involved (2, 4 or 6), at different velocities. In this simulated motion experiment, the sound sources that travelled over a distance of 10° (±5°), 32° (±16°) and 52° (±26°) over the midline were considered as three independent measures of the peri-central field whereas stimuli travelling over the same length on each side were considered as making up the lateral fields.

The subjects were seated in the center of the perimeter facing the sound sources, which were located at ear level. All trials were initiated by a warning buzzer, which served to remind the subjects to fixate a centrally located digital counter. In Experiment 1, the subjects were required to localize the stationary sound source by pointing with the index finger of the hand ipsilateral (i.e. right hand in right hemispace and left hand in left hemispace) to the position where they thought the stimulus had appeared. In Experiment 2, central presentations involved trajectories which began and ended on either side of veridical midline whereas lateral trajectories had a mid-point situated at approximately 52°. The subjects were required to indicate with the index of the hand ipsilateral to the stimulus side where they thought the beginning and the end of the stimulus had occurred. In the case of stimuli crossing the midline the two hands were used, the left to point to that portion of the trajectory present in the left hemispace and the right hand to indicate the position in the right hemispace. The responses were recorded, in degrees, by the experimenter present in the anechoic chamber. In the two experiments, aiming accuracy was assessed by calculating the means and standard deviations of the absolute errors (in degrees of arc) from the perceived and real target positions.

In the first experiment (with a fixed-sound source), the ANOVA indicated that acallosal subjects were significantly less accurate than the controls throughout the field but had no additional impairment in peri-central positions relative to the lateral hemispaces. This result is illustrated in Figure 4. The ANOVA also revealed that sound localization was significantly more precise for targets located near the midline than for laterally presented sounds in both groups.

The first step in the analysis of the second experiment involved comparing the size of the errors made at the beginning with those made at the end of the trajectory. Since no differences were observed, the two measures of accuracy were summed. The resulting total was defined as the global error in estimating the size of the trajectory. The ANOVA carried out on this value demonstrated that performance in acallosals tended to be consistently worse than that of their matched-controls. However, the difference did not reach statistical significance. Moreover, both groups were generally more accurate in pointing to auditory targets in the peri-central than in the lateral positions.

The data obtained in the two experiments clearly indicate that sound localization in all subjects was significantly more precise for targets located in peri-central than in

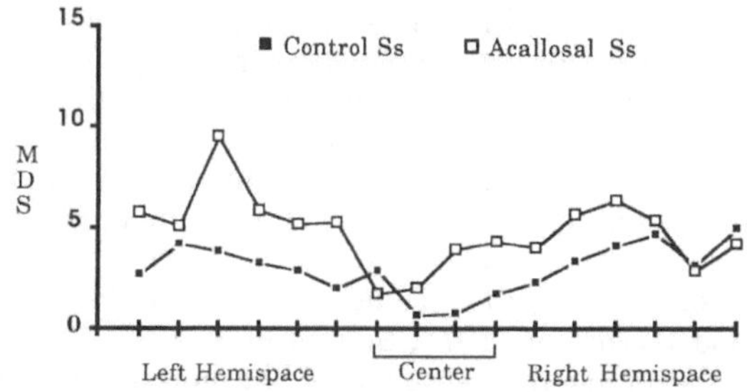

Figure 4. Mean deviation scores (MDS) expressed in degrees of arc obtained in the fixed-sound source experiment.

lateral fields. This result corroborates previous findings in humans (Olfield and Parker, 1984; Makous and Middlebrooks, 1990) and in animals (Strominger, 1969 a,b; Casseday and Neff, 1975; Martin and Webster, 1987) tested under free-field conditions. In both experiments, however, acallosal subjects displayed deficits in localization ability across total auditory field. How does one explain the fact that agenesis affects sound localization but not specifically in midline auditory space, as postulated above? A probable explanation is that the acallosals, having benefited from a long period of cerebral plasticity, have developed compensatory mechanisms involving alternate secondary interhemispheric pathways. The anterior commissure could be a likely candidate for this function since the anatomical studies of Pandya and Seltzer (1986) have shown that in the primate, the superior temporal region sends fibers to the opposite hemisphere both through the anterior commissure and the corpus callosum. Another potential compensatory mechanism that has been advanced mostly within the somatosensory modality would involve an increased use of ipsilateral pathways either by anatomical or functional accentuation (Dennis 1976; see also Jeeves, this volume). In the auditory system, this possibility is even more likely since there are numerous cross-over points at various subcortical levels.

As in previous studies (Lassonde et al., 1988), these experiments suggest that if compensation does occur, it is also subject to some limitations since the acallosal group was significantly less precise in indicating the source of the fixed sound regardless of its position. A consistent feature accompanying callosal agenesis is an impairment in speed to the benefit of performance accuracy (Jeeves, 1969; Lassonde, 1986; Lassonde et al., 1988; 1991). The overall lower levels of performance of acallosals in the fixed-sound condition indicate that these limitations also apply to precision and not just to speed when the latter is not a factor. An explanation for this decrease in accuracy may be attempted if one assumes that the most plausible compensatory mechanism is the enhanced use of ipsilateral auditory pathways. The consequences of such developmental adaptation might be an increased competition between ipsilateral and contralateral pathways, resulting in a generally poorer localization performance. An alternate, possibly contradictory, explanation for this overall decrease in accuracy may be derived from electrophysiological data obtained in our laboratories. In cats, callosal section reduces the number of binaurally-driven cells in the primary auditory cortex, resulting in a greater number of contralaterally-driven monaural cells. Interaural comparisons are thus no longer possible or are at least significantly reduced. It may well be that the congenital absence of the callosum in our human subjects, as is the case with its section in cats, results in a decreased efficiency of the ipsilateral connections thereby leaving the system with an overall reduced capacity for using binaural cues to localize sounds in space.

Although the data show that the congenital absence of the corpus callosum affects the ability to localize a fixed-sound source, and a similar trend was obvious in the simulated motion experiment, the differences between the two groups did not reach statistical significance. This result was somewhat surprising. It may, however, simply be due to the small sample size and the important variability in the results. The difference was in fact very close to significance. It may, alternatively, indicate that processing in the two tasks was not the same. In this regard, studies in the cat have shown that, whereas many auditory neurons in a large number of subcortical structures are tuned to moving sounds (Rauschecker and Harris, 1989; Spitzer and Semple, 1991; Altman et al., 1971), only few cells of the primary auditory area have the same sensitivity (Sovijarvi and Hyvarinen, 1974, Stumpf et al.,1992; Toronchuk et al., 1992). It is conceivable that the congenital absence of the corpus callosum is more easily compensated for by increased interactions between these subcortical structures when moving sounds are used. Stationary sound stimuli, on the other hand, would depend more strongly on cortical mechanisms and hence, show less compensation.

CONCLUSION

The results obtained in the various sensory modalities tested indicate that callosal involvement in midline function is important in the visual and lemniscal somatosensory systems but may be redundant in the auditory and spino-thalamic somatosensory systems. In the latter two systems, receptive fields of individual cells are large in terms of the area of body or the region of space over which they can be driven. Moreover, there is extensive cross-over of information at various levels particularly for the auditory system. Continuity of sensation across hemifields is thus not a problem and the corpus callosum may contribute little to this function. By contrast, the former two systems have cells whose receptive fields are more finely tuned to precise spatial or body coordinates. Moreover, both are completely crossed in nature. In this context, a mechanism ensuring continuity of sensation across sensory space is probably essential.

Thus, in the sensory systems having unilateral projections, callosal absence does affect midline analysis. In fact, the privileged relationship between the corpus callosum and midline integration appears to be genetically determined, confirming its important ecological value. For example, in the visual system, neonate cats have a widespread callosal representation in the cortex. As they mature, there is a gradual withdrawal of the exuberant callosal neurons such that only those cells situated in the region of the vertical midline survive while the others are eliminated through death or withdrawal of their callosal collateral axon (Innocenti, 1986b and this volume). Animals raised without visual experience, such as occurs when the eyelids are sutured, still show the restriction of the callosal cell zone around the midline representation, albeit with some other abnormalities. This confirms that the withdrawal of exuberant neurons is partially independent of experience. More importantly, however, these results as well as ours, indicate that midline integration is a pre-programmed function of the corpus callosum that may be limited in cases of callosal pathology, including the condition in which the callosum is absent from birth. Indeed, our results suggest that the limits of compensation occasionally observed in callosal agenesis with regard to interhemispheric communication (see Lassonde, this volume), may extend to midline analysis. Further studies may, in fact, prove that the callosal role in midline fusion is as crucial as its implication in interhemispheric communication.

REFERENCES

Altman, J.A., Syka, J., and Shmigidina, G.N., 1971, Neuronal activity in the medial geniculate body of the cat during monaural and binaural stimulation, *Exp. Brain Res.* 10: 81-93.

Antonini, A, Berlucchi, G., and Lepore, F., 1983, Physiological organization of callosal connections of visual lateral suprasylvian cortex area in the cat, *J. Neurophysiol.* 48:902-921.

Berlucchi, G., 1972, Anatomical and physiological aspects of visual functions in the corpus callosum, *Brain Res.* 37: 371-392.

Berlucchi, G., 1981, Recent advances in the analysis of the neural substrates of interhemispheric communication, *in:* "Brain Mechanisms of Perceptual Awareness and Purposeful Behavior", O. Pomperano and C.A. Ajemone-Marsan, eds., Raven Press, New York.

Brunk, R., Czihak, E., Opperman, B., and Santibanez, G., 1979, Conditioned audio-visual targeting reflexes in split brain cats, *Acta Neurobiol. Exp.* 39:443-452.

Caminiti, R., Innocenti, G.M., and Manzoni, T., 1979, The anatomical substrate of callosal messages from SI and SII in the cat, *Exp. Brain Res.* 35:295-314.

Casseday, J.H. and Neff, W.D., 1975, Auditory localization: role of auditory pathways in brain stem of the cat, *J. Neurophysiol.* 38:842-858.

Corkin, S., 1964, Somesthetic function after focal cerebral damage in man, Unpublished doctoral thesis, McGill University, Montreal.

Cowey, A. and Wilkinson, F., 1991, The role of the corpus callosum and extra striate visual areas in stereoacuity in macaque monkey, *Neuropsychologia* 29:465-480.

Dennis, M., 1976, Impaired sensory and motor differentiation with corpus callosum agenesis: A lack of callosal inhibition during ontogeny?, *Neuropsychologia* 14:455-469.

Girden, E., 1939, Cerebral determinants of auditory localization, *Am. J. Psychol.* 52:1-15.

Glickstein, M. and Whitteridge, D., 1976, Degeneration of layer III pyramidal cells in area 18 following destruction of callosal input, *Brain Res.* 104:148-151.

Guillemot, J.P., Lepore, F., Prevost, L., Richer, L., and Guilbert, M., 1988, Somatosensory receptive fields of fibers in the rostral corpus callosum of the cat, *Brain Res.* 441:221-232.

Guillemot, J.P., Richer, L., Prevost, L, Ptito, M., and Lepore, F., 1987, Receptive field properties of somatosensory callosal fibers in the monkey, *Brain Res.* 402:293-302.

Guillemot, J.P., Richer, L., Ptito, M., and Lepore, F., 1992, Somatosensory receptive field properties of corpus callosum fibers in the raccoon, *J. Comp. Neurol.* 321:124-132.

Guillemot, J.P., Richer, L., Ptito, M., and Lepore, F., 1993, Disparity coding in the cat: a comparison between areas 17-18 and 19, *Progress Brain Res.* 95:179-187.

Hamilton, C.R. and Vermeire, B.A., 1986, Localization of visual function with partially split-brain monkeys, *in:* "Two Hemispheres-One Brain: Functions of the Corpus Callosum," F. Lepore, M. Ptito, and H.H. Jasper, eds., Alan R. Liss, New York.

Hamilton, C.R., Rodriguez, K.M., and Vermeire, B.A., 1987, The cerebral commissures and midline stereopsis, *Inv. Ophthal. Vis. Sci.* 28:294.

Imig, T.J., Reale, R.A., Brugge, J.F., Morel, A., and Adrian, H.O., 1986, Topography of cortico-cortical connections related to tonotopic and binaural maps of cat auditory cortex, *in:* "Two Hemispheres-One Brain: Functions of the Corpus Callosum," F. Lepore, M. Ptito, and H.H. Jasper, eds., Alan R. Liss, New York.

Innocenti, G., 1986a, What is so special about callosal connections?, *in:* "Two Hemispheres-One Brain: Functions of the Corpus Callosum," F. Lepore, M. Ptito, and H.H. Jasper, eds., Alan R. Liss, New York.

Innocenti, G., 1986b, General organization of callosal connection in the cerebral cortex *in:* "Cerebral Cortex - Sensory-Motor Areas and Aspects of Cortical Connectivity," E.G. Jones and A. Peters, eds., Plenum Press, New York.

Jacobson, S. and Trojanowski, I.Q., 1974, The cells of origin of the corpus callosum in the rat, cat and rhesus monkey, *Brain Res.* 74:149-155.

Jeeves, M.A., 1969, A comparison of interhemispheric transmission time in acallosals and normals, *Psychon. Sci.* 16:245-246.

Jeeves, M.A., 1991, Stereoperception in callosal agenesis and partial callosotomy, *Neuropsychologia* 29:19-34.

Jenkins, W. and Masterton, R.B., 1982, Sound localization: effects of unilateral lesions in central auditory system, *J. Neurophysiol.* 47:987-1016.

Jenkins, W. and Merzenich, M.M., 1984, Role of cat primary auditory cortex for sound-localization behavior, *J. Neurophysiol.* 52:819-847.

Jones, E. and Powell, T.P.S., 1968, The commissural connections of the somatic sensory cortex in the cat, *J. Anat.* 10:433-455.

Kelly, J.P., 1985, Trigeminal system, *in:* "Principles of Neural Science," E.R. Kandel and J.H. Schwartz, eds., Elsevier, New York

Lassonde, M., 1986, The facilitatory influence of the corpus callosum on intrahemispheric processing, *in:* "Two Hemispheres-One Brain: Functions of the Corpus Callosum," F. Lepore, M. Ptito, and H.H. Jasper, eds., Alan R. Liss, New York.

Lassonde, M., Perenin, M.T., Tassinari, G., Corbetta, M., and Cavanagh, P., 1988, Central mechanisms of stereopsis in man, *Adv. Biosci.* 70:95-98.

Lassonde, M., Sauerwein, H., Chicoine, A.J., and Geoffroy, G., 1991, Absence of disconnexion syndrome in callosal agenesis and early callosotomy: Brain reorganization or lack of structural specificity during ontogeny?, *Neuropsychologia* 29:481-495.

Lassonde, M., Sauerwein, H., McCabe, N., Laurencelle, L., and Geoffroy, G., 1988, Extent and limits of cerebral adjustment to early section or congenital absence of the corpus callosum, *Beh. Brain Res.* 30:165-181.

Ledoux, J.E, Dentsoh, G., Wilson, D.H., and Gazzaniga, M.S., 1977, Binocular depth perception and the anterior commissure in man, *Psychologist* 20:55.
Lepore, F. and Guillemot, J.P., 1982, Visual receptive field properties of cells innervated through the corpus callosum in the cat, *Exp. Brain Res.* 46:413-424.
Lepore, F., Ptito, M., and Guillemot, J.P., 1986a, The role of the corpus callosum in midline fusion, *in:* "Two Hemispheres-One Brain: Functions of the Corpus Callosum," F. Lepore, M. Ptito, and H.H. Jasper, eds., Alan R. Liss, New York.
Lepore, F., Ptito, M., and Lassonde, M., 1986b, Stereoperception in cats following section of the corpus callosum and/or the optic chiasm, *Exp. Brain Res.* 61:258-264.
Lepore, F., Samson, A., Paradis, M.C., Ptito, M., and Guillemot, J.P., 1992, Binocular interaction and disparity coding at the 17-18 border: contribution of the corpus callosum, *Exp. Brain Res.* 90:129-140.
Makous, J.C. and Middlebrooks, J.C., 1990, Two-dimensional sound localization by human listeners, *J. Acoustic. Soc. Am.* 87:2188-2200.
Manzoni, T., Barbaresi, P., Bellardinelli, E., and Caminiti, R., 1980, Callosal projections from the two body midlines, *Exp. Brain Res.* 39:1-9.
McKenna, T.M., Whitsel, B.L., Dreyer, D.A., and Metz, C.B., 1981, Organization of cat anterior parietal cortex: Relations among cytoarchitecture, single neuron functional properties, and interhemispheric connectivity, *J. Neurophysiol.* 45:667-697.
Martin, R.L. and Webster, W.R., 1987, The auditory spatial acuity of the cat in the interaural horizontal and median vertical planes, *Hearing Res.* 30(1):239-252.
Middlebrooks, J.C. and Pettigrew, J.D., 1981, Functional classes of neurons in primary auditory cortex of the cat distinguished by the sensibility to sound location, *J. Neurosci.* 1:107-120.
Mitchell, D.E. and Blakemore, C.B., 1970, Binocular depth perception and the corpus callosum, *Vision Res.* 10:49-54.
Moore, C.N., Casseday, J.H., and Neff, W.D., 1974, Sound localization: the role of the commissural pathways of the auditory system of the cat, Brain *Res.* 82:13-26.
Olfield, S.R and Parker, S.P.A., 1984, Acuity of sound localization: a topography of auditory space. I. Normal hearing conditions, *Perception* 13:581-600.
Pandya, D.N. and Seltzer, B., 1986, The topography of commissural fibers, *in:* "Two Hemispheres-One Brain: Functions of the Corpus Callosum," F. Lepore, M. Ptito, and H.H. Jasper, eds., Alan R. Liss, New York.
Payne, B.R., 1986, Role of callosal cells in the functional organization of cat striate cortex, *in:* "Two Hemispheres-One Brain: Functions of the Corpus Callosum," F. Lepore, M. Ptito, and H.H. Jasper, eds., Alan R. Liss, New York.
Petit, D., Lepore, F., Picard, N., and Guillemot, J.P., 1990, Bilateral receptive fields in cortical area SII: Contribution of the corpus callosum and other interhemispheric commissures, *Somatosens. Mot. Res.* 7(2):97-112.
Poirier, P., Richer, L., Lepore, F., and Guillemot, J.P., 1989, Contribution of the corpus callosum to auditory localization in the cat, *Soc. Neurosci. Abstr.* 15:112.
Poirier, P., Lassonde, M., and Lepore, F., 1993, Sound localization in acallosals human listeners, *Brain* 116:53-69.
Ptito, M., Lepore, F., Lassonde, M., Dion, C., and Miceli, D., 1986, Neural mechanisms for stereopsis in cats, *in:* "Two hemispheres-One Brain: Functions of the Corpus Callosum," F. Lepore, M. Ptito, and H.H. Jasper, eds., Alan R. Liss, New York.
Ramchandran, V.S., Cronin-Golomb A., and Myers, J.J., 1986, Perception of apparent motion by commissurotomy patients, *Nature* 320:358-359.
Rauschecker, J.P. and Harris, L.R., 1989, Auditory and visual neurons in the cat's superior colliculus selective for the direction of apparent motion stimuli, *Brain Res.* 490:56-63.
Rivest, J., Cavanagh, P. and Lassonde, M., in press, Interhemispheric depth judgment, *Neuropsychologia.*
Sauerwein, H., Lassonde, M., Cardu, B., and Geoffroy, G. 1981, Interhemispheric integration of sensory and motor functions in agenesis of the corpus callosum, *Neuropsychologia* 19:445-454.
Schiavetto, A., Lepore, F., and Lassonde, M., in press, Somesthetic discrimination thresholds in the absence of the corpus callosum, *Neuropsychologia.*
Shoumura, K., Ando, T., and Kato, K., 1975, Structural organization of callosal OBg in human corpus callosum agenesis, *Brain Res.* 93:241-252.
Sovijarvi, A.R.A. and Hyvarinen, J., 1974, Auditory cortical neurons in the cat sensitive to the direction of sound source movement, *Brain Res.* 73:455-471.
Spitzer, M.W. and Semple, M.N., 1991, Interaural phase coding in auditory midbrain: influence of dynamic stimulus features, *Science* 254:721-723
Strominger, N.L., 1969a, Subdivisions of auditory cortex and their role in localization of sound in space, *Exp. Neurol.* 24:348-362.
Strominger, N.L., 1969b, Localization of sound in space after unilateral and bilateral ablation of auditory cortex, *Exp. Neurol.* 25:521-533.
Stumpf, E., Toronchuk, J.M., and Cynader, M.S., 1992, Neurons in cat primary auditory cortex sensitive to correlates of auditory motion in three dimensional space, *Exp. Brain Res.* 88(1):158-168.
Timney, B, Elberger, A., and Vanderwater, M.L., 1991, Binocular depth perception in the cat following early corpus callosum section, *Exp. Brain Res.* 60:19-26.

Toronchuk, J.M., Stumpf, E., and Cynader, M.S., 1992, Auditory cortex neurons sensitive to correlates of auditory motion - underlying mechanisms, *Exp. Brain Res.* 88:169-180.

Weinstein, S., 1968, Intensive and extensive aspects of tactile sensitivity as a function of body part, sex, and laterality, *in:* "The Skin Senses," D.K. Kenshalo, ed., Charles. C. Thomas, Springfield

VISUAL INTEGRATION IN CALLOSAL AGENESIS

A. David Milner

Psychological Laboratory
University of St Andrews
St Andrews, Fife
Scotland , KY16 9JU

INTRODUCTION

The visual system is wired up in a rather strictly lateralized fashion. Aside from a thin band of the retina (perhaps about 1 deg of visual angle) bordering the vertical meridian, information impinging on the right hemiretina is transmitted faithfully to targets on the right side of the brain, and similarly on the left. These retinofugal neurons synapse upon a number of lateralized structures in the thalamus [especially the pars dorsalis and pars ventralis of the lateral geniculate nucleus (LGNd and LGNv)], and in the midbrain [especially the superior colliculus (SC) and the pretectal nuclei]. From these subcortical structures, the principal route to the cerebral cortex is the direct one from the LGNd to the primary visual cortex (V1), but there are also substantial indirect ones. There are multiple routes from the midbrain visual areas to thalamic nuclei (notably within the pulvinar complex) and from there to secondary and tertiary cortical visual areas.

In the primate brain, visual information appears to remain laterally segregated in all of these routes to the cortex (Rodieck, 1979); thus a visual stimulus presented to the left of a central fixation point, and therefore falling to the right of the fovea, activates a cascade of subcortical visual structures on that right side of the brain, up to and including the right striate cortex (V1). Single neurons recorded physiologically from these structures in monkeys rarely respond to visual stimuli presented on the "wrong" side of the visual field (except within a degree or so of the vertical meridian). Thus it is generally supposed, not unreasonably, that visual integration *between* the two hemifields in man must require interhemispheric transmission through the cerebral commissures, especially the corpus callosum.

This view is directly supported by physiological data from the monkey cortex. Gross and Mishkin (1977) argue persuasively that the neural hardware required to register the identity of a pattern in one hemifield as being the same as one presented in the other hemifield lies in the inferior temporal cortex. Most neurons here have bilateral receptive fields: that is, unlike neurons in earlier visual areas, they respond to a preferred stimulus not just in the contralateral hemifield but also in the "wrong", ipsilateral, hemifield. Experiments in which either the splenium of the corpus callosum and/or the anterior commissure are selectively sectioned in monkeys show that although many bilateral receptive fields survive either one of these operations, none survive both

Callosal Agenesis, Edited by M. Lassonde and
M.A. Jeeves, Plenum Press, New York, 1994

(Gross et al., 1977). Thus, the argument runs, a necessary condition for similar visual patterns presented to different sides of the visual field to be seen as alike is the integrity of the cerebral commissures.

Neurobehavioral data in the monkey also support this view: a monkey shows good transfer of pattern discrimination training between the hemispheres, provided that either the splenium or the anterior commissure is intact; whenever both are cut, no significant transfer is seen (Hamilton, 1982; Hamilton and Vermeire, 1986). In most such experiments investigators have studied monkeys in which the optic chiasm has been sagittally sectioned, in order for visual information to be directed unilaterally in the brain by the simple expedient of blindfolding one eye. But a more recent study has shown similar results in a more naturalistic paradigm where the monkey is trained to fixate a central stimulus and its hemifields independently stimulated with the chiasm intact (Eacott and Gaffan, 1989).

Eacott and Gaffan's technique provides a close parallel to many human commissurotomy studies, where the subject is asked to fixate a central point whilst a visual pattern or patterns are briefly presented in one or both visual hemifields. In humans who have undergone transection of all of the cerebral commissures, a visual disconnection apparently comparable to that seen in the monkey has been repeatedly reported (Sperry, 1968a). Patients are unable to name or describe visual stimuli in the left hemifield, nor are they able to match them as similar to or different from ones on the right.

Tasks of this type have also been used extensively to assay the visual cross-integration abilities of individuals with callosal agenesis (e.g. Sperry, 1968b; Ettlinger et al., 1972; Sauerwein and Lassonde, 1983). Not all patients are able to perform well in perceptual matching tests even when the processing required is restricted to a single hemifield, or in naming tasks when the information is given to the right hemifield (Ettlinger et al., 1972). However, the overwhelming message from studies of this kind is that acallosal subjects are very nearly as good at tasks that would require interhemispheric communication in normal subjects as they are at tasks that would not[1]. In other words, although an acallosal subject may be impaired at naming a left-hemifield drawing or printed word, she is typically no more impaired than would be expected from her performance with right-hemifield stimuli. Similarly, although she may be impaired at left-right matching, she is typically no more so than at up-down matching.

BILATERAL RETINAL PROJECTIONS?

It has often been argued that the preserved left-hemifield naming in callosal agenesis could be due to the presence of bilateral speech mechanisms in these subjects (Sperry, 1968b; Denenberg, 1981). However, as has been argued elsewhere, the independent evidence for such bilaterality (at more than a neurological base-rate frequency) is far from convincing (Milner and Jeeves, 1979, 1981; Chiarello, 1980). More importantly, the speech bilaterality hypothesis cannot explain the ability of acallosal subjects to cross-match between the two hemifields. To subserve this ability, it is difficult to envisage any possible mechanism other than either:

(a) some kind of non-callosal commissural transmission of visual information, or:
(b) enhanced bilateral retinal projections.

Option (b), though rarely considered, cannot be ruled out *a priori*. The strict segregation of retinal fibers to brain targets on their own, ipsilateral, side of the brain could be relaxed in callosal agenesis. It is known, for example, that in hamsters that have received a unilateral lesion of the SC at birth, abnormal retinotectal projections develop from the contralateral eye onto the intact SC on the "wrong" side of the brain, and that these can be functionally active in behavior (Schneider, 1979). Furthermore a greatly increased nasotemporal overlap of retinal projections seems to be present in the optic tract of the opossum, an acallosal marsupial (Volchan et al., 1988). It is therefore not impossible that abnormal bilateral retinofugal projections could be present in callosal agenesis, whether due to generalized midline abnormalities in optic tract ontogenesis or to the absence of inhibitory corticocortical or corticofugal interactions between visual brain structures.

But if the normal nasotemporal overlap (usually estimated at about 1 deg) is increased up to several degrees in agenesis, then one could make a simple prediction. This would be that visual information from both sides of the retina within this enhanced range should arrive on a given side of the brain at much the same instant; so the acallosal subject should be able to make a simple finger reaction equally quickly whichever side of the visual field were stimulated. Using well-lateralized light stimuli, Jeeves (1969) discovered some years ago that this time difference between the reaction time of the ipsilateral hand versus the contralateral hand is typically an order of magnitude larger in acallosal than in normal subjects. Presumably this is because there is no quick route available through the callosum in these subjects, and a more circuitous route has to be used.

Lines (1984) set out to discover what happened to this "crossed - uncrossed difference" (CUD) when an acallosal patient (KC) was tested with stimulus eccentricities much smaller than used by Jeeves. As the data presented in Table 1 show, the CUD averaged out at 20.9 ms overall, a typical value for acallosal patients. Moreover the data clearly show that this large CUD was maintained right down to eccentricities of 1 degree of visual angle; indeed it remained statistically significant even at 0.5 deg. Lines indeed goes so far as to question (on the basis of these and comparable data from normal subjects: Lines and Milner, 1983) whether the generally-supposed naso-temporal overlap is a reality at all in the human retina, or if so whether it has any functional significance.

Table 1. Reaction times (in ms) to a lateralized light stimulus in acallosal patient KC. The effect of stimulus eccentricity.[1]

	Eccentricity			
	0.5 deg	**1 deg**	**2 deg**	**4 deg**
Crossed RTs	321.3	325.7	337.4	352.3
Uncrossed RTs	308.6	305.1	313.4	325.9
Mean CUD	12.7	20.5	24.1	26.3

[1]Data from Lines (1984)

Thus it seems from reaction time evidence that punctate information arriving at one hemiretina probably has no functional access to ipsilateral brain structures via abnormally decussating optic nerve fibers. Accordingly, it seems unlikely that more complex visual information would have direct bilateral access to permit pattern cross-matching without the need for commissural transmission. Confirmatory evidence comes from a recent report that the nasotemporal division is already sharp at the 69th fetal day in the monkey, which is before any appreciable loss of ganglion cells occurs (Lia et al., 1988); this argues against the possibility that extensive "wrongly" projecting optic nerve fibers might be retained in callosal agenesis.

It seems likely, therefore, that some kind of non-callosal commissural transmission of visual information must occur in callosal agenesis. So where in the brain might this information cross over?

CANDIDATE COMMISSURES

One major candidate must be the anterior commissure (AC), which, as indicated above, is known to play an important role in interhemispheric visual integration in the monkey. There is post-mortem and MRI evidence that most acallosal patients possess an AC (Loeser and Alvord, 1968; Lemire et al., 1975; Barkovich and Norman, 1988). It is possible that the AC may even be abnormally enlarged in certain cases, though probably of reduced size in some others (see Rauch and Jinkins, this volume). However there has recently been a report of a sevenfold range in cross-sectional area of the AC even in normal human subjects (Demeter et al., 1988), so that a quantitative population comparison would be needed to substantiate any claim for an abnormal size distribution in acallosals. (It is notable, incidentally, that acallosal mice show no enlargement of the AC, whether the defect is genetic or radiation-induced: Wahlsten and Ozaki, this volume; Schmidt, this volume).

Certainly a number of writers have favored the likely role of the anterior commissure. One major reason is undoubtedly that the complexity of visual information that can evidently be transmitted in acallosal subjects (e.g. simple words, figural drawings, geometrical patterns) would seem to demand cortical participation. It is known that the AC in the monkey heavily interconnects the anterior two-thirds of the two temporal lobes, including those inferior neocortical areas most intimately implicated in pattern processing (Jouandet and Gazzaniga, 1979; Pandya and Seltzer, 1986; Demeter et al., 1990). In contrast there is little evidence for form analysis in subcortical visual structures such as the SC (e.g. Goldberg and Robinson, 1978), though some does seem to be present in its thalamic target structures (Chalupa and Abramson, 1988).

In contrast to monkeys, however, it seems that in normal humans the AC may be comparatively insignificant for visual function: when the AC is spared in commissurotomy operations, there seems to be no better visual integration than when it is cut (McKeever et al., 1981; Holtzman, 1984). Furthermore the AC in humans has a cross-sectional area of only some 1 percent of the CC, in contrast to a ratio of almost 5 percent in monkeys (Foxman et al., 1986; indeed the ratio for numbers of axons approaches 6 percent: Lamantia and Rakic, 1990). It is thus conceivable that the AC does not carry visual information at all in normal humans, as is probably the case in most non-primate species (such as the cat: Hamilton, 1982; Jouandet, 1982).

But there is indirect embryological evidence that in acallosal brains the AC might command a far wider cortical territory than normal. In the normal course of events, the AC develops earlier than the corpus callosum (CC), and Auroux and Roussel (1967) reported that around the 9-10th week of fetal life, prior to the appearance of the CC, the AC supplies a very widespread range of neocortical areas. With the growth of the CC however, most of these "exuberant" connections retract, such that by the 4th month of fetal life the AC has a very much more restricted range, while callosal connections occupy the vacated territory. Clearly, then, if there is a causal link in this chain of events, e.g. due to the operation of synaptic competition between callosal and AC neurons, then a *failure* of the CC to grow could result in a maintenance of the exuberant AC projections. These might be present whether or not the AC is physically enlarged.

The alternative scenario is that mesencephalic commissures might play a crucial role in visual integration in callosal agenesis. Specific candidates would not only be the intertectal (collicular) commissure but also the posterior commissure (whose targets include the posterior pretectal nucleus: Shoumura et al., 1987). Indeed in those few individuals who lack the anterior commissure as well as the CC, any visual cross-integration must of necessity depend on such subcortical pathways. It is true that animal studies indicate that the commissural cells in the superior colliculus lie mainly in the intermediate and deeper layers, as opposed to the more visually responsive superficial laminae (Edwards, 1977). Nonetheless a substantial number of these commissural neurons in the hamster (Rhoades et al., 1986) and the majority of them in the cat (McIlwain, 1991) are visually responsive. It is also quite possible that there might be compensatory or disinhibited over-development of such pathways in acallosal brains.

How do behavioral and electrophysiological studies of acallosal patients bear upon these two broad alternatives? We will consider, in turn, the evidence from reaction times, event-related potentials, and from tachistoscopic perceptual tasks employing lateralized or central stimuli, respectively.

REACTION-TIME DATA

Experiments using simple reaction times to estimate a callosal crossing time, alluded to earlier, have an 80-year history. Poffenberger (1912) argued that the time difference between simple manual reaction times (RTs) to point visual stimuli on the left versus right of fixation should reflect an interhemispheric transmission time (ITT); this should go one way for use of the left hand (controlled by the right hemisphere), and the other way for right-hand RTs. The mean of these 2 differences (the crossed - uncrossed difference, or CUD) should provide an estimate of the ITT. Many investigators since Poffenberger have performed such experiments, and the usual finding is a CUD of around 3 ms in normal subjects (Bashore, 1981).

As reviewed elsewhere (Berlucchi, 1978; Milner and Rugg, 1989), such reaction-time techniques appear to measure non-visual callosal transmission latencies in normal subjects. The main evidence for this is that manipulation of visual stimulus parameters which have a substantial effect on the absolute values of the RTs themselves nonetheless have no detectable effect on the CUDs. Thus a range of stimulus eccentricities from 5 deg to 35 deg (Berlucchi et al., 1971) and a range of stimulus luminances from near-threshold to high intensity (Milner and Lines, 1982) left CUDs invariant in groups of normal subjects.

In contrast, however, there is evidence that such RT techniques may allow the measurement of visual latencies in acallosal subjects. Similar luminance variations to those used by Milner and Lines were used by Milner (1982) and by Milner et al (1985), testing 2 different acallosal subjects, and revealed clear and statistically significant variations in the CUDs measured (see Table 2). As shown in the Table, the range of stimulus luminance was such as to cause mean RTs to vary by a range of 50-60 ms. As would be expected if the cross-midline transfer was visually-coded, it was found that the lower the stimulus energy, the longer the CUD.

Table 2. Reaction times (in ms) to a lateralized light stimulus in two acallosal subjects. The effect of stimulus luminance.

Patient		Luminance Level low (3)	medium (4)	high (7)
KC[1]	Crossed RTs	354.7	322.8	297.4
	Uncrossed RTs	330.4	300.2	283.4
	Mean CUD	24.3	22.6	14.0
BF[2]	Crossed RTs	362.7	-	303.6
	Uncrossed RTs	340.4	-	294.1
	Mean CUD	22.3	-	9.5

[1]Data from Milner (1982); [2]Data from Milner et al. (1985)

There are few comparative data published by others on the effects of stimulus manipulations on CUD in acallosals. A decrease in luminance was studied by Clarke and Zaidel (1989), who found a non-significant increase of the CUD from 18 ms to 22 ms in their acallosal subject MM; however the mean RTs themselves only increased from 333 ms to 361 ms, which may indicate that the luminance difference was insufficient to render the effect on CUD significant. Clarke and Zaidel (1989) also manipulated stimulus eccentricity, using 4 deg and 10 deg presentations. MM again showed a trend in the predicted direction (CUDs were 9.9 ms at 4 deg and 15.1 ms at 10 deg), but again this was non-significant. On the other hand, using a range of eccentricities closer to fixation, Lines (1984) did find a significant effect of eccentricity on CUD in patient KC (see Table 1).

It would seem reasonable to weight positive evidence more heavily than negative in such experiments, particularly since the variance of RTs can be greatly increased in neurological populations (see Milner, 1986, for a discussion). Thus the balance of evidence at present suggests that the CUD in acallosals reflects a visually-coded relay; the question is, which of the possible routes might it take?

There is no direct answer to this question, but there is indirect evidence. In the California patients LB and NG, all of the cerebral commissures were cut (Bogen et al., 1988), so that the AC would not be available for mediating crossed RTs. Yet these patients can perform such reactions, and their CUDs are not longer than those of some acallosals (Sergent and Myers, 1985; Clarke and Zaidel, 1989). But are they using a visual relay? The evidence from stimulus intensity variations is shown in Table 3. For patient LB, an increase in luminance halved his CUD in one experiment (Sergent and Myers) while only decreasing it by 3 ms in the other (Clarke and Zaidel). Surprisingly, the former result missed significance; the reason may be surmised to lie in the small number of RTs measured (100 in total per intensity condition), resulting in high standard errors (see Milner, 1986). For patient NG, a reversed effect of luminance on CUD was recorded in the earlier experiment; but this seems of dubious reliability since NG's mean RTs were very similar in the two luminance conditions (361 ms vs 366 ms), and because in the later study the effect was in the predicted direction and statistically significant (Clarke and Zaidel, 1989).

Table 3. CUD data gathered on commissurotomized patients LB and NG in two published experiments: The effects of stimulus luminance.

Patient		Luminance[3] Low	High
LB	S & M[1]	37.3	20.3
	C & Z[2]	36.7	33.9
NG	S & M[1]	31.7	69.1
	C & Z[2]	50.5	33.9

[1]Data from Sergent and Myers (1985); [2]Data from Clarke and Zaidel (1989)

[3]It should be noted that the luminance values used were not identical across the two experiments; however, the overall mean RTs are roughly comparable, particularly in the "low" luminance conditions.

Thus on balance there is some evidence for an effect of luminance on CUDs in the predicted direction; and the interpretation of this as arguing for a visual relay in these split-brain subjects is bolstered by the finding that 3 out of 4 such patients showed significant stimulus eccentricity effects on the CUD (Clarke and Zaidel, 1989). It seems inevitable that this visual relay could only be subcortical, since there is no AC present; and if this can occur in commissurotomized patients, it could also do so in acallosal subjects.

ELECTROPHYSIOLOGICAL DATA

In a series of studies, it has been demonstrated that interhemispheric transmission times can be measured rather directly by recording the event-related potential (ERPs) from symmetrical points on either side of the scalp (Rugg et al., 1984, 1985a, 1987; Lines et al., 1984; Saron and Davidson, 1989). In these studies a lateralized stimulus is used as the trigger for a manual response, and the latency of early deflections in the recorded waveform [specifically either the positive deflection at around 100 ms (P100) or the negativity at around 160 ms (N160)] is compared between the directly stimulated (contralateral) hemisphere and the indirectly stimulated (ipsilateral) hemisphere.

The ITTs measured by use of these techniques generally average at between 10-15 ms for electrodes placed over occipital regions, and this value varies as a function of stimulus luminance (Lines et al., 1984). In contrast, ITTs estimated by use of centrally placed electrodes, i.e. over sensorimotor cortex, average at only 3-4 ms, and these do not vary with stimulus luminance. These data are consistent with a model postulating the availability of two independent callosal routes in the normal brain during reaction time performance: a posterior one through the splenium, which is visually coded but relatively slow, and a more anterior pathway, perhaps carrying a motor command, which is considerably faster (Milner and Lines, 1982). If information travels both routes, then clearly the "race" to determine the response will almost always be won by the faster anterior pathway, give or take some biological variability. Thus CUDs determined behaviorally from RTs would be expected to be short and stimulus-invariant[2].

We have applied our ERP techniques to the study of the effects of callosal pathology, including the two acallosal subjects KC and BF (Rugg et al., 1985b). We were able to record normal visual ERPs in the contralateral hemisphere in both patients, in that the amplitude and latency of the "direct" N160 was within normal limits whichever visual hemifield was stimulated. However in neither case were we able to detect a corresponding deflection in the waveform recorded over the ipsilateral hemisphere. Similar ERP data have been reported for a different acallosal subject during participation in a more complex behavioral paradigm, by Brown (1991). These data support our interpretation that the ipsilateral response recorded in normal subjects is callosally mediated. However they do not further our understanding of the nature or fate of cross-midline transfer of information in the acallosal brain.

Another ERP approach has been taken in a recent study by Munte and Heinze (1991), who report that semantic information is integrated between the hemispheres in an acallosal subject. They showed this by recording normal N400 responses (a late "endogenous" brain potential) to semantically anomalous terminal words presented to either visual hemifield at the end of a sentence frame presented to one hemifield. Kutas et al. (1988), in contrast, found no such N400 evidence for semantic priming when the terminal word was presented in the left hemifield in callosotomized patients, except in two patients where there exists independent evidence for bilateral language. Kutas et al.'s evidence appears to indicate that a spared AC is not sufficient to permit interhemispheric semantic integration in callosotomized patients, although their paradigm did not test this directly like Munte and Heinze's (1991). If this is so, then there is no reason to assume that the AC could subserve the observed integration in Munte and Heinze's acallosal patient.

VISUAL PROCESSING OF LATERALIZED STIMULI

The starting point for the present chapter was the remarkable absence of a "disconnection syndrome" (Sperry, 1968a) in acallosal patients when tested with lateralized patterns, colors, words, etc, in a tachistoscope (reviewed by Milner and Jeeves, 1979). It was therefore of great interest when a paper by Martin (1985) appeared, which argued cogently for an involvement of the AC in mediating such successful integration. Martin reported that his acallosal patient GB performed well at *identifying* words and abstract shapes presented unilaterally to either visual hemifield; however GB was much poorer at *localizing* stimuli within a spatial frame in his right than in his left hemifield. In interpreting his data, Martin argued that the inferior temporal region of the monkey is involved in visual pattern recognition, but not in spatial functions (cf. Ungerleider and Mishkin, 1982; though see Milner and Goodale, in press, for a contrary perspective). On this assumption it may reasonably be assumed that the AC carries pattern information, but that only the corpus callosum is able to carry spatial information. If, therefore, the right hemisphere is dominant for the performance of visuospatial tasks, then the right hemifield will be selectively disadvantaged for such tasks by the congenital absence of the CC.

Martin's argument is ingenious and seems to provide a convincing case for the AC. We therefore set out to discover whether his data were generalizable to other acallosals. Unfortunately they were not: both our patients KC and BF performed well at right-field localization using a task modeled on one of Martin's (Jeeves and Milner, 1987). The only obvious explanation of this within Martin's theory would be that our patients had an anomalous cerebral dominance for spatial perception, such that either hemisphere could independently perform Martin's task. We therefore tested spatial matching between the hemifields in one of our patients (KC) to provide a more direct test: this task was also performed well, though admittedly not as well as digit matching.

Interestingly, this explicit spatial matching task was failed by two callosotomized patients with intact AC (Holtzman et al., 1981), although the same patients were able to *use* this consciously-inaccessible spatial information controlling attentional and oculomotor shifts (Holtzman et al., 1981; Holtzman, 1984). Like our own findings, these data indicate that spatial information *can* be transferred by non-callosal commissures, though the transmitted information may not then reach awareness. More recent work indicates that even completely commissurotomized subjects can cross-match crude spatial information, and, consistent with Holtzman's findings, that more precise information is transmitted without it attaining explicit awareness (Sergent, 1991). It would seem likely then that the intertectal commissure mediates the spatial integration we have observed in our acallosal subjects, particularly given the spatial coding known to exist within the superior colliculus (Sparks and May, 1990). If so, then Martin (1985) may have been correct in his supposition that the AC cannot transmit spatial information, albeit for the wrong reasons.

Are there any other indications as to the possible limitations upon the transmission capacity of visual commissures in agenesis? There was some early evidence for a selective difficulty when matching was tested using dot patterns of different densities (Ettlinger et al., 1972). Lepore et al. (this volume) have reported a possibly related difficulty: acallosals were impaired at judging the relative depth of dot patterns between the hemifields, whether or not motion parallax cues were available. These deficits could be interpreted as due to an inability to transfer precise spatial frequency information between the hemispheres. But such fine-grain information would be expected to be available at the level of the inferotemporal cortex, while in contrast the spatial filtering characteristics of cells in the SC are known to be coarse. Thus these deficits do not argue preferentially for the involvement of the AC.

More recently, evidence has been reported that the transmission of multiple stimulus features may be impaired in agenesis, while single stimulus features (color, shape, number) can be transmitted successfully (Karnath et al., 1991). This finding might be interpreted as an "attentional" limitation on cross-integration; but it is not clear that such a limitation would be expected for the AC on the basis of what is known of inferotemporal visual processing. On the other hand, if acallosals could be shown to be capable of matching different views of a given object or face across hemifields, then that would provide rather convincing evidence for the integration of high-level "object-centered" coding (see Milner and Goodale, in press) thought to typify the temporal lobe mechanisms that feed the AC in the monkey.

PERCEPTION AT THE VISUAL MIDLINE

It was suggested some years ago (Milner and Jeeves, 1979) that a particular role for the splenial fibers that cross-connect the midline regions of the *early* visual areas of the cortex (as opposed to the association cortex of the temporal lobe) might be to facilitate symmetry perception. A systematic study of two total acallosals (MG and SG) provides some new data on this question (G.K. Humphreys, personal communication). Mirror-symmetric dot patterns were constructed on a computer, and presented briefly either at the fixation point on the screen, or centered at 100 pixels (approximately 3 deg) to right or left. On different occasions, the patterns were oriented either with the axis of symmetry vertical or at 45 or 90 degrees to the vertical. Two matched controls perceived the vertically oriented patterns best when they were presented exactly on the vertical midline of the visual field. However one of the two acallosals (SG) shows exactly the opposite pattern of results. As predicted from the suggestion of Milner and Jeeves, SG was selectively disadvantaged at perceiving symmetry only when the pattern was *both* at the midline *and* vertically oriented. MG did not show this predicted result, but may have fixated less well than the other subjects - accurate fixation would be a necessary condition for the prediction to be made. In the absence of a splenial mechanism specific for the midline, symmetry judgments may have to be based on less immediate perceptual impressions.

The perception of stereoscopic depth at the midline also requires a bilateral integration of visual information, since the disparate images on the two retinae of an object in depth would fall on opposite sides of the vertical meridian. Such midline stereoscopic perception appears to be selectively impaired in splenium-sectioned patients (Mitchell and Blakemore, 1970; Hamilton and Vermeire, 1986; Jeeves, 1991) and in acallosals (including KC: Jeeves, 1979, 1991), *at first testing*. These data would seem to be consistent with the finding that intact splenial fibers are necessary for the preservation of binocular properties in cortical neurons sensitive to retinal stimulation close to the midline (Berlucchi, 1972). However recent retests have found good midline stereopsis in KC (Cooper and Rogers, 1988) and in LB (Hamilton and Vermeire, 1986). It would be interesting to see whether practice would improve the stereoscopic performance reported by Lassonde (1986) who found impairments (unrestricted to the midline) in several acallosal subjects. It is also notable that monkeys are unimpaired at midline stereopsis when callosotomized after lengthy training (Cowey and Wilkinson, 1991). It appears then that midline stereopsis can often be recovered in the absence of the CC, but it is unclear what pathway mediates the compensation. Physiological studies in animals indicate the presence of neurons with binocular properties in the SC as well as in the cortex (Goldberg and Robinson, 1978); in principle, therefore, position information could cross the midline to allow the computation of binocular disparity at either the midbrain or the AC level.

Finally, brief mention should be made of the perception of apparent motion across the vertical meridian, i.e. between stimuli appearing successively in opposite hemifields. This was shown to be possible in some acallosals by Ettlinger et al (1972). Interestingly, it has recently been observed in completely commissurotomized subjects as well (Ramachandran et al., 1988); so this too can plausibly be attributed to subcortical integration, especially since many cells in the SC are motion-selective (Gordon, 1975).

CONCLUSION

Despite the general tendency in recent years to attribute intact bilateral visual integration in agenesis to the functioning of the AC, the evidence reviewed here fails to provide any clear support for that view. The balance of evidence from reaction time studies leans toward a role for the intertectal commissure, since that appears to provide the means by which crossed reactions are achieved following commissurotomy, and the chronometric (CUD) data are broadly similar in the two cases. Secondly, what seemed like persuasive evidence for the use of the AC in tachistoscopic performance (Martin, 1985) has not proved generalizable. It will of course be interesting for the future, with the improving resolution of MRI (see Rauch and Jinkins, this volume) to compare acallosal patients who are known to have, or not to have, an anterior commissure, on a range of these behavioral tasks.

If much of the bilateral integration we see in the visual experience of acallosals operates at the subcortical level, it becomes interesting to speculate whether it also exists in commissurotomized subjects, but cannot in most cases be brought to conscious awareness. The studies of midline stereo vision suggest that even in acallosals a learning process has to operate; this might have enabled KC to bring disparity information to awareness, as apparently she could in the later study of Cooper and Rogers (1988). It may well be that in everyday life an acallosal individual habitually uses binocular information derived from the retinal midline region in order to control ocular vergence movements, but that these are performed at a subcortical level and without any subjective experience of depth.

Of great relevance to this idea is the impressive bilateral visual integration demonstrated in recent years by Sergent (1987, 1990) in the commissurotomized patients LB and NG, and to a lesser extent AA. In these studies, it seems that inter-hemifield integration can guide a unified response without the patient having any acknowledged perceptual awareness of the contents of the left hemifield (cf Holtzman et al., 1981; and see Milner and Rugg, 1992, for other instances of "implicit" visual skills). Interestingly, the visual complexity of the information handled in these integrative processes (including alphanumeric symbols, words, and facial expressions) far exceeds any previously suspected as being within the processing capacities of the visual midbrain. However the cross-integration seems to be served by the transmission of highly abstract "semantic" data, rather than strictly visual information.

It may be that it takes many years of recovery before a commissurotomized patient becomes capable of using subcortically transmitted information in the guidance of behavior in the way described by Sergent. But even then, the evidence suggests that this information is used without its reaching awareness, whereas the acallosal subject seems not to lack such conscious access. Therefore whatever special compensation it is that allows awareness of the transmitted visual information may require not only a long "recovery" period but one that begins at birth, or at least very soon afterwards.

Footnotes

1. There are rare exceptions to this (Donoso and Santander, 1982), whose anatomical correlates are unknown.
2. The Milner/Lines model clearly predicts that partial section of different parts of corpus callosum should have differential effects on the behavioral CUD. It would predict a normal short (3 ms) CUD providing the anterior route remained available, but a longer (10-15 ms) CUD if only the posterior route remained intact. Aglioti et al (1991) have recently reported such RT data on 5 partial callosal-section patients, and found normal short CUDs in all of them. However it is unclear whether the CUD of the patient with only the splenium intact was *significantly* shorter than the predicted larger value. Indeed in another recent paper (DiStefano et al., 1992; Lassonde, this volume) a different callosotomized patient whose surgery spared the splenium is reported as exhibiting a CUD of 58.3 ms. One must surmise that the true CUD values in such patients probably lie somewhere between these two estimates, both of which may have been subject to large standard errors.

REFERENCES

Aglioti, S., Berlucchi, G., Pallini, R., Rossi, G., and Tassinari, G., 1991, Interhemispheric transfer in patients with partial and total callosotomy, *Third IBRO World Congress,* Montreal, Canada (Abstract).

Auroux, M. and Roussel, C., 1967, Relations neocorticales assurées par la commissure blanche antérieure chez le foetus humain. *CR. Assoc. Anat.* 137:152-159.

Barkovich, A.J. and Norman, D., 1988, Anomalies of the corpus callosum: correlation with further anomalies of the brain, *Am. J. Roentgenol.* 151:171-179.

Bashore, T.R., 1981, Vocal and manual reaction time estimates of interhemispheric transmission time, *Psychol. Bull.* 89:352-368.

Berlucchi, G., 1972, Anatomical and physiological aspects of visual functions of corpus callosum, *Brain Res.* 37:371-392.

Berlucchi, G., 1978, Interhemispheric integration of simple visuomotor responses, in: "Brain Mechanisms and Perceptual Awareness," P.A. Buser and A. Rougeul-Buser, eds., Raven, New York.

Berlucchi, G., Heron, W., Hyman, R., Rizzolatti, G., and Umilta, C., 1971, Simple reaction times of ipsilateral and contralateral hand to a lateralized visual stimulus, *Brain* 94:419-430.

Bogen, J.E., Schultz, D.H., and Vogel, P.J., 1988, Completeness of callosotomy shown by magnetic resonance imaging in the long term, *Arch. Neurol.* 45:1203-1205.

Brown, W.S., 1991, Evoked potentials and bilateral field advantage in agenesis of the corpus callosum, *Poster presented at IBRO Satellite Symposium on Callosal Agenesis,* Quebec City, Que., Canada.

Chalupa, L.M. and Abramson, B.P., 1988, Receptive-field properties in the tecto- and striate-recipient zones of the cat's lateral posterior nucleus, *Prog. Brain Res.* 75:85-94.

Chiarello, C. 1980, A house divided? Cognitive functioning with callosal agenesis, *Brain Lang.* 11:128-158.

Clarke, J.M. and Zaidel, E., 1989, Simple reaction times to lateralized light flashes. Varieties of interhemispheric communication routes, *Brain* 112:849-870.

Cooper, A.M. and Rogers, B.J., 1988, Midline stereopsis in agenesis acallosals. *Neurosci. Lett. Suppl.* 32:S112.

Cowey, A. and Wilkinson, F., 1991, The role of the corpus callosum and extra striate visual areas in stereoacuity in macaque monkeys, *Neuropsychologia* 29:465-479.

Demeter, S., Ringo, J.L., and Doty, R.W., 1988, Morphometric analysis of the human corpus callosum and anterior commissure, *Hum. Neurobiol.* 6:219-226.

Demeter, S., Rosene, D.L., and Van Hoesen, G.W., 1990, Fields of origin and pathways of the interhemispheric commissures in the temporal lobe of macaques, *J. Comp. Neurol.* 302:29-53.

Denenberg, V.H., 1981, Hemispheric laterality in animals and the effects of early experience, *Behav. Brain Sci.* 4, 1-49.

DiStefano, M., Sauerwein, H.C., and Lassonde, M., 1992, Influence of anatomical factors and spatial compatibility on the stimulus-response relationship in the absence of the corpus callosum, *Neuropsychologia* 30: 177-185

Donoso, A. and Santander, M., 1982, Hemialexia y afasia hemianoptica en agenesia del cuerpo calloso, *Rev. Chil. Neuro-Psiquiatr.* 20:137-144.

Eacott, M.J. and Gaffan, D., 1989, Interhemispheric transfer of visual learning in monkeys with intact optic chiasm, *Exp. Brain Res.* 74:348-352.

Edwards, S.B., 1977, The commissural projection of the superior colliculus in the cat, *J. Comp. Neurol.* 173:23-40.

Ettlinger, G., Blakemore, C.B., Milner, A.D., and Wilson, J., 1972, Agenesis of the corpus callosum: a behavioural investigation, *Brain* 95:327-346.

Foxman, B.T., Oppenheim, J., Petito, C.K., and Gazzaniga, M.S., 1986, Proportional anterior commissure area in humans and monkeys, *Neurology* 36:1513-1517.

Goldberg, M.E. and Robinson, D.L., 1978, The superior colliculus, *in*: "Handbook of Behavioral Neurobiology. Sensory Integration," Vol. 1., R.B. Masterton, ed., Plenum Press, New York.

Gordon, B., 1975, Superior colliculus: structure, physiology and possible functions, *in:* "Physiology," Series 1, Vol. 3, C.C. Hunt, ed., Butterworth, London.

Gross, C .G. and Mishkin, M., 1977, The neural basis of stimulus equivalence across retinal translation, *in:* "Lateralization in the Nervous System," S. Harnad, R. Doty, J. Jaynes, L. Goldstein, and G. Krauthamer, eds., Academic Press, New York.

Gross, C.G., Bender, D.B., and Mishkin, M., 1977, Contribution of the corpus callosum and the anterior commissure to visual activation of inferior temporal neurons, *Brain Res.* 131:227- 239.

Hamilton, C.R., 1982, Mechanisms of interocular equivalence, *in:* "Advances in the Analysis of Visual Behavior," D.J. Ingle, M.A. Goodale, and R.J.W. Mansfield, eds., MIT Press, Cambridge.

Hamilton, C. R. and Vermeire, B. A., 1986, Localization of visual functions with partially split-brain monkeys, *in:* "Two Hemispheres - One Brain: Functions of the Corpus Callosum," F. Lepore, M. Ptito, and H.H. Jasper, eds., Alan R. Liss, New York.

Holtzman, J.D., 1984, Interactions between cortical and subcortical visual areas: evidence from human commissurotomy patients, *Vision Res.* 24:801-813.

Holtzman, J.D., Sidtis, J.J., Volpe, B.T., Wilson, D.H., and Gazzaniga, M.S., 1981, Dissociation of spatial information for stimulus localization and the control of attention, *Brain* 104:861-872.
Jeeves, M.A., 1969, A comparison of interhemispheric transmission times in acallosals and normals, *Psychon. Sci.* 16:245-246.
Jeeves, M.A., 1979, Some limits to interhemispheric integration in cases of callosal agenesis and partial commissurotomy, *in:* " Structure and Function of the Cerebral Commissures," I.S. Russell, M.W. van Hof, and G. M. Berlucchi, eds., Macmillan, London.
Jeeves, M.A., 1991, Stereo perception in callosal agenesis and partial callosotomy, *Neuropsychologia* 29:19-34.
Jeeves, M.A. and Milner, A.D., 1987, Specificity and plasticity in interhemispheric integration: evidence from callosal agenesis, *in:* "The Duality and Unity of the Brain," D. Ottoson, ed., Macmillan, London.
Jouandet, M.L., 1982, Neocortical and basal telencephalic origins of the anterior commissure of the cat, *Neuroscience* 7:1731-1752.
Jouandet, M.L. and Gazzaniga, M.S., 1979, Cortical field of origin of the anterior commissure of the rhesus monkey, *Exp. Neurol.* 66:381-397.
Karnath, H.O., Schumacher, M., and Wallesch, C.W., 1991, Limitations of interhemispheric extracallosal transfer of visual information in callosal agenesis, *Cortex* 27:345-350.
Kutas, M., Hillyard, S.A., and Gazzaniga, M.S., 1988, Processing of semantic anomaly by right and left hemispheres of commissurotomy patients: evidence from event-related potentials, *Brain* 111:553-576.
Lamantia, A.S. and Rakic, P., 1990, Cytological and quantitative characteristics of four cerebral commissures in the rhesus monkey, *J. Comp. Neurol.* 291:520-537.
Lassonde, M., 1986, The facilitatory influence of the corpus callosum on intrahemispheric processing, *in:* "Two Hemispheres - One Brain: Functions of the Corpus Callosum," F. Lepore, M. Ptito, and H.H. Jasper, eds., Alan R. Liss, New York.
Lemire, R.J., Loeser, J.D., Leech, R.W., and Alvord, E.C., 1975, "Normal and Abnormal Development of the Human Nervous System," Harper & Row, Hagerstown, MD.
Lia, B., Snider, C.J., and Chalupa, L.M., 1988, The nasotemporal division of the retinal ganglion cell decussation pattern in the fetal rhesus monkey. *Soc. Neurosci. Abstr.* 14:458.
Lines, C.R., 1984, Nasotemporal overlap investigated in a case of agenesis of the corpus callosum, *Neuropsychologia* 22:85-90.
Lines, C.R. and Milner, A.D., 1983, Nasotemporal overlap in the human retina investigated by means of simple reaction time to lateralized light flash, *Exp. Brain Res.* 50:166-172.
Lines, C.R., Rugg, M.D., and Milner, A.D., 1984, The effect of stimulus intensity on visual evoked potential estimates of interhemispheric transmission time, *Exp. Brain Res.* 57:89-98.
Loeser, J.D. and Alvord, E.C., 1968, Clinicopathololological correlations in agenesis of the corpus callosum, *Neurology* 18:745-756.
Martin, A., 1985, A qualitative limitation on visual transfer via the anterior commissure, *Brain* 108:43-63.
McIlwain, J.T., 1991, Visual input to commissural neurons of the cat's superior colliculus, *Vis. Neurosci.* 17:389-393.
McKeever, W.F., Sullivan, K.F., Ferguson, S.M., and Rayport, M., 1981, Typical cerebral hemisphere disconnection deficits following corpus callosum section despite sparing of the anterior commissure, *Neuropsychologia* 19:745-755.
Milner, A.D., 1982, Simple reaction times to lateralized visual stimuli in a case of callosal agenesis, *Neuropsychologia* 20:411-419.
Milner, A.D., 1983, Neuropsychological studies of callosal agenesis, *Psychol. Med.* 13:721-725.
Milner, A.D., 1986, Chronometric analysis in neuropsychology, *Neuropsychologia* 24:115-128.
Milner, A.D., and Goodale, M.A., in press, Visual pathways to perception and action, *in:* " The Visually Responsive Neuron: From Basic Neurophysiology to Behavior" T.P. Hicks, S. Molotchnikoff, and T. Ono, eds., Elsevier, Amsterdam..
Milner, A.D. and Jeeves, M.A., 1979, A review of behavioural studies of agenesis of the corpus callosum, *in:* "Structure and Function of the Cerebral Commissures," I.S. Russell, M.W. van Hof, and G. M. Berlucchi, eds., Macmillan, London.
Milner, A.D., and Jeeves, M.A., 1981, The functions of the corpus callosum in infancy and adulthood, *Beh. Brain Sci.* 4:30-31.
Milner, A.D., and Lines, C.R., 1982, Interhemispheric pathways in simple reaction time to lateralized light flash, *Neuropsychologia* 20:171-179.
Milner, A.D. and Rugg, M.D.,1989, Interhemispheric transmission times, *in:* "Developments in Clinical and Experimental Neuropsychology," J.R. Crawford and D.M. Parker, eds., Plenum Press, New York.
Milner, A.D., and Rugg, M.D., 1992, "The Neuropsychology of Consciousness," Academic Press, London.
Milner, A.D., Jeeves, M.A., Silver, P.H., Lines, C.R., and Wilson, J., 1985, Reaction times to lateralized visual stimuli in callosal agenesis: Stimulus and response factors, *Neuropsychologia* 23:323-331.
Mitchell, D.E. and Blakemore, C., 1970, Binocular depth perception and the corpus callosum, *Vision Res.* 10:49-54.
Munte, T.F. and Heinze, H.J., 1991, Corpus-callosum Agenesie. Interhemisphärische Integration semantischer Information, *Nervenarzt* 62: 629-636.

Pandya, D.N. and Seltzer, B., 1986, The topography of commissural fibers, *in:* "Two Hemispheres - One Brain: Functions of the Corpus Callosum," F. Lepore, M. Ptito, and H.H. Jasper, eds., Alan R. Liss, New York.
Poffenberger, A.T., 1912, Reaction time to retinal stimulation with special reference to time lost in conduction through nerve centers, *Arch. Psychol. (NY)* 23:1-73.
Ramachandran, V.S., Cronin-Golomb, A., and Myers, J.J., 1988, Perception of apparent motion by commissurotomy patients, *Nature* 320:358-359.
Rhoades, R.W., Mooney, R.D., Szczepanik, M., and Klein, B.G., 1986, Structural and functional characteristics of commissural neurons in the superior colliculus of the hamster, *J. Comp. Neurol.* 253:197-215.
Rodieck, R.W., 1979, Visual pathways, *Annu. Rev. Neurosci.* 2:193-225.
Rugg, M.D., Lines, C.R., and Milner, A.D., 1984, Visual evoked potentials to lateralized visual stimuli and the measurement of interhemispheric transmission time, *Neuropsychologia* 22:215-225.
Rugg, M.D., Lines, C.R., and Milner, A.D., 1985a, Further investigation of visual evoked potentials elicited by lateralized stimuli: effects of stimulus eccentricity and reference site, *Electroencephalogr. Clin. Neurophysiol.* 62:81-87.
Rugg, M.D., Milner, A.D., and Lines, C.R., 1985b, Visual evoked potentials to lateralized stimuli in two cases of callosal agenesis, *J. Neurol. Neurosurg. Psychiatry* 48:367-373.
Rugg, M.D., Milner, A.D., Lines, C.R., and Phalp, R., 1987, Modulation of visual event-related potentials by spatial and non-spatial visual selective attention, *Neuropsychologia* 25:85-96.
Saron, C.D. and Davidson, R.J., 1989, Visual evoked potential measures of interhemispheric transfer time in humans, *Behav. Neurosci.* 103: 1115-1138.
Sauerwein, H. and Lassonde, M.C., 1983, Intra- and interhemispheric processing of visual information in callosal agenesis, *Neuropsychologia* 21:167-171.
Schneider, G.E., 1979, Is it really better to have your brain lesion early? A revision of the 'Kennard principle', *Neuropsychologia* 17:557-583.
Sergent, J., 1987, A new look at the human split brain, *Brain* 110:1375-1392.
Sergent, J., 1990, Furtive incursions into bicameral minds, *Brain.* 113:537-568.
Sergent, J., 1991, Processing of spatial relations within and between the disconnected cerebral hemispheres, *Brain* 114:1025-1043.
Sergent, J. and Myers, J.J., 1985, Manual, blowing, and verbal simple reactions to lateralized flashes of light in commissurotomized patients, *Percept. Psychophys.* 37:571-578.
Shoumura, K., Imai, H., Kimura, S., Suzuki, T., and Ara, M., 1987, Posterior commmissural connections of area pretectalis and neighbouring structures in cat, with special reference to pupilloconstrictory pathway via posterior commissure, *Jpn. J. Ophthalmo.* 3:289-304.
Sparks, D.L. and May, L.E., 1990, Signal transformations required for the generation of saccadic eye movements, *Annu. Rev. Neurosci.* 13:309-336.
Sperry, R.W., 1968a, Hemisphere disconnection and unity in conscious awareness, *Am. Psychol.* 23:723-733.
Sperry, R.W., 1968b, Plasticity of neural maturation, *Dev. Biol. Suppl.* 2:306-327.
Ungerleider, L.G. and Mishkin, M., 1982, Two cortical visual systems, *in:* "Advances in the Analysis of Visual Behavior," D.J. Ingle, M.A. Goodale, and R.J.W. Mansfield, eds., MIT Press, Cambridge.
Volchan, E., Bernardes, R.F., Rocha-Miranda, C.E., Gleiser, L., and Gawryszewski, L.G., 1988, The ipsilateral field representation in the striate cortex of the opossum, *Exp. Brain Res.* 73:297-304.

THE DEVELOPMENT OF INTERHEMISPHERIC TRANSFER OF TACTILE INFORMATION IN CASES OF CALLOSAL AGENESIS

Gina M. Geffen,[1] Judith Nilsson,[2]
Donald A. Simpson,[3] and Malcolm A. Jeeves[4]

[1]Cognitive Psychophysiology Laboratory
Psychology Department and Clinical Studies Unit
University of Queensland and
Queensland Health Department
Brisbane, Australia, 4072
[2]South Australia Head Injury Service, Julia Farr Centre
Adelaide, Australia, 5063
[3]NH and MRC Road Accident Research Unit
University of Adelaide and Department of Neurosurgery
Adelaide Children's Hospital, Adelaide, Australia, 5001
[4]Psychological Laboratory
University of St. Andrews
St. Andrews, Fife, Scotland, KY16 9JU

INTRODUCTION

Congenital absence of corpus callosum in relatively asymptomatic individuals raises the question of mechanisms responsible for interhemispheric transfer of information. Possible alternative methods of communication between the cerebral hemispheres involve the anterior commissure, subcortical commissural pathways, bilateral representation of functions that are normally predominantly under unilateral control, enhanced transmission of information on the ipsilateral sensory and motor pathways, and behavioral cross-cueing (Jeeves, 1990). Before it is possible to pinpoint a particular mechanism, it is necessary to determine whether subjects with congenital absence of the corpus callosum (ACC) differ from normal individuals when performing tasks that require interhemispheric transfer, and whether transfer deficits, where present, are similar in magnitude to those shown by patients with surgical section (partial or complete) of the corpus callosum. We report a study of seven cases of callosal agenesis using a tactile finger localization task previously shown to be sensitive to extent and location of acquired lesions of the corpus callosum (Geffen et al., 1985). Developmental improvements in interhemispheric transfer in normal children were also elicited using this task (Quinn and Geffen, 1986). However, the proposal that this improvement reflects the concomitant myelination of the corpus callosum between ages 5-11 years has been questioned (Pipe, 1991).

Previous studies of interhemispheric transfer of tactile information have not shown a transfer deficit or tactile disconnection in ACC cases (e.g. Lassonde et al., 1988). The task selected, however, should exclude finger and hand movements during initial stimulation in order to lateralize the input. Proprioceptive information, e.g. heaviness and position sense, is available bilaterally in adult commissure sectioned patients, presumably due to midline transfer at a subcortical level (Gandevia, 1978; McCloskey, 1973). Movement of the fingers during exploration of shapes or textures provides position information to both cerebral hemispheres. Reports of ACC patients performing tactile finger or body-part localization are also contradictory. These studies are examined in detail below.

The patient, PH, a subject in the present study, was reported to show no evidence of deficit in finger cross-localization at ages 6 and 18 years (Jeeves and Rajalakshmi, 1964; Jeeves, 1965, 1979). Similarly, two acallosal subjects (a 19-year-old female and a similarly aged male) showed highly accurate cross-localization of fingers and finger segments (Saul and Gott, 1973; Saul and Sperry, 1968; Sperry, 1968). Two four-year-old female acallosal patients of borderline intelligence could correctly localize touch on ipsi- and contralateral sides of their bodies (Field et al., 1978). Normal cross-localization of touched fingers by showing the corresponding finger on the non-stimulated hand was inferred from a report of two ACC patients aged 9, and 17 years (Sauerwein et al., 1981).

In contrast to the above studies in which the six patients varied in age from 4 to 19 years, with different intelligence quotients (IQs) and different clinical pathologies, there are several studies in which deficits of cross-localization of fingers have been reported (Dennis, 1976; Ettlinger et al., 1972; Gazzaniga, 1970; Reynolds and Jeeves, 1977).

When the patients in the above reports are combined, 54% of them showed cross-localization deficits. Chiarello (1980) reviewed 29 reported cases of callosal agenesis and concluded that 41% of these subjects showed a deficit on this task. It was the only tactile task to reveal cross-localization impairments. The impairment was attributed to loss of spatial information during interhemispheric transfer, rather than the conversion of the spatial code to a motor response after interhemispheric transfer (Chiarello, 1980). It is possible that differences in administration of the finger localization task (e.g. extent of movement of the touched finger that is allowed) as well as differences between patients produce contradictory findings. One advantage of the task used by Geffen et. al (1985) involved the stimulation of a sequence of fingers in one condition. Cases with partial section of the corpus callosum were more likely to show a cross-localization impairment when they had to localize a sequence of two or three fingers rather than a single finger. Similarly, normal adults and 11 year olds required three finger sequences to elicit worse cross than ipsilateral localization differences (Geffen et al., 1985; Quinn and Geffen, 1986). The more difficult conditions (2 and 3 finger localization) were expected to produce cross-localization deficits in our adult ACC subjects. However, we expected the children to show deficient one finger cross-localization.

The development of interhemispheric transfer in the absence of the corpus callosum may be inferred by comparing the cross-localization performances of ACC subjects of different ages (5 to 26 years). We also report longitudinal data from two male subjects (MR and AS) to complement the cross-sectional information.

METHOD

Subjects

Neurological. Seven subjects with agenesis of the corpus callosum were tested. Their radiological evidence, intelligence test results, handedness and age tested are shown in Table 1.

Table 1. Profile of callosal agenesis cases.

Initials	Sex	Verbal IQ	Performance IQ	Preferred Hand	Callosal Agenesis	Age Tested
MR	M	94	82	Left	Complete [CT (9 months) MR (age 13)]	5, 7, 8, 13, 15
DC	M	112	108	Right	Complete [Ultrasound (3 days) MR (age 10)]	8
AS	M	83	77	Left	Partial [CT (age 10) MR (age 21)]	13, 22
PH	M	92	69	Right	Complete [Air encephalo-gram (age 6) MR (age 36)]	26
AC	F	102	73	Right	Complete [CT (age 8)]	10
DH	F	62	68	Left	Complete [Air encephalo-gram (2 months)]	11
JW	F	81	65	Left	Complete [CT (age 14)]	15

CT=Computed tomography; MR=Magnetic resonance scan

The case histories and MR scans of MR, DC, AS and PH are described in Geffen, Forrester, Jones and Simpson (this volume). DH has been previously described (Field et al., 1976).

AC was delivered close to term with normal birth weight. Clinical findings were normal apart from a coarse lateral nystagmus to both sides but investigations revealed nothing amiss. At one month, the infant was noted to roll her head rhythmically from side to side with deviation of the eyes. Her skull x-ray and urinary chromatography for amino acids showed no abnormalities. The child showed a mild hypotonia with slow developmental features, although intellectual performance was within normal limits. She was placed in the general category of a possible hypotonic cerebral palsy and received physiotherapy from the Spastic Centre. In early 1980 she was investigated for a large head with a circumference of more than two standard deviations above the mean, bilateral pale discs and abnormally brisk deep tendon jerks in the lower limbs, possible evidence of hydrocephalus. The CT scan showed evidence of agenesis of the corpus callosum. In 1980 the child was transferred to a small state school which could provide her with remedial teaching.

JW's mother suffered persistent vomiting attacks throughout her entire pregnancy. Labor was induced in the 38th week because of mother's high blood pressure. At two weeks postnatal, the baby had a noticeably large head. Milestones of sitting and walking were late. At 17 months she was admitted to hospital for investigation of a large head and possible febrile convulsions shown by right side twitching. EEG showed ill-defined epileptic spike and slow wave activity bilaterally but focal features were absent. An angiogram showed evidence of agenesis of the corpus callosum. At 13 years 6 months CT scan confirmed agenesis of the corpus callosum with severe bilateral porencephalic dilation of occipital horns of the lateral ventricle. Occipital cortices were markedly thinned. Her VIQ was low average and PIQ in the mentally deficient range. She is left-handed. JW has had considerable learning difficulties throughout her schooling and became increasingly short sighted. Seizures have been fairly well controlled by medication. She presented as an attractive, friendly young woman but with limited potential to cope with outside employment.

Controls. The controls comprised seven different age groups. These consisted of 5-year-olds (N = 16, range 5.3 to 6 years, mean 5.7 years); 7-year-olds (N = 16, range 7.1 to 7.9 years, mean 7.3 years); 9-year-olds (N = 16, range 9.3 to 9.9 years, mean 9.7 years); 11-year-olds (N = 18, range 11.3 to 12.2 years, mean 11.7 years including a pair of 11-year-old twins which were dizygotic twin controls for DH and her neurologically normal dizygotic twin sister, LH); 13-year-olds (N = 20, range 12.9 to 13.7 years, mean 13.3 years); 15-year-olds (N = 11, range 14.9 to 15.8 years, mean 15.3 years); and an adult student group (N = 16, range 19.4 to 28.3 years, mean 22.5 years). The 5, 7, 9, 11, 15-year-old and adult groups contained equal numbers of males and females, whereas the 13-year-olds consisted of 13 males (4 left handed) and 7 females.

Apparatus and Materials

The six primary questions of the Annett (1970) handedness questionnaire were used to establish hand preference. A right (left) response to all six questions classified the child as right- (left-) handed, while a mixture of 'right', 'left' or 'either' responses provided a mixed classification.

Subjects sat at a table with their arms extended and resting comfortably on a cushion on top of the table. During testing, a sleep mask was worn to exclude vision. The point of a Lafayette aesthesiometer was used to stimulate the tips of fingers. A protocol with one, two and three finger random sequences in sets of 16 trials for each condition was used. Each finger (excluding the thumb) was represented within each sequence of trials the same number of times. Within each two and three finger trial, each finger was stimulated only once, i.e. two or three different fingers were touched.

Procedure

The procedure described by Quinn and Geffen (1986) was used, and is summarized below. Subjects were tested individually, in one session lasting between 35 and 60 minutes.

The task involved brief, light stimulation to the tip of one finger (or a sequence of fingers) using the point of the aesthesiometer. Movement of the stimulated finger either by the experimenter or the subject was minimized, and the palm of the stimulated hand faced downwards. The subject indicated which finger had been stimulated by touching it with the thumb of the same hand in the same (within) hand condition, or touching the corresponding finger on the opposite hand with the thumb of that hand in the opposite (between) hand condition. There were two between-hand response conditions: one involved holding the hand in the same orientation (palm-down) as the stimulated hand (mirror- image match). In the second condition, the response hand was held palm-up (left-to-right match), i.e. in a different orientation from the stimulated hand. Subjects were instructed to respond immediately following stimulation.

Localization of a sequence of fingers involved stimulating two or three fingers sequentially before a response was required. For a trial to be scored correct, subjects not only had to touch the correct fingers but to touch them in the same order as stimulated.

Subjects were initially familiarized with the task using one-finger stimulation, and with vision. Twelve practice trials (four within-hand and eight between-hand) were given in the pre-testing with vision, and then a further 12 trials with vision excluded. The experimenter demonstrated how the response was to be made, and if a subject failed to understand the task requirements following the pre-testing with vision, further demonstration was provided before proceeding to the practice trials with vision excluded. For the subjects who were also tested with two- or three-finger sequences, further practice trials were given with the appropriate number of fingers stimulated.

There were six conditions, the type of matching response (same hand; other hand, same palm orientation; other hand, different palm orientation) and the stimulated hand (left or right). The order of same and other hand testing was counterbalanced in an ABBA design, with each testing session beginning with same hand responses. All other conditions were also counterbalanced to control for practice and fatigue effects. For each type of matching response, the left hand was stimulated first for half the subjects, and the right hand stimulated first for the remainder of the subjects. There were 16 trials per condition, yielding a total of 96 trials with one finger stimulated per trial, 96 trials with two fingers stimulated per trial, and 96 trials with three fingers stimulated per trial.

RESULTS

The dependent variable was number of trials correct, corrected for chance, yielding corrected proportion of trials (P(C) - C /I-C, where P(C) = obtained proportion correct and C = chance probability of a correct response. These probabilities were .25 for one, .083 for two, and .042 for three-finger stimulation).

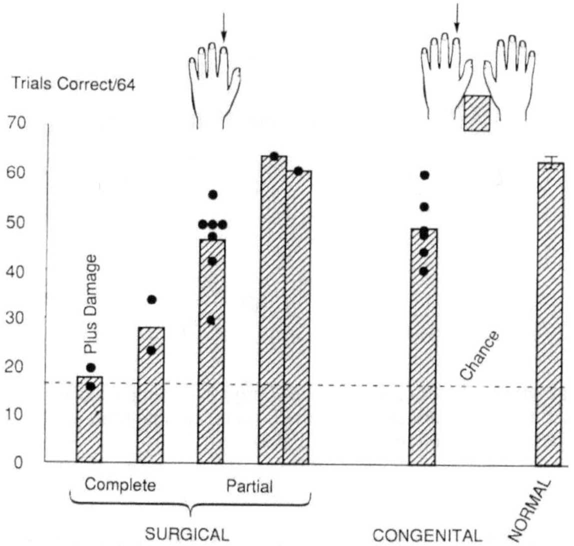

Figure 1. Single finger localization performance responding with the same hand as stimulated (correct range shown, stippled grey) or with the other hand (diagonally striped histograms). From left to right the histograms depict mean number of trials correct for two subjects with commissurotomies to relieve intractable epilepsy (AA and RY) who also had extra-callosal cortical lesions; two subjects (LB & NG) with complete commissurotomies and minimal extra-callosal damage; seven subjects with partial surgical section involving the trunk and anterior splenium of the corpus callosum to remove midline abnormalities; one subject (WY), with section of the genu of the corpus callosum.. As shown, his pre- and post-operative cross-localization performance was normal; seven congenitally acallosal subjects; sixteen normal adults.

Figure 1 shows the mean number of trials correct with ipsi- and contralateral responses for the ACC subjects compared to four subjects with complete commissurotomies and nine subjects with partial section of the corpus callosum (from Geffen et al., 1985). Single finger localization using the same hand as stimulated was highly accurate in all subjects. Cross-localization was similarly accurate in normal subjects, in one ACC (PH, see Figure 2), and in the case with pre- and post-operative section of the genu of the corpus callosum. Varying levels of inaccurate cross-localization performance were observed. Complete section of the corpus callosum caused most disruption to cross-localization. Performance of the ACC cases was equivalent to that after partial surgical section of the corpus callosum.

Tactile cross-localization improves between 5 and 11 years of age in normal children (Geffen et al., 1985). Figure 2 shows the performance of ipsi- and cross-localization of the seven acallosal subjects according to age. While their age-matched controls demonstrated the expected improvement in cross-localization, the effects of age (7 to 11 years) on this ability in ACC cases were not evident. Close-to-normal performance was demonstrated by AS (13 years), but MR scanning revealed an intact genu that presumably enabled interhemispheric transfer of information. The only adult, PH, showed normal single finger cross-localization. The mechanism that enables transfer of simple tactile information between the cerebral hemispheres appears to develop after age 15 in cases of complete absence of the corpus callosum.

Indication of a sequence of two or three fingers touched was a more difficult task than localizing a single finger, with correspondingly lower levels of accuracy achieved by both control and acallosal subjects (see Figure 3). The very low level of within-hand accuracy shown by AC and DH indicates that their knowledge of finger topography and/or ability to encode, store and reproduce a sequence of tactile events was well below normal. Since the baseline for inferring deficient cross-localization is adequate ipsilateral localization, the transfer deficit in AC and DH could not be estimated.

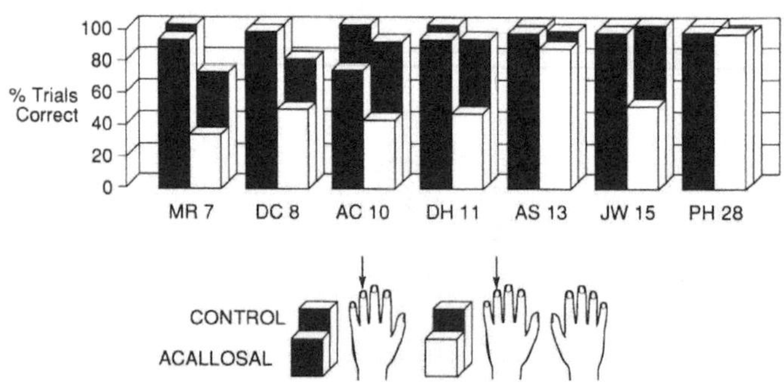

Figure 2. The individual within- (black) and between- (grey) hand finger localization performance with a *single* finger stimulated of acallosal subjects (age 7 to 28 years) compared to age matched controls (shown behind acallosal results). Chance correction has been applied.

Both AS (age 13, genu of corpus callosum present) and PH (28 years) had shown normal cross-localization performance with one finger stimulated. Cross-localization of two fingers was 28% worse than ipsilateral localization (AS), while PH showed 12% poorer cross than ipsilateral finger sequence performance. Reference to Table 2 indicates that these differences are roughly twice as large as the corresponding differences in normal subjects of 13 and adult ages. AS appeared to be cross-localizing at roughly a 7-year-old level, and PH at the level of an 11-year-old. The deficit shown was similar to that of partial section of the corpus callosum.

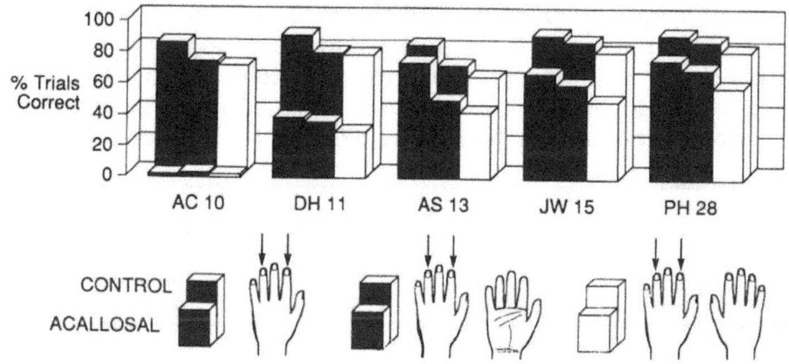

Figure 3. Within- (black) and between-hand localization of a sequence of fingers of the five older subjects with congenital absence of the corpus callosum compared to age matched controls (shown behind). Chance correction has been applied. Cross localization responses were made with the responding hand held palm up (grey) or palm down (white). The stimulated hand was always palm down. PH had three fingers stimulated per trial, while all other subjects had two fingers stimulated per trial.

Table 2. Age effects on cross localization of a stimulated finger, or sequence of fingers. Scores are mean (SD) percentage between same (within) - other (between) hand responding.

	WITHIN - BETWEEN HAND RESPONSE	
Age (Years)	**One Finger**	**Two / Three Fingers**
5	47 (14)	
7	30 (23)	26 (7)
9	12 (10)	21 (5)
11	9 (8)	12 (6)
13	6 (5)	15 (6)
15	0.25 (0.3)	4 (2)*
Adult	1 (1)	7 (7)*
Partial Section (including trunk)	28 (11)	18 (8)
Complete Section	82 (11)	31 (12)

* Three fingers stimulated

In summary, the analysis of cross-sectional age differences in the ability of cases of callosal agenesis to transfer tactile information indicates that all seven cases showed some deficit of cross-localization but normal ipsilateral localization of a single finger. In five cases, the cross-localization deficit was apparent using single finger localization. The adult case of complete agenesis (PH) and the 13-year-old with partial absence of the corpus callosum (AS) required the more difficult task of localizing a sequence of fingers to exhibit their cross-localization deficits. The deficits were similar in magnitude to those shown by partial section involving the trunk of the corpus callosum.

Table 3. Acallosal subjects, MR (complete absence) and AS (genu evident): mean percentage trials correct of single finger localization responding with the same (within) or the other hand (between). Chance correction has been applied. There were 16 trials per condition, however for between hands, trials have been averaged across same vs different orientation of responses hand making 32 trials for this condition.

			TYPE OF MATCHING RESPONSE		
Subject	**Age (Years)**	**Hand Stimulated**	**Within Hand**	**Between Hands**	**Within - Between**
MR	5	Left	87	28	59
		Right	87	*	
		Mean	87		
	7	Left	96	37	59
		Right	90	32	58
		Mean	93	34	59
	8	Left	100	50	50
		Right	100	29	71
		Mean	100	40	60
	13	Left	100	58	42
		Right	100	33	66
		Mean	100	46	54
	15	Left	100	71	29
		Right	100	46	54
		Mean	100	58	42
AS	13	Left	96	91	5
		Right	100	86	14
		Mean	98	89	9
	22	Left	100	92	8
		Right	100	87	13
		Mean	100	90	10

* Incomplete test session

Table 4. Controls. Accuracy of single finger localization. Scores are mean percentage (SD) of trials correct averaged across stimulated hand yielding 32 trials per within hand condition. Between Hands measures were obtained by averaging across hand stimulated and orientation of response hand making 64 trials per condition.

	TYPE OF MATCHING RESPONSE		
Age (Years)	**Within Hand**	**Between Hands**	**Within - Between**
5	97 (4)	50 (14)	47
7	98 (4)	68 (26)	30
8*	98 (4)	78 (21)	20
13	100 (0)	94 (6)	6
15	100 (0)	100 (0)	0
Adult	100 (0)	99 (9)	1

* 8-year -olds' data interpolated between 7 and 9

The comparison of single cases of different ages with a common congenital disorder assumes a similarity between them that their case histories and IQ levels indicate may not exist. A description of longitudinal study of two cases follows. The first, MR, was studied between 5 and 15 years of age covering the period of childhood to adolescence and the second, AS was studied at ages 13 and 22, providing some indication of development between adolescence and adulthood.

Single finger localization is shown in Table 3 for the acallosal subjects and Table 4 for controls. While MR's within-hand finger localization at age 5 was about 10% below normal, by age 7 his left hand responses were within the normal range and right hand at a borderline level. In contrast, his cross-localization responses remained roughly one and a half standard deviations below normal from ages 5-15 years. By age 15 his ability to transfer tactile information between the hemispheres resembled that of normal 5-year-olds and patients with partial section of the trunk of the corpus callosum, since the difference between ipsi- and cross-localization accuracy was 42%. In addition, from age 7 onwards, he was consistently better at transferring information from his right hemisphere (left hand stimulated) to his left (right hand response) than in the opposite direction. Hand differences and direction of transfer were not significant effects in control subjects.

Two possible interpretations are 1) the ipsilateral sensory pathways from MR's left hand to his left hemisphere transmit finger localization information better than those from his right hand to his right hemisphere. This would enable more accurate left hemisphere control of the contralateral right hand response, since the right hemisphere would have impoverished information to encode a contralateral left hand response. Alternatively, 2) right hemisphere motor control of left hand responses could be worse than left hemisphere/right hand. The second alternative seems unlikely because MR is left handed. However, with two fingers stimulated, MR showed a relative right hand advantage. This is discussed further below.

The condition of sequential finger localization within each hand provided a sensitive indicator of the relative integrity of the left and right parietal cortex. Control subjects showed no hand differences in ipsilateral localization of a sequence of fingers. Their results, combined across hand stimulated, are shown in Table 6. Reference to Table 5 indicates that MR was better with his right than left hand from ages 7 to 13, and then at age 15 was slightly superior with his left hand stimulated. Yet at this age (15 years), transfer of single finger localization was still 35% better with his left hand stimulated (that is, from his right to his left hemisphere than vice-versa, see Table 3).

More striking was MR's very depressed performance on the task of sequential finger localization within each hand and which did not require interhemispheric transfer. This result implies that MR may have inefficient parietal lobe function that is more marked in the right than the left half of the brain.

The performance of AS was similar in many respects to that of MR. His ability to localize a single finger on the same hand as stimulated was as accurate as control subjects. He showed better transfer with his left hand stimulated and right hand responding than in the opposite direction (at least one s.d. below controls, see tables 3 and 4). These effects were consistent between 13 and 22 years of age. AS' ability to reproduce a sequence of stimulated fingers using the same hand as touched was also asymmetrical: at 13 and 22 years his right hand was better than his left, with normal right hand performance when 22. Ipsilateral localization of finger sequences reached control levels of accuracy when AS was 22, and then only with his right hand stimulated (and responding). AS, like MR, is also left handed, yet showed a right hand advantage for within hand localization of finger sequences.

Transfer of finger sequences was performed more accurately from left to right hand than in the opposite direction. However, left to right transfer showed no improvement between ages 13 and 22. In contrast, AS showed a marked gain in accuracy between these ages with the right hand stimulated and the left responding. As mentioned when discussing his results in relation to the other acallosal subjects, he showed a cross-localization deficit equivalent to that after partial section involving the trunk of the

Table 5. Acallosals: Accuracy (mean percentage of trials correct; 16 trials per condition) of finger localization while responding with the same hand (within) or the other hand (between). Two/three fingers stimulated.

				TYPE OF MATCHING RESPONSE			
				Between Hands			
Subject	Age (Yrs)	Hand Stimulated	Within Hand	Same Orientation (S)	Different Orientation (D)	S+D / 2	Within-Between
MR	7	Left	9				
	(2) #	Right	18				
		Mean	14	*			
	8	Left	5				
	(2)	Right	18				
		Mean	11				
	13	Left	18				
	(2)	Right	37				
		Mean	29				
	15	Left	59				
	(2)	Right	53				
		Mean	56				
AS	13	Left	59	52	56	54	5
	(2)	Right	86	32	42	37	29
		Mean	73	42	49	45	28
	22	Left	74	41	61	51	23
	(3)	Right	94	48	80	64	30
		Mean	84	45	71	58	26

* Level of performance within hand (same matching response) was too poor to measure matching responses with the other hand. # Numbers in parentheses refer to the number of fingers stimulated per trial.

Table 6. Controls: Accuracy of finger localization [mean percentage (SD) of trials correct averaged across stimulated hand, 32 trials per condition] while responding with the same hand as stimulated (within) or with the other hand (between). Adults and 15 years old had three fingers stimulated per trial. The other age groups had two fingers stimulated.

		TYPE OF MATCHING RESPONSE			
		Between hands			
Age (Yrs)	Within Hand	Same Orientation (S)	Different Orientation (D)	S+D / 2	Within-Between
7	57 (20)	30 (26)	32 (21)	31 (23)	26
8*	68 (19)	45 (24)	48 (23)	47 (23)	21
9	78 (19)	60 (23)	65 (25)	62 (24)	16
13	79 (18)	59 (13)	66 (15)	63 (13)	15
15	96 (4)	89 (11)	93 (4)	92 (5)	4
Adult	87 (9)	77 (18)	83 (16)	80 (14)	7

* 8-year -olds' data interpolated between 7 and 9

corpus callosum especially when required to transfer finger sequences from his right to his left hand.

DISCUSSION

The analysis of longitudinal data of two cases tends to confirm the findings of the cross-sectional results on acallosal subjects. All subjects showed a cross-localization deficit. The developmental lag in ability to transfer tactile information between the cerebral hemispheres was substantial (greater than seven years), with normal performance on the simplest task being shown by one adult subject (PH, 28) and in one condition (left hand touched, right response) by the other adult (AS, 22). All subjects showed a deficit in ipsilateral localization of finger sequences, indicative of compromised intrahemispheric functions (cf. Lassonde, 1986). The longitudinal data have also revealed a dissociation of development of the left and right hands, and by implication, the right and left cerebral hemispheres. In both left-handed male subjects, ipsilateral right hand localization of finger sequences was more accurate than that of the left hand. Transfer of information of which single finger was touched was superior from left to right hands than in the opposite direction. When tested at age 22, AS showed better transfer of sequences from right to left hand but still maintained the left to right hand advantage in the transfer of single finger information. Normal subjects (including left handers) showed neither hand asymmetry nor directional transfer differences.

The cross-sectional and longitudinal results of ipsi- and cross-localization of fingers indicate that ACC cases do show deficits in transferring tactile information. The size of the transfer deficit is equivalent to partial section of the corpus callosum involving the trunk. Patients with complete commissurotomy show a much larger cross-localization deficit than those with partial sections (Geffen et al., 1985). The mechanism used by ACC subjects to transfer tactile information is therefore at least as accurate as the use of a portion of the corpus callosum. It is notable that the one case, AS, with an incomplete corpus callosum (genu present), could transfer single finger information normally, provided this went from his left to his right hand.

Since the corpus callosum is absent in six of seven cases, each of the alternative mechanisms for interhemispheric transfer of tactile information will be considered in turn. Although the anterior commissure could be visualized in the four cases with MR scans, the structure was reported as shrunken rather than enlarged. Moreover, interconnections between the somatosensory areas of the left and right cortices do not traverse the anterior commissure (Pandya and Seltzer, 1986). Subcortical commissural pathways are a possible mechanism that cannot be ruled out. Behavioral cross-cueing was controlled by disallowing movement of the stimulated fingers. However, very small movements might have occurred.

Bilateral representation of touch at the fingertips could occur if ipsilateral sensory pathways transmitted touch information. Similarly, the contralateral hemisphere to the hand stimulated could employ ipsilateral motor pathways to control the other, responding hand. Enhancement of ipsilateral transmission has been the favored alternative mechanism by authors reporting behavioral (Dennis, 1976; Lassonde et al., 1986; Sauerwein et al., 1981) and electrophysiological evidence (Laget, et al., 1977; Vanasse et al., this volume) in ACC cases. Further research that records electromyographic response to detect finger movements and uses a task that separates stimulation from response components (e.g. tactile simultaneity judgments, Clark and Geffen, 1990) is required. This would enable isolation of the mechanism(s) responsible for transfer.

CONCLUSION

The dissociation of development of the left and right hands in ipsilateral localization of finger sequences, as well as the asymmetry that was observed in direction of transfer, could only have been uncovered by longitudinal study. Such dissociations and asymmetrical transfer have not been noted in the literature on normal development and were not observed in the cross-sectional comparisons of the different age groups. One function of the corpus callosum during normal development, not discovered previously, may be to enable parallel development of sensorimotor functions in each half of the brain.

In conclusion, the interhemispheric transfer of tactile information in congenital acallosals was found to be inefficient, asymmetrical, and developed (in accuracy and symmetry) very slowly indeed. It will be interesting to test MR at 21 years, as well as more right handed male and female acallosal subjects of a variety of ages.

Acknowledgments

This work was supported by grants from the Australian Research Council and Julia Farr Centre Foundation Inc.

REFERENCES

Annett, M., 1970, A classification of hand preference by association analysis, *Br. J. Psychol.* 61:303-321.

Chiarello, C., 1980, A house divided? Cognitive functioning with callosal agenesis, *Brain Lang.* 11:128-158.

Clark, C.R. and Geffen, G.M., 1990, Hemispheric equivalence for simultaneity judgements of somatosensory stimuli, *Cog. Neuropsychol.* 7:311-327.

Dennis, M., 1976, Impaired sensory and motor differentiation with corpus callosum agenesis: A lack of callosal inhibition during ontogeny?, *Neuropsychologia* 14:455-469.

Ettlinger, G., Blakemore, C.B., Milner, A.D., and Wilson, J., 1972, Agenesis of the corpus callosum: a behavioural investigation, *Brain* 95:327-346.

Field, M., Ashton, R., and White, K., 1978, Agenesis of the corpus callosum: report on two pre-school children and review of the literature, *Dev. Med. Child Neurol.* 20:47-61.

Gandevia, S.C., 1978, The sensation of heaviness after surgical disconnection of the cerebral hemispheres in man, *Brain* 101:295-305.

Gazzaniga, M. S., 1970, "The Bisected Brain," Appleton-Century-Crofts, New York.

Geffen, G., Nilsson, J., Quinn, K., and Teng, E.L., 1985, The effect of lesions of the corpus callosum on finger localization, *Neuropsychologia* 23:497-514.

Jeeves, M.A., 1965, Psychological studies of three cases of congenital agenesis of the corpus callosum, *in:* "Functions of the Corpus Callosum," E. G. Ettlinger, ed., Churchill Livingstone, London.

Jeeves, M.A., 1979, Some limits to interhemispheric integration in cases of callosal agenesis and partial commissurotomy, *in:* "Structure and Function of the Cerebral Commissures," I. S. Russell, M. W. van Hof and G. Berlucchi, eds., University Park Press, Baltimore, Md.

Jeeves, M.A., 1990, Agenesis of the corpus callosum, *in:* "Handbook of Neuropsychology," Vol. 4, F. Boller and J. Grafman, eds., Elsevier, North Holland, Amsterdam.

Jeeves, M.A. and Rajalakshmi, R., 1964, Psychological studies of a case of congenital agenesis of the corpus callosum, *Neuropsychologia* 2:247-252.

Laget, P., D'Allest, A.M., Fihey, R., and Lortholary, O., 1977, L'intérêt des potentials évoqués somesthésiques homolatéraux dans les agénésies du corps calleux, *Rev. E.E.G. Neurophysiol. Clin.* 7:498-502.

Lassonde, M., Sauerwein, H., McCabe, N., Laurencelle, L., and Geoffroy, G., 1988, Extent and limits of cerebral adjustment to early section of the corpus callosum, *Behav. Brain Res.* 30:165-181.

McCloskey, D.I., 1973, Position sense after surgical disconnexion of the cerebral hemispheres in man, *Brain* 96:269-276.

Pandya, D. P. and Seltzer, B., 1986, The topography of commissural fibers, *in*: "Two Hemispheres-One Brain: Functions of the Corpus Callosum," F. Lepore, M. Ptito, and H. H. Jasper, eds., Alan R. Liss, New York.

Pipe, M., 1991, Developmental changes in finger localization, *Neuropsychologia* 29:339-342.

Quinn, K. and Geffen, G., 1986, The development of tactile transfer of information, *Neuropsychologia* 24:793-804.

Reynolds, D. and Jeeves, M.A., 1977, Further studies of tactile perception and motor co-ordination in agenesis of the corpus callosum, *Cortex* 13:257-272.

Sauerwein, H.C., Lassonde, M.C., Cardu, B., and Geoffroy, G., 1981, Interhemispheric integration of sensory and motor functions in agenesis of the corpus callosum, *Neuropsychologia* 19:445-454.

Saul, R.E. and Gott, P.S., 1973, Compensatory mechanisms in agenesis of the corpus callosum, *Neurology* 18:307.

Saul, R. E. and Sperry, R. W., 1968, Absence of commissurotomy symptoms with agenesis of the corpus callosum, *Neurology 18*:307.

Sperry, R.W., 1968, Plasticity of neural maturation, *Dev. Biol. Suppl.* 2: 306-327.

SHORT- AND MIDDLE-LATENCY SOMATOSENSORY EVOKED POTENTIALS IN CALLOSAL AGENESIS

Michel Vanasse,[1] Luc Forest,[2] and Maryse Lassonde[2]

[1]Service de Neurologie, Département de Pédiatrie
Hôpital Sainte-Justine pour les Enfants,
Université de Montréal
3175, Chemin Côte Sainte-Catherine, Montréal
Qué., H3T 1C5, Canada.
[2]Groupe de Recherche en Neuropsychologie Expérimentale
Département de Psychologie
Université de Montréal, C.P. 6128, Succ. A
Montréal, Qué., H3C 3J7, Canada

INTRODUCTION

Experimental transection of the corpus callosum in animals and callosotomy practiced in human adults for therapeutic purposes have led to the description of the disconnection syndrome. Typically, adult patients or animals who undergo a callosal section show a permanent deficit in interhemispheric transfer and integration of sensory and motor functions (for reviews, see Sperry et al., 1969; Gazzaniga, 1970). By contrast, few of these deficits are seen in individuals with agenesis of the corpus callosum when there are tested in certain tasks requiring intermanual comparisons. For example, whilst callosotomized patients are unable to cross-match objects held in both hands out of view, acallosal subjects have no difficulties in performing this task although their response times may be longer than in callosally-intact subjects (Ettlinger et al., 1972, 1974; Saul and Gott, 1973; Jeeves, 1979; Sauerwein et al., 1981). The same is true when callosotomy is performed before puberty (Lassonde et al., 1986, 1991).

The mechanisms used by acallosal subjects to compensate for the absence of the callosal commissure are still a matter of speculation. Several hypotheses have been proposed, including bilateral speech representation (Sperry, 1968; Ettlinger, 1972, 1974; Saul and Gott, 1973) or the use of behavioral strategies (Gazzaniga, 1970). Reviews by Chiarello (1980) and Sauerwein et al. (this volume) tend not to confirm these two propositions. In fact, it has often been suggested that an enhanced proficiency of the ipsilateral components of the somatosensory system is a more likely explanation for the absence of disconnection syndrome in acallosal subjects, at least in the tactuomotor modality (Nebes and Sperry, 1971; Dennis, 1976; Gazzaniga and Ledoux, 1978; Sauerwein et al., 1981). Hence, by relying on information provided by both ipsi- and contralateral somesthetic pathways, acallosal subjects would perform a bimanual matching task using an *intra-*, rather than an *inter-*, hemispheric comparison process.

Callosal Agenesis, Edited by M. Lassonde and M.A. Jeeves, Plenum Press, New York, 1994

This hypothesis has received some support from electrophysiological studies carried out in both hemispherectomized and acallosal patients. For instance, Arnott et al. (1982) reported that stimulation of the hand contralateral to the decorticated hemisphere could still evoke ipsilateral somatosensory potentials in hemispherectomized patients. Similarly, by recording evoked responses to wrist stimulation, Laget et al. (1977) demonstrated a greater responsiveness of the ipsilateral pathway in acallosal subjects as compared with normals where the ipsilateral component was negligible or totally absent. The latter results were not replicated, however, when using vibratory stimulation in an acallosal subject (Gott et al., 1985). Although a difference in stimulation site and procedure might explain the discrepancies, other technical problems such as the use of a cephalic reference and/or a limited amount of trials, make all of these studies difficult to interpret. Clearly, better methodological controls are required in order to confirm the existence of bilateral cortical connections of the lemniscal pathway in acallosal subjects.

TECHNICAL FACTORS IN SOMATOSENSORY EVOKED RECORDINGS

Sensory or mixed nerve (such as the median nerve) stimulation produces or evokes a series of potentials along the lemniscal sensory pathway. Several technical factors can influence the recording of somatosensory evoked potentials (SEPs) and therefore their latency and morphology (Chiappa et al., 1980; Mauguiere et al., 1981). Active and reference electrode locations, low and high frequency filters, time of analysis (or sweep time) and number of stimulations are probably the most important of these factors. When stimulating the median nerve, one records potentials originating from the roots of the brachial plexus (at Erb's point), cervical cord, brainstem and cerebral cortex. These potentials are designated according to their polarity and mean latency in normal adults. For instance, N20 is a negative potential with a mean latency of 20 msec in normal adults (Donchin et al., 1977). SEPs are further subdivided into short, middle- and long-latency potentials. Short-latency potentials are those generated within the first 40 msec after the stimulation while middle- and long-latency responses are recorded between 40 and 100 msec, and after 100 msec after stimulation, respectively.

SEPs are potentials of low amplitude which can only be extracted from baseline cortical activity by averaging a large number of responses to the electrical stimulation. Each stimulation produces a series of potentials which have the same latency on each occasion. By electronically averaging or adding these potentials, it becomes possible to extract these fixed-latency responses from random cortical activity. The number of nerve stimulations is thus extremely important and in most laboratories at least 500, and frequently 1000 to 2000, stimulations are recorded (Mauguiere et al., 1981). High and low frequency filters are used to eliminate different artifacts. However, the filter bandpass may also have an influence on the morphology of evoked potentials, especially when it is too narrow (Desmedt et al., 1974).

Finally, in most laboratories, SEP recording is done using an algebraic summation of the difference in potential between the active and reference electrodes. It is therefore important that the reference electrode be as neutral as possible. A frontal or cephalic reference electrode is frequently used but it has the disadvantage of picking up, at least partially, the potentials recorded by the active electrode located over the central or parietal head regions. This technique can thus create a distortion of the recorded potentials (Desmedt and Cherron, 1982). Following Desmedt's and Mauguiere's technique (Desmedt and Cherron, 1981, 1982; Mauguiere, 1983), we have used a non-cephalic reference electrode, located over the shoulder or on the ear lobe contralateral to the stimulated limb. Only when using this technique can two types of potentials be distinguished: the far- and near-field potentials (Desmedt and Brunko, 1980).

As their name indicates, far-field potentials are generated within the peripheral or central nervous system and can be recorded over all head regions. They originate from the brachial plexus (P9), the cervical cord (P11 and P13) and the brainstem (P14 and

N18). These potentials are probably generated by a change in volume conduction and are recorded bilaterally, which is an important factor to consider in a study such as the present one (Desmedt and Cherron, 1981, 1982; Mauguiere, 1983; Mauguiere et al., 1983). By contrast, near-field potentials are much more localized and they have to be picked by an electrode close to the potential's source. Thus, for short-latency SEPs, N20s and P27s are best recorded with an active electrode located over the parietal head region whereas P22s and N30s are better seen over the central head region. Middle-latency SEPs may also best be visualized when recorded over the central region although in some individuals, N60s are better seen over the frontal head regions (Bastuji, 1988).

It has been shown conclusively that N20s and P22s are the potentials generated by the arrival of the electrical volley in the primary sensory and motor cortex, respectively. As yet, the generators of subsequent potentials are not known precisely (Desmedt et al., 1987; Bastuji, 1988). These technical factors are of utmost importance especially in a study aimed at verifying the bilaterality of cortical evoked responses to a unilateral stimulation.

THE STUDY

Subjects

The sample consisted of four acallosal subjects. The first two subjects, a 21-year-old left-handed subject (M.G.) and his 29-year-old right-handed sister have been extensively described elsewhere (e.g. Lassonde et al., 1988). Briefly, both subjects were diagnosed as having complete callosal agenesis, first by pneumoencephalography and later by CT scan. In L.G., the diagnosis was further confirmed by MR imaging, which also indicated the presence of an intact anterior commissure in this subject. Both siblings have a global IQ of 78.

The third subject (S.G.) is the elder sister of the two subjects described above. Callosal agenesis was diagnosed in this subject after she and her parents volunteered to have a CT scan examination in order to assess the distribution of the congenital abnormality in the family. At the time of testing, S.G., a right-handed subject, was 30 years old. She has a global IQ of 84 and has no other known neurological pathology.

The fourth subject (S.B.), a right-handed 16-year-old male, started to have absence seizures at the age of six. A CT scan taken at the time revealed complete callosal agenesis. The seizures are presently well controlled with Carbamazepine. S.B. obtains a global IQ of 68.

The control group consisted of ten adult subjects without any neurological pathology.

Methodology

Two SEP studies were conducted. In the first one, far-field and short-latency (N20 and P27) SEPs were recorded in the four acallosal subjects and in 10 normal controls of similar age. In the second study, short- (N20, P22, P27 and N30) and middle- (P45 and N60) latency SEPs were recorded in three of these four acallosal subjects (M.G., S.G., S.B.) and in four normal adults.

Both studies involved a stimulation of the median nerve at the wrist level. Electrical stimulations of 0.1msec duration at the rate of 5.1/sec were first applied on the left median nerve and then on the right side. Stimulus intensity was adjusted to produce a minimal thumb movement. Evoked potentials were picked up by surface electrodes placed at different levels. In the first experiment, the four active electrodes were placed at Erb's point, over the cervical spine (at the level of C2) and over both parietal head regions (C'3 and C'4, 2cm behind C3 and C4 respectively, as defined by the 10-20 International System). The reference electrode was located on the shoulder contralateral to the stimulation site. The time of analysis was 40 msec. In the second study, the active electrodes were located at F3 and F4 and C3 and C4 (again according to the 10-20 International System) and the reference electrode was placed on the ear lobe contralateral to the stimulated side. The time of analysis was 100 msec.

Electrode impedance was kept inferior to 5 kW. The grounding electrode was placed between the stimulation and recording electrodes. The electrical activity was collected, filtered (5-3000 Hz), amplified and averaged with 0.2 msec resolution. In both experiments, two series of 1000 sweeps were recorded for each stimulation site and analysis was made by comparing the outlines of the waves.

Results

Recorded potentials were labeled by visual inspection according to the nomenclature suggested by Donchin et al. (1977). Absolute latencies were measured from the onset of the stimulus to the peak of the corresponding waves. The following latencies were analyzed: N20, P22, P27, N30, P45 and N60. Central conduction time (P14-N20 interpeak latency) was also measured as well as interside differences for all SEPs.

Statistical analysis for interside differences was carried out using a two-tail t-test. Since no statistical differences were observed between the two sides, results from left and right mėdian nerve stimulations were combined for further analysis. The ipsi- and contra-lateral waves thus obtained were superimposed with the early components of both waves (P9, P11 and P14) having a perfect fit (see Figure 1).

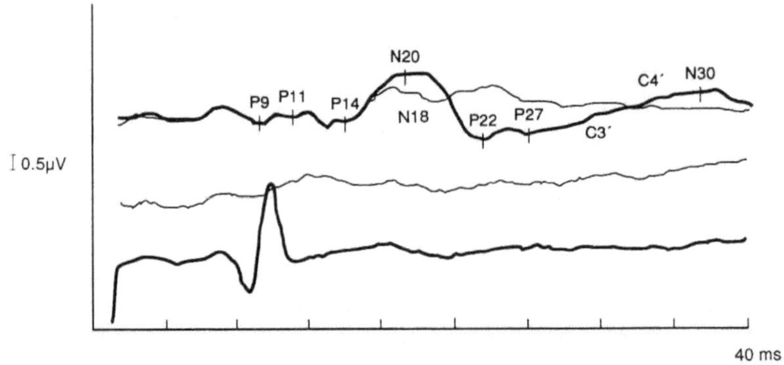

Figure 1. Far-field and short-latency SEPs obtained from a normal individual after stimulation of the right median nerve. The ipsilateral and contralateral (bold) cortical responses have been superimposed in order to illustrate the measurement of the N20 and P22 potentials' amplitude in relation to N18. The two lower traces are recorded at the cervical level (N13) and at Erb's point, respectively.

Being interested in a possible bilateral representation of SEPs, we chose to measure the amplitude of the primary cortical components (N20 and P27) by subtracting the responses recorded over the parietal head region ipsilateral to the stimulated limb from the ones recorded contralaterally. With this approach, the amplitude of the "genuine" N20 (N20 - N18) and P27 (P27 - N18) could be evaluated. Other amplitude components (P22, N30, P45 and N60) were measured from baseline to peak. Table 1 presents the mean amplitude obtained by the acallosal and control groups for each of the wave components.

On visual inspection, none of the acallosal subjects obtained clear bilateral SEPs after median nerve stimulation (see Figures 2 and 3). Paired t-tests, however, revealed that the amplitude of the N20 component was smaller in acallosal subjects as compared to normal controls. No such differences were seen for any of the other components.

Table 1. SEP amplitude in μVolts.

Groups	N20	P22	P27	N30	P45	N60
Acallosal subjects	(4)	(4)	(3)	(3)	(3)	(3)
Mean	0.73	1.41	2.28	1.47	1.05	2.05
Standard Deviation	0.54	0.26	1.04	0.63	0.55	0.77
Controls	(10)	(10)	(4)	(4)	(4)	(4)
Mean	1.45	0.89	1.64	0.85	1.00	2.19
Standard Deviation	0.73	0.93	0.50	0.52	0.40	0.77
p Value	<0.05	N.S.	N.S.	N.S.	N.S.	N.S.

Figures in brackets indicate the number of subjects recorded in each group. Statistical analysis was performed using a two-tailed test.

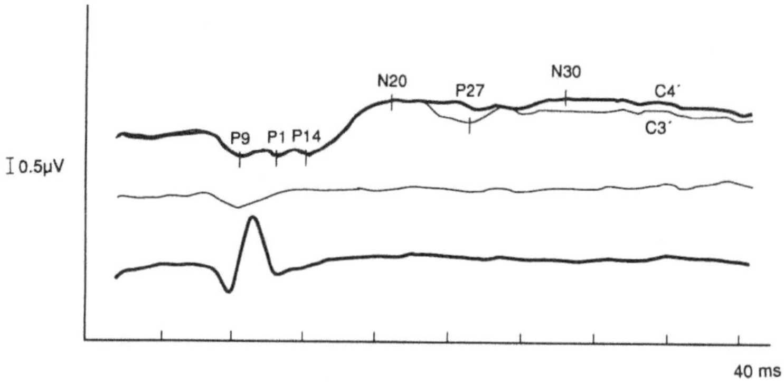

Figure 2. Far-field and short-latency SEPs obtained from an acallosal subject after stimulation of the left median nerve. The ipsilateral and contralateral (bold) cortical responses have been superimposed in order to illustrate the reduction in amplitude of the N20 potential. The two lower traces are recorded at the cervical level (N13) and at Erb's point, respectively.

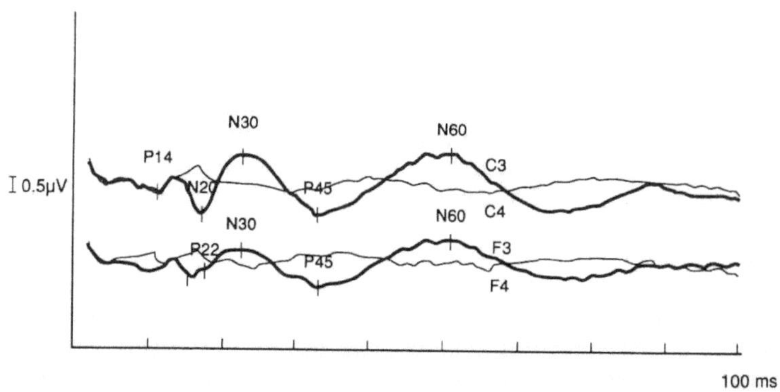

Figure 3. Short- and middle-latency SEPs obtained from an acallosal subject after stimulation of the right median nerve. The central (C3 and C4) and frontal (F3 and F4) ipsilateral and contralateral (bold) cortical responses have been superimposed. Both types of potentials are better visualized over the central head regions and are recorded only contralaterally to the stimulated side.

When latencies were compared, it appeared that the N20 and P27 components were shorter in our experimental subjects than in the normal controls. No significant differences were seen for the other SEP latencies (P22, N30, P45 and N60) or for central conduction time (P14-N20 interpeak latency).

Discussion

Two of our findings clearly dissociated the acallosals' SEPs from those of the control subjects. Both of these findings relate to the first cortical component observed in SEPs. First, N20 and P27 latencies were significantly shorter in acallosals subjects as compared to normal controls. The signification of this finding is unclear, however, particularly in view of the fact that the central conduction time (P14- N20) was normal. Second, we observed a reduction in amplitude of the contralateral N20 potentials in the acallosals. Since this amplitude was measured by subtracting the ipsilateral response from the contralateral one, this finding could support, at least partially, the hypothesis of a bilateral cortical representation of the lemniscal pathway in these subjects. The reduction in amplitude could indeed be explained by the fact that in acallosal subjects, median nerve stimulation evokes a bilateral N20 whereas in normal individuals, this stimulation only generates sensory cortical potentials contralaterally to the stimulated side.

It is important to point out, however, that the ipsilateral response was certainly of lower magnitude than the contralateral one. Considering the way we measured the N20 amplitude, if the ipsilateral and contralateral N20s had been of the same magnitude, the calculation would have yielded a zero value. Furthermore, the fact that no such reduction in amplitude was observed for other short- (P22, P27 and N30) or middle-latency (P45 and N60) SEPs also suggests that ipsilateral lemniscal connections are not as functional as the contralateral ones in acallosal subjects. If such were the case, all SEP components should have been clearly visible both ipsi- and contra-laterally to the stimulated side.

These results suggest that an increased proficiency of ipsilateral *lemniscal* connections is not the only compensatory mechanism used by acallosals to overcome disconnections deficits. This interpretation should not be restrictive, however: other sensory ipsilateral connections may play an important role in these subjects. In fact, it has often been proposed that uncrossed *spinothalamic* pathways may be enhanced in acallosal subjects (Dennis, 1976; Sauerwein et al., 1981). In this context, it has been shown that ventral spinothalamic connections are sufficient to sustain the slow reacquisition of a size discrimination lost after transection of the dorsolateral columns and hence, of the lemniscal impulses (Mountcastle, 1984). Because the ventral spinothalamic pathway contains axons of smaller diameter than the lemniscal fibers and because some of these axons are unmyelinated, sensory information transmitted by this pathway cannot be recorded by standard SEP techniques, a fact which could explain why no bilateral responses were not clearly observed in our subjects. It should also be remembered that nerve transmission is slower in the ventral spinothalamic tract than in the lemniscal pathway. This would be consistent with the finding that, behaviorally, response times are often longer in acallosals subjects and in young callosotomized patients as compared to IQ-matched controls (Jeeves and Rajalakshmi, 1964; Jeeves, 1979; Sauerwein et al., 1981; Lassonde et al.,1988, 1991).

CONCLUSION

Our results suggest that ipsilateral lemniscal connections can only partially explain the absence of a disconnection syndrome in acallosal subjects. It is quite likely, however, that the persistence of ipsilateral ventral spinothalamic connections plays a more important role as a compensatory mechanism for the absence of the largest interhemispheric commissure in these subjects.

REFERENCES

Bastuji, H., 1988, Etude de la composante Evoquée Somesthésique N60 chez l' Homme, Unpublished doctoral thesis, University of Lyon.

Chiappa, K.H., Choi, S.K., and Young, R.R., 1980, Short-latency somatosensory evoked potentials following median nerve stimulation in patients with neurological lesions, *in*: " Progress in Clinical Neurophysiology," J.E. Desmedt, ed., Karger, Basel.

Chiarello, C., 1980, A house divided? Cognitive functioning with callosal agenesis, *Brain Lang.* 11:128-153.

Dennis, M., 1976, Impaired sensory and motor differentiation with corpus callosum agenesis: A lack of callosal inhibition during ontogeny? *Neuropsychologia* 14:455-469.

Desmedt, J.E. and Brunko, E., 1980, Functional organization of far field and cortical components of sensory evoked potentials in normal adults, *in:* "Progress in Clinical Neurophysiology," J.E. Desmedt, ed., Karger, Basel.

Desmedt, J.E., Brunko, E., Debecker, J., and Carmeliet, J., 1974, The system bandpass required to avoid distortion of early components when averaging somatosensory evoked potential, *Electroencephalogr. Clin. Neurophysiol.* 37:407-410.

Desmedt, J.E. and Cherron, G., 1981, Non-cephalic reference recording of early somatosensory potentials to finger stimulation in adult or aging normal man: differentiation of widespread N18 and contralateral N20 from the prerolandic P22 and N30 components, *Electroencephalogr. Clin. Neurophysiol.* 52:257-275.

Desmedt, J.E. and Cherron, G., 1982, Recent progress in the understanding of subcortical somatosensory evoked potentials, *in:* "Clinical Applications of Evoked Potentails in Neurology," J. Courjon, F. Mauguière, and D. Revol, eds., Raven Press, New York.

Desmedt, J.E., Nguyen, T.H., and Bourguet, M., 1987, Bit-mapped color imaging of human evoked potentials with reference to the N20, P22, P27 and N30 somatosensory responses, *Electroencephalogr. Clin. Neurophysiol.* 68:1-19.

Ettlinger, G., Blakemore, C.B., Milner, A.D., and Wilson, J., 1972, Agenesis of the corpus callosum: a behavioral investigation, *Brain* 95:327-346.

Ettlinger, G., Blakemore, C.B., Milner, A.D., and Wilson, J., 1974, Agenesis of the corpus callosum: a further behavioral investigation, *Brain* 97:255-234.

Gazzaniga, M.S., 1970, "The Bisected Brain," Appleton Century Crofts, New York.

Gazzaniga, M.S. and Ledoux, J.E., 1978, "The Integrated Mind," Plenum Press, New York.

Geffen, G., Nilsson, J., Quinn, K., and Teng, E.L., 1985, The effect of lesions of the corpus callosum on finger localization, *Neuropsychologia* 23:497-514.

Gott, P.S., Hughes, E.C., and Binggeli, R.L., 1985, Interhemispheric transmissions: Assessment with vibratory somatosensory evoked potentials, *Int. J. Neurosci.* 27:121-130.

Jeeves, M.A., 1979, Some limits to interhemispheric integration in cases of callosal agenesis and partial commissurotomy, *in:* "Structure and Function of Cerebral Commissures," S.I. Russel, M.W. van Hof, and G. Berlucchi, eds., University Park Press, Baltimore.

Jeeves, M.A. and Rajalakshmi, R., 1964, Psychological studies of a case of congenital agenesis of the corpus callosum, *Neuropsychologia* 2:247-252.

Laget, P, D'Allest, A.M. Fihey, R., and Lortholary, O.L., 1977, L'intérêt des potentiels évoqués somesthésiques homolatéraux dans les agénésies du corps calleux, *Rev. E.E.G. Neurophysiol. Clin.* 7:498-502.

Lassonde, M., Lortie, V., Ptito, M., and Geoffroy, G., 1981, Hemispheric asymmetry in callosal agenesis as revealed by dichotic listening performance, *Neuropsychologia* 19:455-458.

Lassonde, M., Ptito, M., and Laurencelle, L., 1984, L Étude tachistoscopique de la spécialisation hémisphérique chez l' agénésique du corps calleux, *Rev. Can. Psychol.* 38:527-536.

Lassonde, M., Sauerwein, H., Chicoine, A.J., and Geoffroy G., 1991, Absence of disconnexion syndrome in callosal agenesis and early callosotomy: Brain reorganization or lack of structural specificity during ontogeny?, *Neuropsychologia* 29:481-495.

Lassonde, M., Sauerwein, H., Geoffroy, G., and Décarie, M., 1986, Effects of early and late transection of the corpus callosum in children, *Brain* 109:953-967.

Lepore, F., Ptito, M., and Guillemot, J.P., 1986, The role of the corpus callosum in midline fusion, *in:* "Two hemispheres- One Brain: Functions of the Corpus Callosum," F. Lepore, M. Ptito, and H.H. Jasper, eds., Alan R. Liss, New York.

Mauguière, F., 1983, Les potentiels évoqués somesthésiques cervicaux chez le sujet normal. Analyse des aspects obtenus selon le siège de l' électrode de référence, *Rev. E.E.G. Neurophysiol. Clin.* 13:259-272.

Mauguière, F., Brunon, A.-M., Echallier, J.-F., and Courjon, J., 1981, Intérêts des potentiels évoqués précoces dans l'exploration des voies de la sensibilité lemniscale, *Rev. Neurol.* 137(1):1-19.

Mauguière, F., Desmedt, J.E., and Courjon, J., 1983b, Neural generators of N18 and P14 farfield somatosensory evoked potentials studied in patients with lesion of thalamus or thalamo-cortical radiations, *Electroencephalogr. Clin. Neurophysiol.* 56:283-292.

Mountcastle, V.B., 1984, Central nervous system mechanisms in mechanoreceptive sensibility, *in:* "Handbook of Physiology: The Nervous System," Vol. III, Sec.1, J.M. Brookhart and V.B. Mountcastle, eds., American Physiological Society, Bethesda, Maryland.

Nebes, R.D. and Sperry, R.W., 1971, Hemispheric deconnection syndrome with cerebral birth injury in the dominant arm area, *Neuropsychologia* 9:247-259.

Sauerwein, H., Lassonde, M., Cardu, B., and Geoffroy, G., 1981, Interhemispheric integration of sensory and motor functions in agenesis of the corpus callosum, *Neuropsychologia* 19(3):449-454.

Saul, R.E. and Gott, P.S., 1973, Compensatory mechanisms in agenesis of the corpus callosum, *Neurology* 23:443.

Sperry, R.W., 1968, Hemisphere deconnection and unity in conscious awareness, *Amer. Physiol.* 23:723-733.

Sperry, R.W., Gazzaniga, M.S., and Bogen, J.E., 1969, Interhemispheric relationships: The neocortical commissures; syndromes of hemisphere disconnection, *in:* "Handbook of Clinical Neurology, " P.J. Vinken and G.W. Bruyn, eds., Elsevier, North Holland, Amsterdam.

Yakolev, P.I. and Lecours, A.-R., 1967, The myelogenetic cycles of regional maturation in the brain, *in:* "Regional Development of the Brain in Early Life," A. Minkowski, ed., Blackwell, Oxford.

MOTOR COORDINATION IN CALLOSAL AGENESIS

Priscilla H. Silver and Malcolm A. Jeeves

Psychological Laboratory
University of St. Andrews
St.Andrews, Fife
Scotland, KY16 9JU

INTRODUCTION

In the early 1960s we reported deficits shown by acallosals when performing bimanual coordination tasks under a speed stress (Jeeves, 1965; Jeeves and Rajalakshmi, 1964). These early findings were confirmed by Chiarello in her 1980 review paper on cognitive functioning in callosal agenesis (Chiarello, 1980). It was evident from her review of studies of twenty-nine acallosals that problems with bimanual coordination were one of the most commonly reported features of these patients' performance. A nineteen-year follow-up showed that this deficit does not disappear in the course of development (Jeeves, 1979).

Building on the work of Zaidel and Sperry (1974) and Preilowski (Preilowski, 1972, 1975) who studied bimanual coordination in total commissurotomy patients, we carried out detailed investigations of bimanual coordination in two adult acallosal patients (Jeeves et al., 1988a). The task used was that devised by Bruno Preilowski (1972, 1975). His results made it clear that normally the anterior forebrain commissures are important for the maintenance of rapid and precise bilateral motor coordination. In this paper we summarize the major findings from our study of bimanual coordination in acallosals using Preilowski's methods.

Our second line of investigation noted the results of developmental studies of motor coordination in normal children. Denkla (1974) had pointed out that there is a coincidence between the leveling off of motor performance at eight years old and a similar leveling off in aspects of neural maturation and brain growth. She inferred from this that performance on simple coordination tasks may reflect intrinsic brain maturation. Developing this view, the importance of interhemispheric connections for the efficient performance of tasks involved in bimanual coordination has been linked with neural maturation data.

With these leads in mind we suggest that the evident deficits in bimanual coordination in the absence or dysgenesis of the corpus callosum are consistent with evidence pointing to improvement during normal development in motor coordination paralleling the normal neural maturation of the major forebrain commissure (Jeeves et al., 1988b). Studies illustrative of this link are also summarized below.

Speculating about compensatory mechanisms which may occur when the callosum is absent from birth, several investigators have suggested the likelihood of the enhanced

Callosal Agenesis, Edited by M. Lassonde and
M.A. Jeeves, Plenum Press, New York, 1994

development of the ipsilateral or uncrossed sensory and motor pathways (e.g. Dennis, 1976). This alerts us to another factor that may affect bimanual coordination in acallosals. Specifically, it has been suggested that in the absence of the callosum, the normal inhibition of the ipsilateral, uncrossed, pathways does not occur and as a result the output from such uncrossed pathways may compete with that from crossed pathways in a way that does not occur in normals. In the latter, it is argued that the uncrossed outputs are inhibited by the corpus callosum (Jeeves, 1991). To investigate this we studied the reaching and grasping behavior of the same two acallosals who took part in our bimanual coordination study (Jeeves and Silver, 1988). The relevant part of the results from our study will be summarized in considering the factors which may be at work in producing the motor coordination deficits of acallosals.

BIMANUAL COORDINATION IN CALLOSAL AGENESIS

Method

The subjects were two acallosal patients, KC and BF. Their performance was compared to that of 10 adult normal age-, gender-, and handedness-matched controls. Extensive reports of other investigations on KC and BF have appeared elsewhere (Jeeves et al., 1988a). Detailed information about them appears in that publication (Jacobson and Jeeves, this volume).

Figure 1. Apparatus.

An adaptation of Preilowski's apparatus was used. Figure 1 illustrates how the subject turned two crank-handles in order to control a pen which drew a line between guidelines on the screen, in front of the subject. Each handle was turned anti clockwise. One complete turn of the right handle moved the pen 100 mm from right to left and one turn of the left handle moved the pen 100 mm upwards. The pen was always placed at the lower right hand end of the parallel lines. To draw a line at 90° only the left hand handle should be turned, to draw a line at 180° only the right hand handle. The remaining three angles (Figure 2) were to be drawn between guidelines at 135° or 157.5° or 112.5°. To draw a line at 135° both handles should be turned at the same speed. To draw a line at 157.5° the right hand handle would have to be turned faster than the left and vice versa to draw 112.5°. The task given to the subjects was to trace the four lines on one chart all at the same angle before being presented with another chart with lines to be drawn at another angle.

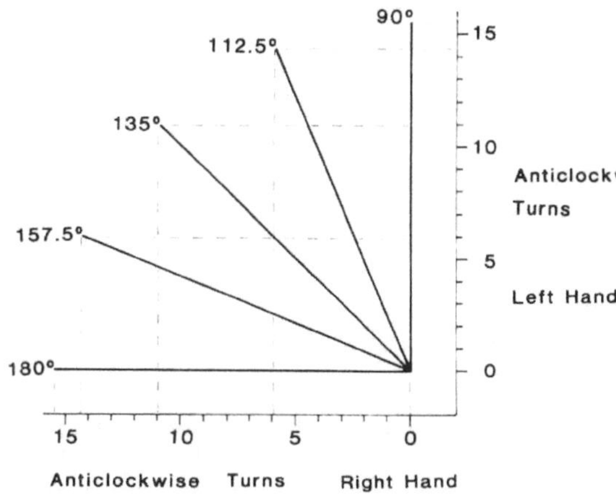

Figure 2. The task.

Each session started with the two charts of one-hand lines followed by twelve charts of two-hand lines. All were given in balanced sequence followed again by the two one-hand lines. Both acallosal patients and control subjects completed nine sessions in all. During these nine sessions visual feedback, in the sense that the paper with parallel lines and the tracing pen were in full vision, was given to all subjects. Thus the subjects were very highly practiced. Preilowski had introduced further sessions in which halfway through the drawing of each line, he required the subject to close his eyes so that the task had to be completed without visual feedback. In our experiment we introduced a screen for the condition without visual feedback to make it easier for those subjects who found it difficult to keep their eyes closed.

In the St. Andrews version of the Preilowski apparatus provision was made for adjusting the separation of the handles so that the task would be equally amenable to young children. One other variation in our procedure compared with Preilowski's was that he did not introduce the condition without visual feedback until completion of nine sessions with feedback. We introduced a half-unseen condition at the end of each session from the third to the ninth.

The principal measures taken of the subjects' performances were the speed of completion of drawing a line together with the number of times that the pen crossed the parallel guidelines. We also measured the deviation from the correct orientation of the line when visual feedback was withdrawn.

Results

Figure 3 taken from Preilowski's paper (1972) gives a typical example of the performance of a patient with anterior callosal section when visual feedback was withdrawn as compared with that of a normal control. Looking at the data from our studies, Figures 4 and 5 summarize the results from the 'seen' condition for lines

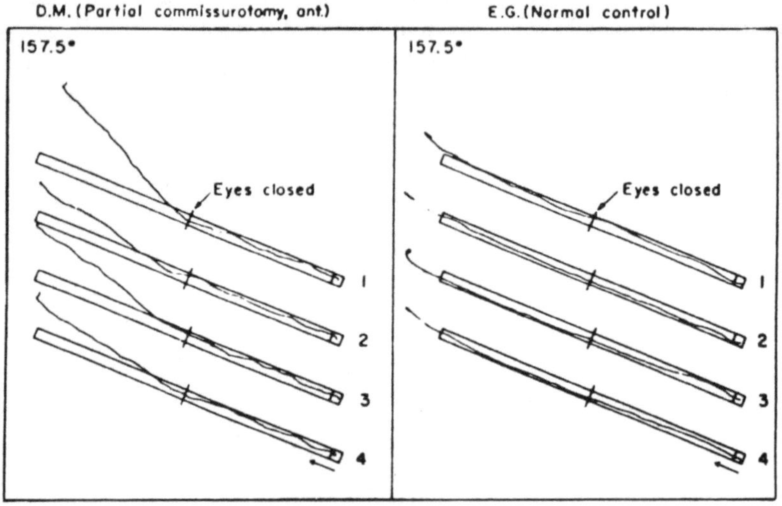

Figure 3. Performance sample of Preilowski's partial commissurotomy patient (D.M.) and a normal control (E.G.), when feedback was eliminated after crossing the halfway mark. The numbers indicate consecutive trials.

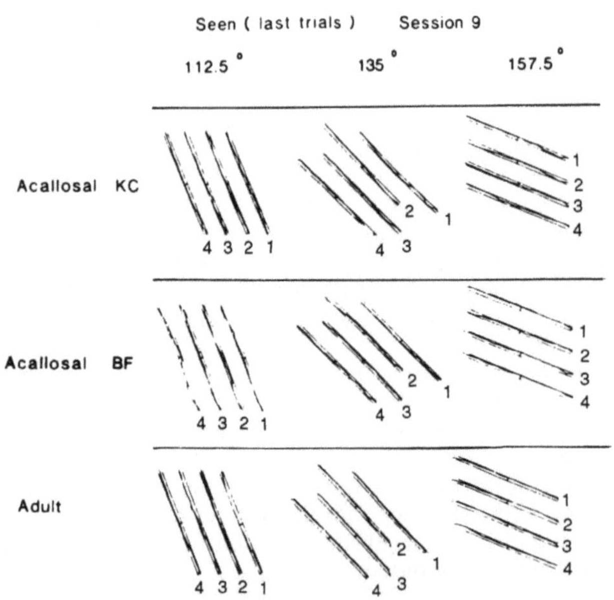

Figure 4. Examples of performance by two acallosals and a typical adult control, with visual feedback throughout.

requiring both hands to participate. Figure 4 indicates the accuracy of the line drawing at the end of the ninth session for two acallosal patients and for one normal control. In Figure 5 for each angle of two-hand line, the time taken by each of these adult acallosals averaged over the last two sessions is compared with the mean and range for the ten controls; one acallosal subject's times were near the upper 95% limit of the normal range, the other's slightly longer times were outside the range. Figures 6 and 7 refer to the 'half-unseen' condition.

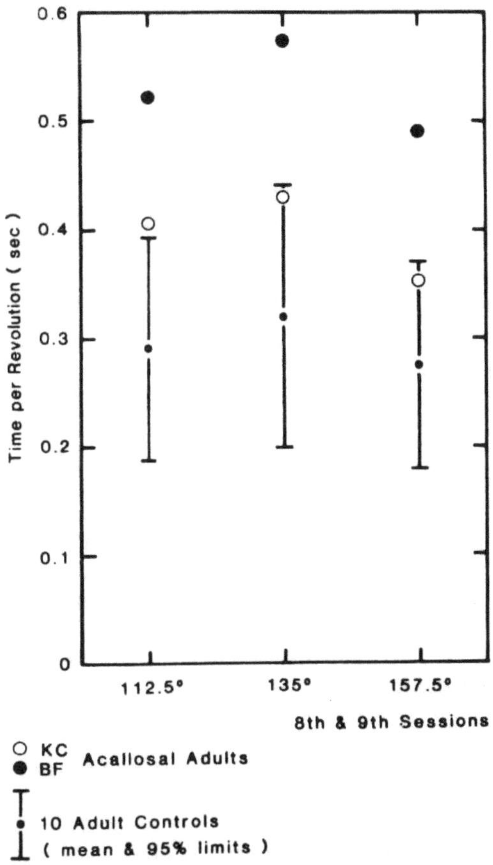

Figure 5. Time taken per revolution of the time-determining hand by acallosals and by control adults during the last two sessions. Angles 112.5o, 135o, 157.5o require hand-input L>R, L=R, and L<R respectively.

In Figure 6 we see the actual two-hand lines drawn by the patients and the typical control drawn at the end of the ninth session. The first half of each line was in the 'seen' condition but the second half was without visual feedback - the 'unseen' condition. Figure 7 shows the magnitude of deviation from the correct direction under the unseen condition, averaged over the last two sessions. The two acallosal subjects are compared with the mean and range of the ten normal controls.

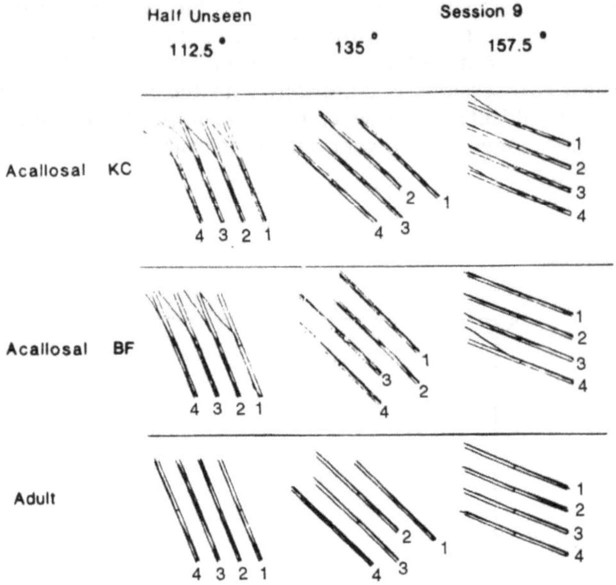

Figure 6. Examples of performance by two acallosals and a typical adult control, visual feedback withdrawn at the mid-point.

Briefly, the major findings were (1) the two acallosal subjects are slower in peak performance achieved after prolonged practice compared with the controls; (2) when visual feedback is withdrawn acallosal subjects are unable to maintain the correct direction of the line they were drawing prior to the withdrawal of visual feedback (Figure 7 summarizes this result); (3) it is noteworthy that the acallosals, both of whom were right-handed, deviated more when drawing lines at 112.5°. It is under this condition that the output of the right hand must be inhibited or restrained more than the left to continue drawing the line at 112.5°.

THE FOREBRAIN COMMISSURES AND DEVELOPMENTAL ASPECTS OF BIMANUAL MOTOR COORDINATION

The coincidence between the leveling off of motor performance of eight-year-old children and the structural maturation of callosal fibers noted by Denckla (1974) raises the question of whether younger children, say six-year-olds, manifest any performance characteristics on our bimanual task reminiscent of aspects of the performance of acallosals. This study on neurologically normal subjects explores that issue.

Method

The method and procedure used in this study were the same as described above except that the separation of the crank handles was adjusted to provide a comfortable position for young children.

The subjects were three groups, each of ten subjects whose mean ages were 20 years, 10 years and 6 years. The younger children, with suitable encouragement and refreshment breaks, were able to perform the task at a high level.

Results

Having in mind the specific issue raised in the introduction concerning the comparable effects on bimanual coordination of dysgenesis of the corpus callosum and the putative immaturity of the callosal pathway in young children, we shall focus on the comparative behavior of the 6-year-old, 10-year-old and adult subjects *when visual*

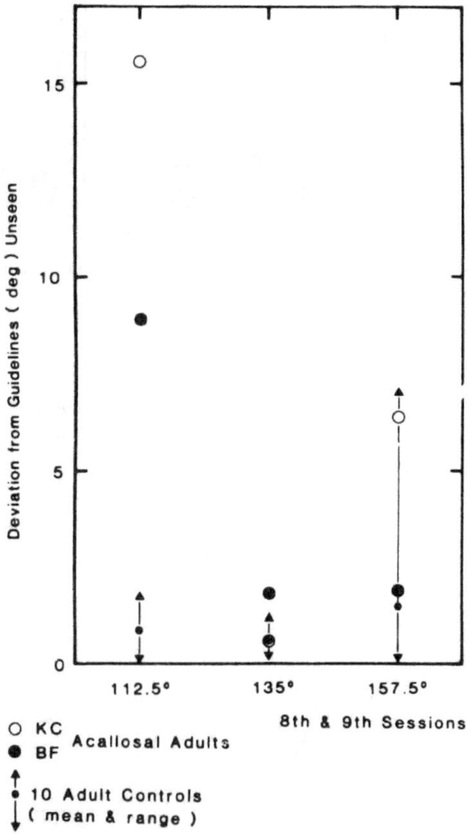

Figure 7. Magnitude of deviation from guidelines after withdrawal of visual feedback, during the last two sessions (8 and 9), by acallosals and by control adults. Angles 112.5o, 135o, and 157.5o require hand-input L>R, L=R, and L<R respectively.

feedback was withdrawn. Figure 8 summarizes the results of this aspect of the study. It is evident that when the visual feedback is withdrawn the 6-year-old children find it extremely difficult to maintain the direction of line drawing prior to withdrawal of visual feedback. Where an asymmetrical performance is called for, as in the case of the lines being drawn at 112.5o and 157.5o, their performance drops away markedly. Of special note is how the performances of the youngest children are so similar to the performances of the acallosal adults. Figure 9 illustrates this finding.

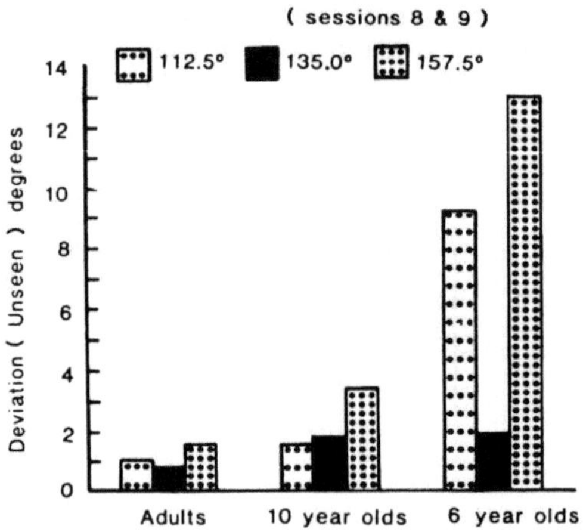

Figure 8. Deviation (unseen). Final performance by three age groups at the angles 112.5o, 135o, and 157.5o.

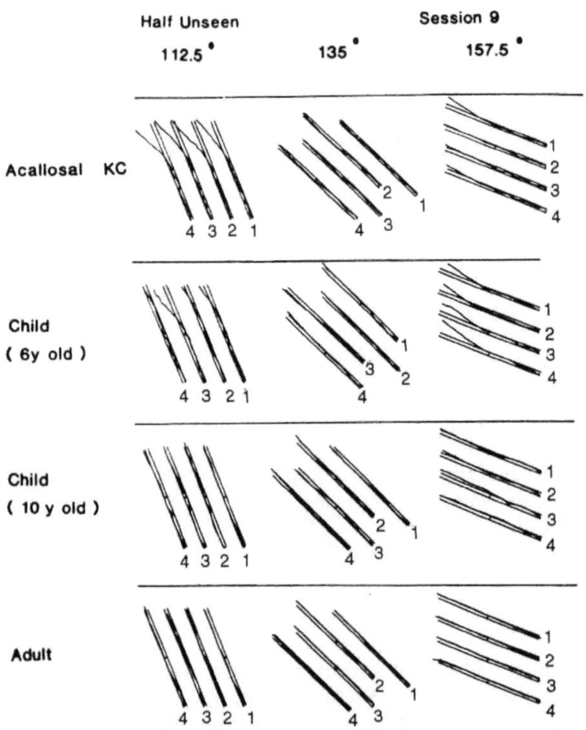

Figure 9. Examples of performance by an acallosal (K.C.) and by a typical normal participant from each of three age groups, visual feedback withdrawn at the mid-point.

It is interesting to note that when this same task was given to a 10-year-old girl in whom the middle two-thirds of the callosum had been sectioned to enable the surgeon to remove a benign tumor in the third ventricle, her performance was very similar to that of normal controls of her age, rather than of an acallosal subject of the same age. This is not surprising since the work of Preilowski made it clear that it is the *anterior* parts of the callosum which are essential for the maintenance of efficient bimanual motor coordination with and without visual feedback and these were intact in this surgical patient.

REACHING AND GRASPING BEHAVIOR IN ACALLOSAL PATIENTS

Jeannerod (1986a, 1986b) demonstrated that prehension movements directed at visual objects involved two clearly different, although superimposed, segmental components. One corresponds to the transportation of the hand as a whole to the vicinity of the target object (the reaching component), the other corresponds to the detailed adjustment of the hand posture and the position of the fingers (the grasping component). The first involves proximal joints and muscle groups whereas the second involves distal joints and muscles. Jeannerod reported that in patients with unilateral lesions to the motor cortex, the grip formation of the hand contralateral to the lesion is severely affected. Specifically, the fingers and, in particular, the index finger tended to remain outstretched until contact with the object had been made. Jeannerod interpreted the results as demonstrating the role of the cerebral cortex in the control of finger grip formation. He concluded that "integration at cortical level of visual and somatosensory cues from the involved hand is a necessary condition for grip formation to be adapted to the grasp".

Method

The task given to our two patients KC and BF is illustrated in Figure 10 (Jeeves and Silver, 1988). They were required to lift the hand from a start pad which initiated a timer and to grasp an object briefly displayed for less than 200 msecs in front of them, to the left or to the right of a fixation point. They were instructed to do this as quickly as possible. A digit from 0 to 9 was displayed at the fixation point. This had to be correctly reported to confirm that correct fixation had been maintained. The whole performance was recorded on video tape on two cameras, one from above and one from

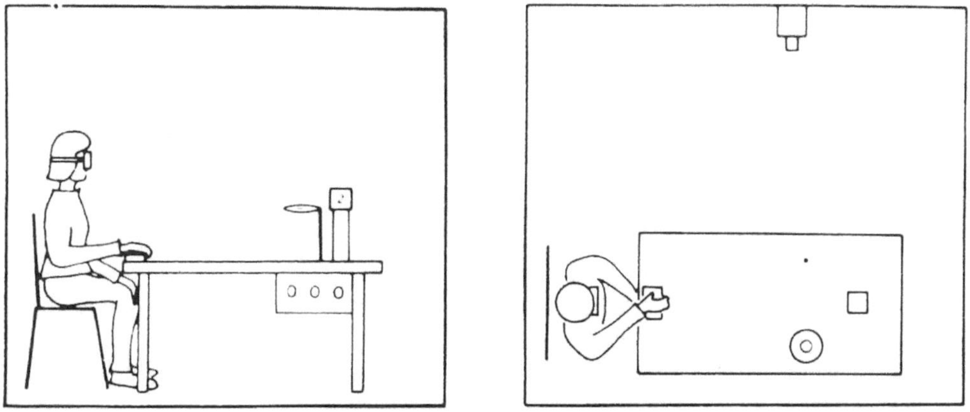

Figure 10. Subject ready to grasp saucer. Side view, as video record. Top view, showing camera.

the side (see Figure). In addition, a digital timer which started when the hand was lifted and stopped when the object was grasped gave a record of the timing of the movement. The subjects wore a mask with a green filter and the background lighting in the room was red. The subjects therefore sat in darkness except for the brief periods when white light lit up the room for the subjects to see the objects to be grasped. Full details of this study are given elsewhere (Jeeves and Silver 1988).

Results

The main points from the results are illustrated in Figure 11 (Jeeves and Silver, 1988). Here it can be seen that where finger and thumb separation has been measured from the video records and plotted against time, the hand opens and closes in normals in such a way that at the time of arriving at the object the fingers have closed on to the shape and size of the object. This is shown in the inverted triangle on the upper part of Figure 11. Examination of the lower part of Figure 11 indicates that in the case of the acallosals the hand opens but remains in a gaping posture until it hits the object and only then do the fingers begin to close in an appropriate configuration on the object. It was possible to calculate the rate of change of closure of the hand. This showed that the acallosal patient was not only outside the range of the controls, but differed significantly from the mean of the controls; the rate of closing in the acallosals is significantly slower than that in the normals.

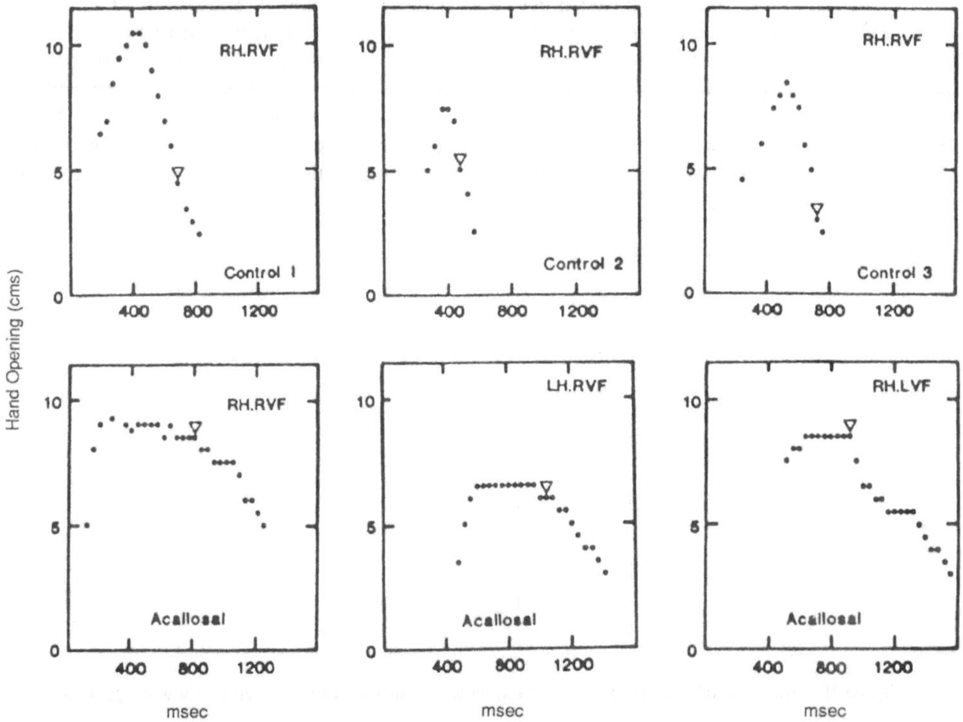

Figure 11. Finger-thumb separation (measured from record) plotted against time. Upper three graphs: control subjects. Lower three graphs: acallosal (K.C.).

CONCLUSION

It is evident from the results of the studies summarized above that the behavior of the acallosals differs from that of the normal matched controls. It is also evident that in the bimanual motor coordination task the 6-year-old children performed in a way that approximated more to that of the acallosal adults than to normal 10-year -old or 20-year-old subjects.

On Preilowski's bimanual coordination task one of the two acallosal patients tested failed to attain, even after extended practice, the level of performance achieved by normal controls. This applies to both the quality and speed of performance and is consistent with earlier reported studies of acallosals. The second adult acallosal, whilst performing at the lower end of the range of the normals, is significantly slower than the normals when drawing lines at 112.5^{o}. When the same task was given to an acallosal boy (Jeeves et al., 1988b) for whom the clinical report had identified early signs of praxic difficulties accounting for clumsiness and poor coordination, we found that these difficulties had diminished markedly. We tested him when 11 years old and whilst performing like normal matched controls on measures of speed, he was nevertheless inferior to normals on measures of accuracy. Taken together, these results suggest that whatever the alternative pathways available to acallosals, they set a ceiling to the performances achievable. It is noteworthy that the partial commissurotomy patient tested at age 10 years and in whom the anterior part of the callosum was intact, after extended practice, reached levels of performance comparable with her age-matched controls.

Earlier studies had shown that bimanual coordination can be maintained by using visual and proprioceptive feedback, as long as performance is slow. However, for performance to remain efficient at high speeds it is assumed that a faster, more direct and finer bimanual control is necessary. Such would be provided by the interhemispheric exchange of corollary or feed-forward discharges arising directly from the motor outflow. Such motor corollary discharges have been proposed by Sperry (1950), von Holtz and Mittelstaedt (1950) and Teuber (1964). Further evidence for the importance of such direct interaction comes from the effects on bimanual performance of removing visual feedback. Unlike normals, the acallosals do not, with practice, become independent of visual information.

As an aside, but relevant to other discussions in this book, is the finding that the effect of taking away visual feedback was not symmetrical. Thus, when the right hand was required to slow its rate of responding relative to the left in order to draw a line at 112.5^{o} this produced significantly greater unintended deviation than in the condition of drawing a line at 157.5^{o}. This finding is consistent with the view that the brains of these acallosals are probably normally lateralized for movement control (Milner and Jeeves, 1979). According to Preilowski it was the lack of inhibitory control which gave rise to the poor performance of his partial commissurotomy patients, especially when drawing lines at 112.5^{o} rather than at 157.5^{o}. This would further support an important role for the callosum in inhibition (Chiarello, 1980; Dennis, 1976; Jeeves, 1986).

To sum up, the evidence from the study of bimanual coordination in these acallosal patients, (1) supports Preilowski's view that for fast, coordinated, highly skilled bimanual performance an intact functioning corpus callosum is necessary; (2) is consistent with there being an asymmetry of motor control in acallosals; and (3) lends further support to the view that the inhibitory role of the callosum is crucial in the normal efficient control of bimanual motor output in the absence of visual feedback.

The conclusions from the study of the adult acallosals that normally an intact functioning corpus callosum plays a crucial part in bimanual coordination receives confirmation from the findings of the study in which we compared the performance of 6-year-olds, 10-year-olds and adults. Denkla (1974) has argued that in normal children the level of performance on such tasks reflects intrinsic brain maturation. Of particular interest are the findings in the studies above where we saw that the deviation errors made by the 6-year-olds, when external visual feedback was withdrawn, were significantly greater than those in the 10-year-olds or the adults. The largest deviations were observed when lines which required asymmetrical movements of the two hands were being drawn. Under this condition we may presume that the commissural interchange of information about the activity of the two hemispheres was most important. It is under this condition that it is necessary to slow down one or the other hand with respect to the other in order to draw the lines at 112.5^{o} and 157.5^{o}. Although the major focus of interest in this study was to explore the likelihood of a reduced interhemispheric interchange of feed-forward information through the corpus callosum in very young children we may also note the possibility of a similar reduction in its facilitatory function, as argued by Lassonde (1986) and Payne (1986).

The final study summarized above was of the reaching and grasping behavior of the acallosal patients. One way of interpreting their performance is to draw attention again to the likelihood of enhanced development of the uncrossed ipsilateral pathways in acallosals. This, in the absence of the supposed inhibitory role of the callosum, may have resulted in greater competition between the uncrossed ipsilateral pathways and the crossed contralateral pathways so as to produce the observed deficit in grasping behavior. Such a view would receive further support from the evoked potential studies of acallosals of Laget et al. (1976, 1977). They reported that while in normals ipsilateral evoked potentials from wrist stimulation were negligible or totally absent, in acallosals they observed a greater responsiveness from the ipsilateral pathways. Lassonde et al. (1986) have argued that in this respect they are more like young children who, they believe, rely more heavily on uncrossed connections before the age of 10. Such a suggestion would certainly be consistent with the observation that, for example, infants cannot form a prehension grip before the age of 36-52 weeks although this probably depends, more importantly, on the level of maturational development of the cortex at these ages. Whether it applies also until maturation of the callosum is complete, around age 10, is a matter for empirical investigation. Certainly, the report by von Hofsten (1979) that younger infants show only crude prehensile movements of the fingers after an object has been touched sounds very similar to the findings above with the prehensile movements of the acallosal patients observed here.

If the visuo-motor apparatus for reaching is, as has been suggested, located at brain stem level (Glickstein and May, 1982; Jeannerod, 1986a, 1986b) it will not be affected by absence of the corpus callosum. By contrast, if the control of prehension movement required in grasping is located at the hemispheric level, then it would be susceptible, amongst other factors, to any departure from normal function in the corpus callosum. The absence of the hypothesized cortical facilitatory role of the corpus callosum could be an additional factor accounting for the slower performance of the acallosal in grasping when compared with the normal controls. The failure to show the normal prehensile movements, however, may be due to the absence of the normal inhibitory action of the callosum in suppressing the ipsilateral motor output, which may consequently compete with the contralateral output controlling the fingers.

REFERENCES

Chiarello, C., 1980, A house divided? Cognitive functioning with callosal agenesis, *Brain Lang.* 11:128-158.

Denckla, M., 1974, Development of motor coordination in normal children, *Dev. Med. Child Neurol.* 16:729-741.

Dennis, M., 1976, Impaired sensory and motor differentiation with corpus callosum agenesis: a lack of callosal inhibition during ontogeny?, *Neuropsychologia* 14:455-469.

Glickstein, M. and May, J.G. III, 1982, Visual control of movement: the circuits which link visual to motor areas of the brain with special reference to the visual input of the pons and cerebellum, *in:* "Contributions to Sensory Physiology," W.D. Ness ed., Academic Press, London.

Hofsten, C. von, 1979, Development of visually directed reaching: the approach phase, *J. Hum. Mov. Stud.* 5:160-178.

Holst, E. von and Mittelstaedt, H., 1950, Das Reafferenzprinzip (Wechselwirkungen zwischen Zentralnervensystem und Peripherie), *Naturwissenschaften* 37:464-476.

Jeannerod, M., 1986a, The formation of finger grip during prehension. A cortically mediated visuomotor pattern, *Behav. Brain Res.* 19:99-116.
Jeannerod, M., 1986b, The role of vision in the patterning of prehension movements, *in:* "Two Hemispheres - One Brain: Functions of the Corpus Callosum," F. Leporé, M. Ptito, and H.H. Jasper, eds., Alan R. Liss, New York.
Jeeves, M.A., 1965, Psychological studies of three cases of congenital agenesis of the corpus callosum, *in:* "Functions of the Corpus Callosum," E.G. Ettlinger ed., Churchill, London.
Jeeves, M.A., 1979, Some limits to interhemispheric integration in cases of callosal agenesis and partial commissurotomy, *in:* "Structure and Function of the Cerebral Commissures," I.S. Russell, M.W. van Hof, and G. Berlucchi, eds., Macmillan, London.
Jeeves, M.A., 1986, Callosal agenesis: neuronal and developmental adaptations, *in:* "Two Hemispheres - One Brain: Functions of the Corpus Callosum," F. Lepore, M. Ptito, and H.H. Jasper, eds., Alan R. Liss, New York.
Jeeves, M.A., 1991, Hemispheric interactions, *in:* "Encyclopedia of Human Biology," Vol. 4, Academic Press, New York.
Jeeves, M.A. and Rajalakshmi, R., 1964, Psychological studies of a case of congenital agenesis of the corpus callosum, *Neuropsychologia* 2:247-252.
Jeeves, M.A. and Silver, P.H., 1988, The formation of finger grip during prehension in an acallosal patient, *Neuropsychologia* 26:153-159.
Jeeves, M.A., Silver, P.H., and Jacobson, I., 1988a, Bimanual co-ordination in callosal agenesis and partial commissurotomy, *Neuropsychologia* 6:833-850.
Jeeves, M.A., Silver, P.H. and Milne, A.B., 1988b, Role of the corpus callosum in the development of a bimanual skill, *Dev. Neuropsychol.* 4:305-323.
Laget, P., D'Allest, A.M., Fihey, R., and Lortholary, O., 1977, L'intérêt des potentiels évoqués somesthésiques homolatéraux dans les agénésies du corps calleux, *Rev.E.E.G. Neurophysiol. Clin.* 7:498-502.
Laget, P., Raimbault, J., d'Allest, A.M., Flores-Guevara, R., Mariani, J. and Thierot-Prevost, G., 1976, La maturation des potentiels évoqués somesthésiques (PES) chez l'homme, *Electroencephalogr. Clin. Neurophysiol.* 40:499-515.
Lassonde, M., 1986, The facilitatory influence of the corpus callosum on intrahemispheric processing, *in:* "Two Hemispheres - One Brain: Functions of the Corpus Callosum," F. Lepore, M. Ptito, and H.H. Jasper, eds., Alan R. Liss, New York.
Lassonde, M., Sauerwein, H., Geoffroy, G., and Décarie, M., 1986, Effects of early and late transection of the corpus callosum in children, *Brain* 109:953-967.
Milner, A.D., 1982, Simple reaction times to lateralized visual stimuli in a case of callosal agenesis, *Neuropsychologia* 20:411-419.
Milner, A.D. and Jeeves, M.A., 1979, A review of behavioural studies of agenesis of the corpus callosum, *in:* "Structure and Function of the Cerebral Commissures," I.S. Russell, M.W. van Hof, and G. Berlucchi, eds., Macmillan, London.
Payne, B.R., 1986, Role of callosal cells in the functional organization of cat striate cortex, *in:* "Two Hemispheres - One Brain: Functions of the Corpus Callosum," F. Lepore, M. Ptito, and H.H. Jasper, eds., Alan R. Liss, New York.
Preilowski, B.F.B., 1972, Possible contribution of the anterior forebrain commissures to bilateral motor coordination, *Neuropsychologia* 10:267-277.
Preilowski, B.F.B., 1975, Bilateral motor interaction: perceptual-motor performance of partial and complete 'split-brain' patients, *in:* "Cerebral Localization," K.J. Zülch, O. Creutzfeld, and G.C. Galbraith, eds., Springer, Berlin.
Sperry, R.W., 1950, Neural basis of the spontaneous optokinetic response produced by visual inversion, *J. Comp. Physiol. Psychol.* 43:482-489.
Teuber, H.L., 1964, The riddle of frontal lobe function in man, *in:* "The Frontal Granular Cortex and Behavior," J.M. Warren and R. Akert, eds., McGraw-Hill, New York.
Zaidel, D. and Sperry, R.W., 1974, Memory impairment after commissurotomy in man, *Brain* 97:263-272.

COGNITIVE FUNCTIONING IN CALLOSAL AGENESIS

Hannelore C. Sauerwein,[1] Pierre Nolin,[2] and Maryse Lassonde[1]

[1]Groupe de Recherche en Neuropsychologie Expérimentale
Département de Psychologie
Université de Montréal
C.P. 6128, Succ. A
Montréal, Qué., H3C 3J7, Canada
[2] Département de Psychologie
Université du Québec à Trois-Rivières
Trois-Rivières, Qué., G9A 5H7, Canada

INTRODUCTION

Agenesis of the corpus callosum results from partial or total failure of callosal axons to cross the midline during the first trimester of fetal development (Rakic and Yakovlev, 1968; Rauch and Jinkins, this volume). Although the resulting structural modifications of the brain are shared by all cases, the clinical expression of this malformation seems to depend on the extent of concomitant pathology of the nervous system (e.g. O'Brien, this volume; Wisniewski and Jeret, this volume). In most cases some neurological dysfunction can be observed. At the cognitive level there seems to exist, as would be expected, a strong relationship between the severity of co-existing CNS defects and level of functioning. Especially affected are those with multiple malformations, such as in Aicardi syndrome (Aicardi and Chevrié, this volume) and other syndromes described in this volume (see Andermann and Andermann; Geoffroy, Wisniewski and Jeret). Recent MRI studies by de Léon et al. (1987) and Atlas et al. (1986) have revealed extensive abnormalities of limbic structures, including absence of the hippocampal formation in some cases, and it has been suggested that these subcortical malformations may be at the basis of the mental deficiencies found in so many syndromes that include callosal agenesis (de Léon et al., 1987).

On the other hand, the introduction of such non invasive neuroradiological procedures as computerized tomography and magnetic resonance imaging has also led to the detection of an increasing number of cases without gross neurological dysfunction and with normal, or next to normal, mental status. In her 1980 review of the literature, Chiarello (1980) compiled 29 cases of callosal agenesis with an IQ ranging from 71 to 115 by excluding the more severely retarded cases with lipoma or multiple malformations. Using the same criteria, our own updated search has yielded 17 new cases (Table 1). Additional cases with normal IQ are reported in this volume (e.g. Geffen et al.; Temple and Ilsley). These individuals are of particular interest because

Callosal Agenesis, Edited by M. Lassonde and
M.A. Jeeves, Plenum Press, New York, 1994

Table 1. Background data of cases of callosal agenesis reported between 1980 and 1991.

Subject	Age	Gender	Laterality	IQ			Ear Advantage for verbal stimuli	Source
				V	NV	FS		
GL	60 yrs	M	right	107	122	110	NR	Pirozzolo et al. (1979)
case 1	11 yrs	M	right	NR	NR	88	NR	Lynn et al. (1980)
case 2	35 yrs	M	right	NR	NR	(96)[1]	NR	idem
MG	9 yrs	M	left	63	85	71	left	Sauerwein et al. (1981)
LG	17 yrs	F	right	61	97	75	right	idem
1 case	8 yrs	F	left	NR	NR	104	NR	Meerwaldt (1983)
GB	22 yrs	M	right	93	99	95	right	Martin (1985)
BF	20 yrs	M	right	85	94	86	NR	Milner et al. (1985)
SB	10 yrs	F	right	105	104	104	NR	Temple et al (1987)
case 2	12 yrs	F	right	119	52	(85)[2]	NR	Bigler et al. (1988)
MJ	11 yrs	M	right	106	75	90	NR	Jeeves et al. (1988)
NL	17 yrs	F	left	73	80	74	right	Lassonde et al. (1988, 1990)
CF	6 yrs	F	NR	91	70	(79)[2]	NR	Sanders (1989)
MM	8 yrs	M	right	95	98	96	NR	Clark & Zaidel (1989)
KW	14 yrs	M	left	103	106	105	NR	Temple et al. (1989, 1990)
ML	5 yrs	F	right	NR	NR	85	NR	Lassonde et al. (1991)
SG	29 yrs	F	right	88	82	84	NR	idem

F=Female; M=Male; V=Verbal; P=Performance; FS=Full Scale; NR=Not reported; [1]Estimated from description; [2]Extrapolated from WISC-R IQ Tables.

they provide the opportunity to investigate the extent and limits of compensation in a nervous system that has evolved without the largest interhemispheric pathway.

REVIEW OF PREVIOUS RESEARCH

Earlier cognitive studies have mainly concentrated on the attempt to answer the question of how language is organized in the acallosal brain (i.e. Dennis, 1977, 1981; Sperry, 1968). This question builds on the theory expounded by Selnes (1974), Trevarthen (1974), Moscovitch (1977) and others (see Jeeves, 1986) that the corpus callosum is crucial for the development of hemispheric specialization because it is via this commissure that the hemisphere inherently predisposed to excel in a certain function inhibits simultaneous activation of functionally equivalent neural units in its counterpart. If this were true, it was argued, functions that are usually lateralized in the normal brain would be bilaterally and redundantly represented in the acallosal brain. However, an analysis of the reviewed cases (Table 2) clearly shows that this hypothesis is not supported by empirical evidence.

For one, most acallosals show clear hand-preference and this manual dominance is also evident on objective tests of skilled motor activity (i.e. Jeeves, 1986; Jeeves et al., 1988).

Second, dichotic listening research has yielded small but reliable ear asymmetries to verbal stimuli, although the side of the preferred ear is less consistent than in normals. According to Chirarello's data, summarized in Table 2, there appears to exist a greater incidence of left-ear advantage (LEA) in the acallosal population. It should, however, be pointed out that one subject in each of the three categories (right, left or no EA) in Chiarello's review reversed the direction of ear advantage when retested. The author further mentions that LEA for speech is found in about 20% of normal right-handers when tested with dichotic listening paradigms, a proportion that grossly overestimates the incidence of right-hemisphere language representation derived from sodium amytal studies (Milner et al., 1964; Gruber and Segalowitz, 1977).

Table 2. Background data, hand preference and ear advantage of acallosal cases from review studies.

Description	Chiarello (1980) N=29	Recent Cases N=17
Total ACC	17	16
Partial ACC	10	-
Uncertain (probably total)	2	1
Concomitant CNS pathology	25	12
Neurological Symptoms	28	11
Gender: Male	20	9
Female	9	8
Age at first testing	3-60 yrs (median: 21 yrs)	5-60 yrs (median: 12 yrs)
Hand used for writing:		
Right	22 (2 ambidextr)	12
Left	7 (2 ambidextr)	4
NR	-	1
Ear advantage for verbal stimuli:		
Right	3	4
Left	5	2
None	2	-
NR	19	11

NR=Not reported

In a more recent study, Lassonde et al. (1990) used Bryden and Sprott's (1981) Laterality Index lambda to compare relative accuracy for the two ears. The stimuli were CV pairs, some of which contained a target syllable that had to be recognized by the subject. Four of the six acallosal subjects showed REA (see Table 2: recent cases) and two LEA. Interestingly, with respect to the magnitude of asymmetry, the acallosal group ranked between the normal and low-IQ group by displaying smaller asymmetries than the high-IQ but larger asymmetries than the low-IQ group. The authors interpreted this finding as suggesting that the corpus callosum, by allowing constant exchange of information between the hemispheres, may in fact *reduce* initially existing asymmetries. Consequently, the absence of interhemispheric "cross-talk" in the acallosal brain would favor the full expression of such asymmetries.

A third reason for rejecting the hypothesis of bilateral language representation in the acallosal brain is based on tachistoscopic studies that have demonstrated consistent visual field asymmetries for verbal and non-verbal (visuo-spatial) stimuli (Jeeves and Milner, 1987; Lassonde et al., 1984; Martin, 1981; Meerwaldt, 1983; Milner, 1982; Sauerwein and Lassonde, 1983). However, once more, there appears to be a greater variability among acallosal subjects as to which hemisphere is dominant for verbal processing.

Finally, of eight cases that have been tested with the Wada method, evidence of normal lateralization of speech in the left hemisphere has been found in all but two of the acallosal subjects (Gazzaniga, 1970; McGlone, 1985; Gott and Saul, 1978, Jeeves, personal communication). Both were left-handed. Besides, as Milner (1983) has pointed out, one documented case of unilateral representation of speech suffices to discredit the hypothesis of bilateral organization of linguistic functions in acallosals.

In order to verify the claim made by Sperry (1970a, 1970b) that verbal skills in acallosals develop at the expense of perceptual skills, Dennis (1977) tested two acallosal subjects with average intelligence. She chose syntactic comprehension and visually-guided route finding as verbal and visuo-spatial tasks, respectively, since these functions had previously been found to be affected in acallosals. The results showed that there is no consistent imbalance in favor of language functions in acallosals. In fact, the two subjects showed opposite patterns of functioning, one being impaired exclusively on the verbal and the other on the perceptual task. The author concluded that, while the nature of the deficit (verbal or visuo-spatial) is unpredictable, lateralized functions are compromised in callosal agenesis. She further predicted that no case should be proficient in both skills.

However, an analysis of the IQs obtained from the reviewed cases (Table 3) reveals that there is no basis to this claim. When a discrepancy of 10 points is used to distinguish verbal and perceptual advantages, the proportion of cases that show no imbalance in these skills is in fact slightly higher than that showing an advantage for either function. It is also noteworthy that the distribution of IQ scores is similar in Chiarello's and our reviews in that the largest proportion of the acallosals fall in the low average range of cognitive functioning. One has, however, to bear in mind that these subjects were diagnosed because of some neurological complaint and that asymptomatic cases with greater intellectual capacities may rarely, if ever, have neuropsychological exams. Nevertheless, the fact that a small proportion of acallosals in the two surveys fall in the high average range of mental abilities suggests that agenesis of the corpus callosum *per se* is not the cause of the cognitive deficits so often found in this population. The data also infirm the hypothesis put forth by Cook (1984, 1986) that absence or reduction of interhemispheric "cross-talk" invariably results in lower intelligence.

Table 3. IQ analysis of review cases.

Range	Number of cases	V>P[1]	P>V[1]	V=P[2]	NR[3]
Chiarello (1980)					
70-79	6	3	-	3	-
80-89	13	2	3	5	3
90-109	7	2	3	2	-
110-119	3	1	1	-	1
Mean: 89	29	8	7	10	4
Recent Cases					
70-79	4	1	2	1	-
80-89	5	1	-	2	2
90-109	7	1	-	4	2
110-119	1	-	1	-	-
Mean: 88	17	3	3	7	4

V=Verbal; P=Performance; [1]10 points or more; [2]Less than 10 points difference; [3]Not reported

PERSONAL INVESTIGATIONS

In view of the discordance as to which function, if any, may be selectively affected in callosal agenesis and because of the high incidence of this malformation in the province of Quebec and the frequent letters and calls we have received and continue to receive from distressed parents whose children have been recently diagnosed, our group decided in the early eighties to submit acallosal and two matched control groups to a complete neuropsychological battery and to conduct discriminant analysis (SPSS, 1982) in order to identify the cognitive pattern specific to each group (Lassonde et al., 1982). We were also mindful of the fact that the identification of the strengths and weaknesses in the acallosal group could be useful for their rehabilitation as well as for their educational and vocational orientation.

Subjects

Selection criteria for all groups were absence of hydrocephalus, epilepsy or focal brain damage. From an original group of 10, eight patients with total agenesis of the corpus callosum as diagnosed by CT scan and MRI (in one case) were retained for this study. Descriptive data of the sample are presented in Table 4. For detailed case reports we refer to Nolin (1984) and Lassonde et al. (1982, 1988). Cases 1 to 6 were from the Saguenay region in Quebec and suffered from the Andermann syndrome (see Andermann and Andermann, this volume; Carpenter, this volume). Subjects 7 and 8 were siblings also originating from the province of Quebec but unrelated to the Saguenay cohort. Except for LG, who is gainfully employed, all subjects were enrolled in special schools according to the level of their mental development.

More recently, we also had the opportunity to test an asymptomatic case of callosal agenesis, SG (Lassonde et al., 1991), who, along with her parents, volunteered to have a CAT scan because two of her three siblings (cases 7 and 8) have the malformation. Unlike her acallosal siblings, SG had a normal development, has finished high-school and is presently working at a home for the elderly. The data of this subject are not included in the analysis but are reported in the tables as a basis for comparison.

Two control groups were employed. One group, arbitrarily called low-IQ group, was composed of nine (9) students who attended the same or similar special classes as the acallosal subjects and who functioned approximately at the same level as the acallosals. Another group, called normal-IQ subjects, consisted of 10 students who were recruited from regular schools and junior colleges in the Saguenay district. Both control groups were matched with the acallosal subjects on the basis of age, gender and handedness.

Table 4. Background data of acallosal sample.

Case	Age	Gender	Handedness	V IQ	P IQ	FS IQ	Test
1 MJ	10 yrs	F	right	59	71	63	WISC-R
2 NL	13 yrs	F	left	73	80	74	WISC-R
3 AM	13 yrs	F	left	67	51	55	WISC-R
4 HB	13 yrs	F	right	54	77	64	WISC-R
5 MT	15 yrs	M	left	68	71	68	WISC-R
6 EC	24 yrs	F	left	55	63	55	O-W
7 MG	13 yrs	M	left	63	85	71	O-W
8 LG	21 yrs	F	right	81	81	78	O-W
SG	29 yrs	F	right	88	82	84	WAIS-R

F=Female; M=Male; V=Verbal; P=Performance; FS=Full Scale; O-W= Ottawa-Wechsler

Methods

The tests employed in the evaluation were the WISC-R (Wechsler, 1974), the WAIS-R (Wechsler, 1981) or the Ottawa-Wechsler, a French version of the WAIS

(Gagnon, 1953), the Michigan Neuropsychological Test Battery (MNTB: Smith, 1981; see also Lassonde et al., 1982) and a selection of standardized tests from the Halstead-Reitan Battery (Reitan and Davison, 1974). The functions evaluated were: intellectual capacities, attention and concentration, memory, language, perceptual abilities, motor skills, perceptuo-motor integration and somatosensory functions.

Results

An IQ analysis of the acallosal sample is presented in Table 5. Inspection of Tables 4 and 5 reveals that most of the acallosal subjects, especially those suffering from Andermann syndrome, have lower intellectual capacities than the cases reviewed above (Tables 3). Nevertheless, the analysis of their IQ resembles that of these normal-IQ cases: the number of patients who show no discrepancy between verbal and performance IQ is equal to the number showing an imbalance in favor of one of the skills. The same holds true for the asymptomatic case (Table 4) whose performance on the two scales is not suggestive of an imbalance between verbal and visuo-spatial functions. The performance of this subject is further compatible to that of the reviewed cases in the sense that it places her in the low average (dull) range of cognitive abilities with the largest proportion of the cases in both reviews. It should, however, be mentioned that the IQ of her father, who, like the mother, has an intact corpus callosum, also falls in the low average range of mental abilities (Lassonde and Bryden, 1990). It can also be noted from Table 4 that there is a larger proportion of left-handers in the sample, as one might expect in a clinical population.

Table 5. IQ analysis of acallosal sample.

Range	Nr/Cases	V>P[1]	P>V[1]	V=P[2]
55-69	5	1	2	2
70-79	3	-	1	2
Mean: 66	8	1	3	4

V=Verbal; P=Performance; [1]10 points or more; [2]Less than 10 points difference

The group means for all neuropsychological tests and group comparisons (Scheffé, see Winer, 1962) for the corresponding tests are presented in Table 6. The results of the asymptomatic subject, SG, also shown in Table 6, are reported in terms of raw or scaled scores and compared to standardized norms for her age group (Spreen and Strauss, 1991). Several subtests of the Wechsler Intelligence Scale appear under different headings according to the function they purport to measure. It should be noted that the subtest "Vocabulary" from the Wechsler Intelligence Scale was excluded from the analysis because no standardized French Canadian translation of this subtest existed at the time of testing. Furthermore, the data of the naming task from the MNBT was not submitted to statistical analysis because all subjects obtained perfect scores. The results of all remaining tests were comprised in the analysis.

Intellectual Capacities. It is evident from Table 6, that the acallosals obtained subnormal scores on all cognitive measures (Raven, Wechsler Intelligence Scale), except for the subtest Digit Span, tapping short-term auditory memory and attention. Not surprisingly, the low-IQ group was equally impaired with respect to the normal sample. Comparison between low-IQ and acallosal subjects showed that the former did more poorly than the acallosal group on a verbal tests measuring concept formation and

Table 6. Results of neuropsychological evaluation of acallosal, low-IQ and normal-IQ subjects.

		Group Means			Score	F Values			Centile
Function/Test		**Acallosal**	**Low-IQ**	**Normal**	**SG**	**AvsIQ**	**AvsN**	**IQvsN**	**SG**
Intellectual Capacities									
Information		3.13	3.00	10.70	6	0.0	51.4**	56.7**	9
Digit Span		6.13	3.56	8.40	11	4.3	3.6	17.2**	63
Arithmetic		2.75	4.67	10.00	11	4.0	60.0**	34.6**	63
Comprehension		5.38	3.11	10.90	6	3.1	19.2**	40.7**	9
Similarities		5.88	2.11	10.40	6	9.9*	14.9**	53.4**	9
Picture Completion		7.88	4.59	12.80	8	6.6	18.8**	51.8**	25
Picture Arrangement		5.13	3.00	10.70	8	2.0	14.6**	29.7**	25
Block Design		6.38	5.67	11.20	9	0.3	13.1**	18.4**	37
Object Assembly		6.63	6.67	10.60	6	0.0	7.3*	7.6*	9
Digit Symbol		2.13	3.56	10.90	6	1.9	36.0**	17.2**	9
Raven SPM (Percentile)		34.13	55.67	62.00	57	6.5	6.6*	28.6**	57
Attention/Concentration									
Arithmetic (Wechsler)		2.75	4.67	10.00	11	4.0	60.0**	34.6**	63
Digit Span (Wechsler)		6.13	3.56	8.40	11	4.3	3.6	17.2**	63
Digit Symbol (Wechsler)		2.13	3.56	10.90	6	1.9	36.0**	17.2**	9
Symbol Digit	written	16.00	22.44	44.90	42	2.1	44.7**	28.8**	7
	oral	21.25	28.44	52.40	55	1.4	28.0**	17.7**	43
Trail Making	A	84.38	26.11	17.60	52	15.5**	30.7**	2.4	<10
	B	133.71	75.88	38.60	65	8.1*	27.9**	5.9	50
Verbal Orders		6.63	5.56	10.00	10	1.9	20.2**	37.4**	normal
Memory									
Arithmetic (Wechsler)		2.75	4.67	10.00	11	4.0	60.0**	34.6**	63
Digit Span (Wechsler)		6.13	3.56	8.40	11	4.3	3.6	17.2**	63
Unrelated Sentences		17.75	16.89	27.30	32	0.1	13.5**	17.5**	normal
Tabula Rasa		1.88	1.61	2.65	3	0.5	4.5	8.6*	normal
TPT: Memory		3.13	2.78	5.80	10	0.1	8.9*	12.1	96
Visual Retention (Benton)		2.75	3.89	7.00	8	1.3	18.8**	10.7*	67
Language									
Peabody Vocabulary (IQ)		103.00	73.22	137.80	126	30.7**	44.0**	161.4**	100
Perceptual Abilities									
Facial Recognition (Benton)		32.88	5.67	54.00	52	7.8*	4.9	27.6**	97
Picture Completion (Wechsler)		7.88	4.59	12.80	8	6.6	18.8**	51.8**	25
Visual Organization (Hooper)		22.38	19.22	27.60	28	4.4	12.8**	35.0**	74
Motor Skills									
Finger Tapping	PH	36.13	47.44	53.30	44	6.5	15.6**	1.9	48
	NPH	33.25	45.44	49.20	38	6.6	11.9**	0.7	29
Dynamometer	PH	17.50	23.10	31.10	30	3.2	20.0**	7.4*	42
	NPH	15.63	22.45	28.90	30	4.2	16.7**	4.2	60
Purdue Pegboard	PH	8.77	22.54	28.83	29	24.8**	47.5**	3.4	normal
(60 sec)	NPH	6.90	19.02	26.86	23	22.7**	52.0**	5.8	normal
	BM	4.37	16.58	23.14	21	26.6**	52.9**	4.2	normal
Perceptuo-Motor Integration									
Block Design (Wechsler)		6.38	5.67	11.20	9	0.3	13.1**	18.4**	37
Digit Symbol (Wechsler)		2.13	3.56	10.90	6	1.9	36.0**	17.2**	9
Object Assembly (Wechsler)		6.63	6.67	10.60	6	0.0	7.3*	7.6*	9
Symbol Digit (written)		16.00	22.44	44.90	42	2.1	44.7**	28.8**	7
TPT - Formboard	PH	700.88	288.10	260.22	480	7.9*	11.0*	0.2	3-9
(seconds)	NPH	713.50	260.22	139.10	830	4.8	12.5**	2.2	<1
	BM	547.38	162.11	75.40	235	8.1*	14.1**	0.8	<1
Trail Making	A	84.38	26.11	17.60	52	15.5**	30.7**	2.4	<10
	B	133.71	75.88	38.60	65	8.1*	27.9**	5.9	50
Visual Retention (copy)		5.25	6.56	9.50	10	1.3	13.9**	7.1*	60
Somatosensory Functions									
Face-Hand Test (SDSS)		16.63	20.00	20.00	20	29.5**	31.0**	0.0	78

*p<0.05; **p<0.01; A=Acallosal subjects; IQ=Low-IQ subjects; N= Normal-IQ subjects; PH=Preferred hand; NDH=Non-preferred hand; BM=Bimanual level of abstract thinking (Similarities). The asymptomatic acallosal subject also scored highest on short-term memory and showed excellent ability of verbal (Arithmetic) and non-verbal reasoning (Raven) whereas formal acquisitions (Information), concept formation (Similarities) and integration of social norms (Comprehension), as well as perceptuo-motor integration (Object Assembly and Coding) were at the lower limit of the normal range (9th percentile).

Attention and Concentration. The ability to focus and sustain attention on a given task was somewhat impaired in both acallosal and low-IQ subjects in comparison to normal controls. However, auditory attention (Arithmetic, Digit Span) was less, and visual attention (Trail A) more, affected in the acallosal sample. In fact, the Trail Making Test (Form A) yielded the only significant difference between acallosal and low-IQ subjects. This test was also impaired in the asymptomatic subject, while auditory attention was normal on all measures. However, in the absence of any deficit on the more demanding Form B of the same test, the impaired performance of the acallosals can scarcely be regarded as a serious attentional deficit in this population. It is more plausible that the longer performance time of the acallosal subjects was attributable to an initial slowness in ocular scanning which quickly improved with practice.

Memory. Auditory short-term memory was normal in the acallosal and low-IQ groups as long as the material to be memorized consisted of single units such as digits (Digit Span) and names (Tabula Rasa) but it became deficient when imbued with semantic context (Arithmetic, Unrelated Sentences). No significant differences were found between acallosal and low-IQ subjects. The latter were impaired on all measures except tactual memory. The asymptomatic subject performed in the high average range on all mnemonic tasks. She was given two additional tasks: Rey's Auditory Verbal Learning Test (see Lezak, 1983), on which she scored at the 60th percentile, and the Rey-Osterrieth Figure (Lezak, 1983; Spreen and Strauss, 1991) measuring visuo-spatial memory, which placed her in the average range (30th percentile).

Language. Expressive language testing was limited to observation of spontaneous speech and performance on the color naming task from the MNTB. On the latter, all groups performed at ceiling level. Spontaneous speech was less elaborate in the acallosal and low-IQ groups than in the normal group on all levels (grammatical, lexical and semantic). With regard to receptive language (Peabody Picture Vocabulary Test), the performance of the acallosals was significantly superior to that of their low-IQ pairs. SG also obtained excellent results on the PPVT (100th percentile). Her abilities in the expressive domain, judged on the basis of her discourse and her verbal responses on the WAIS-R, were in the low normal range which is in line with her overall level of functioning.

Perceptual Abilities. Although the callosal agenesis subjects displayed the same difficulties in visual analysis (Picture Completion) and synthesis (Visual Organization) of complex perceptual information as their low-IQ school mates, their performance was normal and superior to that of the low-IQ controls on the particularly difficult test of Facial Recognition. The asymptomatic subject also excelled on this task (97th percentile). Her performance on the other perceptual tasks was well within normal limits.

Motor Skills. Motor skills have a place in the assessment of cognitive functioning inasmuch as they are crucial for adequate performance on all tasks that require manipulation under time constraints (i.e. most tasks of the Wechsler performance subscale). Furthermore, this is the domain which has yielded the most consistent deficits in acallosals throughout the literature (for review, see Jeeves, 1979, 1986 and Jeeves and Milner, 1987). As can be seen in Table 6, the acallosal group was significantly impaired in all motor functions, whereas the low-IQ group performed normally on these tasks. When low-IQ and acallosals were compared, the latter showed specific deficits in both unimanual dexterity and bimanual coordination (Purdue Pegboard), whereas tapping speed and muscle force were less affected. Not surprisingly, the largest deficits were found in the Andermann patients. The asymptomatic subject, on the other hand, exhibited no particular difficulties in this domain. However, reports of her developmental history indicate that SG had been a clumsy child not unlike her other two acallosal siblings.

Perceptuo-motor Integration. This is another area in which deficits have been consistently reported for callosal agenesis patients (Chiarello, 1980; Jeeves, 1979; Temple and Ilsley, this volume) and our sample was no exception. A clear dissociation between the acallosal and the low-IQ groups could be observed in that the former were impaired on all perceptuo-motor tasks whereas the latter obtained subnormal scores only on those tasks that had a strong cognitive component. The test most indicative of perceptuo-motor deficits in the acallosal group was the Tactual Performance Test (TPT). It can be seen that initial learning with the dominant hand as well as intermanual transfer and bimanual execution of the task were impaired in the acallosal group, and this applied not only to the patients suffering from Andermann syndrome but also to the two siblings unrelated to the Andermann cohort and their asymptomatic sister SG. In fact, the latter required the longest time of all acallosal subjects to perform the task and showed no transfer of tactuo-motor learning between the hands. It may be of interest, that her siblings had not shown deficits during an earlier testing session with an easier formboard (see Sauerwein et al., 1981) and that their retest results with the adult board vary greatly to the effect that their performance has become slower with time. Although these results may question the test-retest reliability of this task, they also show that practice does not come into play and that this deficit is not easily compensated by alternative neural and/or behavioral mechanisms. It must, however, be emphasized that the deficits seen in the asymptomatic case concern exclusively those tasks that involve timed performance and interhemispheric transfer. Her performance on other complex tasks of perceptuo-motor integration such as the copy of the designs of the Benton Visual Retention Test (100th percentile) and the Rey-Osterrieth Figure (60th percentile) was strictly normal.

Somatosensory Functions. It is not surprising that somatosensory deficits, in terms of extinctions of the distal parts (hands) during simultaneous face-hand stimulation, were observed in the acallosal patients, given the high proportion of subjects with peripheral neuropathy in our sample. Inspection of the raw data revealed that the deficit was limited to these patients. The other three cases (7, 8, and SG) obtained perfect or near-perfect scores. The deficit can therefore not be considered a characteristic feature of callosal agenesis.

Stepwise Discriminant Analysis. In order to determine which group of tests would best characterize each of the three groups, the group means of all tests were submitted to "Stepwise" Discriminant Analysis (SPSS, 1982). This analysis permits comparison of all variables and consecutive exclusion of any variable that does not further enhance the contrast between the groups. The five tests that yielded maximum contrast between the three groups were the Face-Hand Test, the Peabody Picture Vocabulary Test (IQ), Raven's Standard Progressive Matrices (percentile) as well as the two subtests "Information" and "Picture Arrangement" from the Wechsler Intelligence Scale. The results of the analysis are depicted in Figure 1.

It is evident from Figure 1 that all three groups were correctly classified by these five variables. The Face-Hand Test was expected to be one of the factors that would distinguish the acallosal patients of the Saguenay cohort from both normal and low-IQ controls in view of the specific neurosensory deficits that are associated with the Andermann syndrome. It may also be recalled that the acallosals were superior to the low-IQ controls on the Peabody Vocabulary Test, although they did not attain the same level of competence on this task as the normal-IQ controls, or, for that matter, the asymptomatic acallosal subject. Both the acallosal patients and the low-IQ group appeared to be equally deficient relative to the normal-IQ group with respect to the acquisition of general knowledge (see Information, Table 6). The weaker performance of the acallosals on the non-verbal reasoning task (Raven) as well as the inferior scores of the low-IQ subjects on measures of perceptual organization of sequential information (Picture Arrangement) further differentiated them from the normal-IQ group.

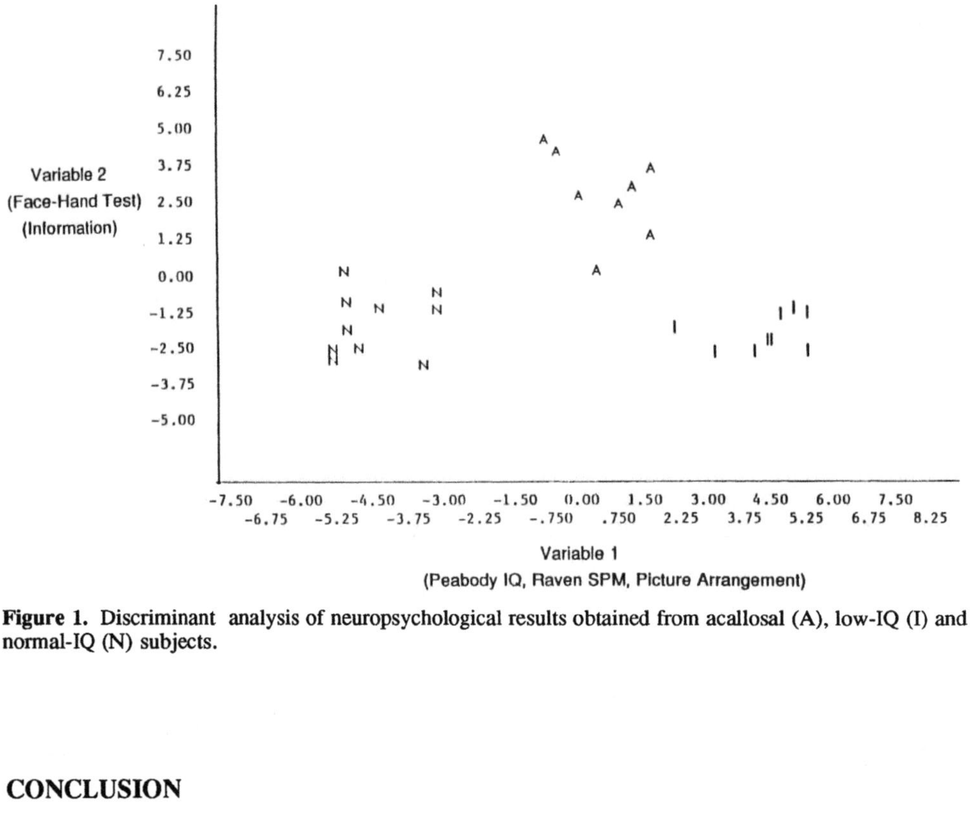

Figure 1. Discriminant analysis of neuropsychological results obtained from acallosal (A), low-IQ (I) and normal-IQ (N) subjects.

CONCLUSION

Taken together, the results of these and the reviewed cases suggest that individuals born without corpus callosum can have normal cognitive abilities but that they tend to function at the lower end of the normal range. In fact, the general level of functioning may be still lower when other important structural abnormalities of the CNS co-exist. However, it should be kept in mind that most samples reported in the literature, including ours, are biased towards a neurological population. Cases with normal or superior mental abilities may elude scientific inquiry because the absence of neurological and cognitive dysfunction does not warrant a neuroradiological or neuropsychological investigation. Some of these inconspicuous cases are occasionally detected at autopsy. Others may be discovered accidentally when a neuroradiological exam for an acute condition is indicated. In this context, a brilliant professional with a transient cerebrovascular accident was recently diagnosed and examined by S.L. Schmidt (personal communication). It should also be pointed out that pre- and postnatal sonography has helped to identify an increasing number of isolated, neurologically normal cases of callosal agenesis (Cioni et al., this volume; Vergani and Strobelt, this volume).

The question as to whether lateralized functions are invariably compromised in acallosals cannot be conclusively answered without testing a larger number of asymptomatic cases. In a clinical population an imbalance is often related to a mental deficit, however mild, which requires special education with limited exposure to certain types of information and environmental stimuli. The fact that our asymptomatic subject, like the cases reviewed above (Table 3), and half of the cases in our clinical sample (Table 5), did *not* show a significant discrepancy between verbal and perceptual functioning would suggest that the imbalance seen in some of the patients described in the literature may be attributable to concomitant cerebral pathology or other variables not directly related to the absence of the corpus callosum.

When compared to adequately matched controls with supposedly intact corpora callosa, acallosals do not show particular cognitive deficits that could be attributed specifically to the absence of the corpus callosum. On the contrary, they may be superior to these subjects on several cognitive measures. They do, however, differ from their low-IQ peers with respect to sensorimotor learning and transfer to the effect that even the asymptomatic subject manifested deficits in this domain. Furthermore, it appears that these deficits, also reported in the literature (see Chiarello, 1980; Jeeves, 1986), cannot easily be surmounted by practice or other forms of compensation. It should, however, be emphasized that such deficits may become evident mainly under particular testing conditions. The same holds true for cognitive subskills that have been the subject of more recent studies (Jeeves and Temple, 1987; Sanders, 1989; see also Geffen et al., this volume; Temple and Ilsley, this volume) Here again, subtle deficits may be revealed that are not evident from the individual's overall level of functioning. This is exemplified by Temple et al. (1990) who found specific impairments in reading subskills in two acallosal children. Yet, the subjects had normal reading levels and were not known as being developmental dyslexics.

With regard to motor functions, impairments in speed, dexterity and/or bimanual coordination have been consistently found in acallosals (Jeeves, 1979, 1986; Milner and Jeeves, 1987; O'Brien, this volume). Some of these deficits were also present in our sample. However, there seems to exist a large individual variability, and asymptomatic cases may show few, if any, visible deficits at adulthood, unless they are submitted to sophisticated experimental tasks involving complex bimanual operations such as described by Silver and Jeeves (this volume).

Thus, it seems reasonable to conclude, as has been suggested before (see Jeeves and Milner, 1987; Temple et al., 1990), that the absence of the corpus callosum may prevent the "fine-tuning" of certain cognitive and motor processes without necessarily affecting the general level of functioning. The apparent absence of observable cognitive deficits in some individuals lacking the corpus callosum is consistent with the findings that callosal defects are often clinically silent (Kretschmer, 1974) and that callosotomized patients without post-operative complications generally do not show changes in their overall functioning after surgery (Akelaitis, 1941; Gazzaniga et al, 1965; Geoffroy et al, 1983; Lassonde et al, 1991b).

We therefore feel that there is no need to unduly alarm the already distressed parents of newly diagnosed children. Rather, each case should be considered individually with regard to the presence or absence of concomitant CNS pathology, and even then the prognosis should be made prudently in the light of acallosal individuals with multiple malformations detected at autopsy that reportedly lived normal and productive lives (i.e. Harcourt-Webster and Rack, 1965). We have also found that manual dexterity and motor coordination in the young child can be greatly and more quickly improved with the help of physiotherapy and occupational therapy. Children tend to eventually outgrow most of the motor deficits, probably due to compensatory mechanisms that come into play with practice. However, therapy in this area may help them to maximize these mechanisms at a time when plasticity is presumably greatest.

REFERENCES

Akelaitis, A.J., 1941, Psychobiological studies following section of the corpus callosum, *AM. J. Psychitatr.* 97:1147-1158.

Atlas, S.W., Zimmerman, R.A., Bilaniuk, L.T., et al., 1986, Corpus callosum and limbic system: Neuroanatomic MR evaluation of developmental anomalies, *Radiology* 160(2):355-362.

Bigler, E.D., Rosenstein, L.D., Roman, M., and Nussbaum, N.L., 1988, The clinical significance of congenital agenesis of the corpus callosum, *Arch. Clin. Neurospychol.* 3:189-200.

Bryden, M.P. and Sprott, D.A.,1981, Statistical determination of degree of laterality, *Neuropsychologia* 19:571-581.

Chiarello, C., 1980, A house divided? Cognitive functioning with callosal agenesis, *Brain Lang.* 11:128-158.

Clarke, J.M. and Zaidel, E., 1989, Simple reaction times to lateralized light flashes, *Brain* 112:849-870.

Cook, N.D., 1984, Callosal inhibition: the key to the brain code, *Behav. Sci.* 29:98-110.

Cook, N.D., 1986, "The Brain Code: Mechanisms of Information Transfer and the Role of the Corpus Callosum," Methuen, London.

De Léon, G.A., De Léon G., Grover, W.D., Zaeri, N., and Alburger, P.D., 1987, Agenesis of the corpus callosum and limbic malformation in Alpert syndrome Type I Acrocephalosyndactyly, *Arch. Neurol.* 44:979-982.

Dennis, M., 1977, Cerebral dominance in three forms of early brain disorder, *in:* "Topics in Child Neurology," M.E. Blaw, I. Rapin, and M. Kinsbourne, eds., Spectrum Pub., New York.

Dennis, M., 1981, Language in an congenitally acallosal brain, *Brain Lang.* 12:33-53.

Ettlinger, G., Blakemore, C.B., Milner, A.D., and Wilson, J., 1974, Agenesis of the corpus callosum: a further behavioural investigation, *Brain* 95:327-346.

Gagnon, M., 1953, "L'Échelle d'intelligence Ottawa-Wechsler," Les Éditions de l'Université d'Ottawa, Ottawa.

Gazzaniga. M.S., 1970, "The Bisected Brain," Appleton- Century- Crofts, New York.

Gazzaniga, M.S., Bogen, J.E., and Sperry, R.W., 1965, Observation on visual perception after disconnexion of the cerebral hemispheres in man, *Brain* 88(2):221-235.

Geoffroy, G., Lassonde, M, Delisle, F., and Décarie, M., Corpus callosotomy for control of intractable epilepsy in children, *Neurology* 33(7):891-897.

Gott, P.S. and Saul, R.E., 1978, Agenesis of the corpus callosum: limits of functional compensation, *Neurology* 28:1272-1279.

Gruber, F.A. and Segalowitz, S. J., 1977, Some issues and methods in the neuropsychology of language, *in:* "Language Development and Neurological Theory," S.J. Segalowitz and F. A. Gruber, eds., Academic Press, New York.

Harcourt-Webster, J.N. and Rack, J.H., 1965, Agenesis of the corpus callosum, *Postgrad. Med. J.* 41:73-79.

Jeeves, M.A., 1986, Callosal agenesis: neuronal and developmental adaptations, *in:* "Two Hemispheres-One Brain: Functions of the Corpus Callosum," F. Lepore, M. Ptito, and H.H. Jasper, eds., Alan R. Liss, New York.

Jeeves, M.A. and Milner, A.D., 1987, Specificity and plasticity in interhemispheric integration: evidence from callosal agenesis, *in:* "The Duality and Unity of the Brain: Unified Functioning and Specialization of the Hemispheres," Wenner-Gren International Symposium Series, Vol. 47, D. Ottoson, ed., Macmillan, Basingstoke.

Jeeves, M.A., Silver, P.H., and Jacobson, I., 1988, Bimanual coordination in callosal agenesis and partial commissurotomy, *Neuropsychologia* 26(6):833-850.

Jeeves, M.A. and Temple C.M., 1987, A further study of language function in callosal agenesis, *Brain Lang.* 32:325-335.

Kretschmer, H., 1974, Callosal tumors, *in:* "Handbook of Clinical Neurology," P.J. Vinken and G.W. Bruyn, eds., Elsevier, North Holland, Amsterdam.

Lassonde, M. and Bryden, M.P., 1990, Dichotic listening, callosal agenesis and cerebral laterality, *Brain Lang.* 9:475-481.

Lassonde, M., Bryden, M.P., and Demers, P., 1990, The corpus callosum and cerebral speech lateralization, *Brain Lang.* 38:195-206.

Lassonde, M., Nolin P., Laurencelle, L., and Geoffroy, G., 1982, Évaluation neuropsychologique d'enfants atteints d'agénésie du corps calleux, *Apprent. Social.* 5(2):115-124.

Lassonde, M., Ptito, M., and Laurencelle, L., 1984, Étude tachistoscopique de la spécialisation hémisphérique chez l'agénésie du corps calleux, *Rev. Can. Psych.* 38(4):527-536.

Lassonde, M., Sauerwein, H., Chicoine, A.J., and Geoffroy, G., 1991a, Absence of disconnexion syndrome in callosal agenesis and early callosotomy: Brain reorganization or lack of structural specificity during ontogeny?, *Neuropsychologia* 29(6):481-495.

Lassonde, M., Sauerwein, H., Geoffroy, G., and Décarie, M., 1991b, Long-term neuropsychological effects of corpus callosotomy in children, *J. Epilepsy (Suppl.)* 3:277-284.

Lezak, M., 1983, "Neuropsychological Assessment," 2nd Edition, Oxford University Press, New York.

Lynn, R.B., Buchanan, D.C., Fenichel, G.M., and Freemon, F.R., 1980, Agenesis of the corpus callosum, *Arch. Neurol.* 37:444-445.

Martin, A., 1985, A qualitative limitation on visual transfer via the anterior commissure, *Brain* 108:43-63.
McGlone, J., 1985, Congenital absence of the corpus callosum: neuropsychological and sodium amytal findings. Paper presented at *INS Annual Meeting*, San Diego, CA.
Meerwaldt, J.D., 1983, Disturbances of spatial perception in a patient with agenesis of the corpus callosum, *Neurosychologia* 21:161-165.
Milner, A.D., 1982, Simple reaction times to lateralized visual stimuli in a case of callosal agenesis, *Neuropsychologia* 20:411-419.
Milner, A.D., 1983, Neuropsychological studies of callosal agenesis, *Psychol. Med.* 13:721-725.
Milner, A.D., Jeeves, M.A., Silver, P.H., Lines, C.R., and Wilson, J., 1985, Reaction times to lateralized visual stimuli in callosal agenesis: stimulus and response factors, *Neuropsychologia* 23:323-331.
Milner, B., Branch, C., and Rasmussen, T., 1964, Observations on cerebral dominance in some non-right-handers, *Trans. Am. Neurol. Ass.* 91:306-308.
Moscowitch, M., 1977, The development of lateralization of language functions and its relation to cognitive and linguistic development. A review and some theoretical speculations, *in:* "Language Development and Neurological Theory," S.J. Segalowitz and F.A. Gruber, eds., Academic Press, New York.
Nolin, P., 1984, Évaluation neuropsychologique de l'agénésie du corps calleux, Unpublished Master's thesis, Université du Québec à Trois-Rivières, Trois-Rivières, Qué., Canada.
Pirozzolo, F.J., Pirozollo, P.H., and Zinman, R.B., 1979, Neuropsychological assessment of callosal agenesis: report of a case with normal intelligence and absence of disconnexion syndrome, *Clin. Neuropsychol.* 1:13-16.
Rakic, P. and Yakovlev, P.I., 1968, Development of the corpus callosum and cavum septi in man, *J. Comp. Neurol.* 132:45-72.
Reitan, R.M. and Davison, L.A., 1974, "Clinical Neuropsychology: Current Status and Applications," Hemisphere, New York.
Sanders, R.J., 1989, Sentence comprehension following agenesis of the corpus callosum, *Brain Lang.* 37:59-72.
Sauerwein, H.C. and Lassonde, M., 1983, Intra- and interhemispheric processing of visual information in callosal agenesis, *Neuropsychologia* 21:167-171.
Sauerwein, H.C., Lassonde, M., Cardu, B., and Geoffroy, G., 1981, Interhemispheric integration of sensory and motor functions in agenesis of the corpus callosum, *Neuropsychologia* 19(3):445-454.
Selnes, O.A., 1974, The corpus callosum: Some anatomical and functional considerations with special reference to language, *Brain Lang.* 1:111-139.
Smith, A., 1981, Principles underlying human brain functions in neuropsychological sequelae of different neuropathological processes, *in:* "Handbook of Clinical Neuropsychology," S.B. Filskov and T.J. Boll, eds., John Wiley, New York.
Sperry, R.W., 1968, Plasticity of neural maturation, *Dev. Biol. (Suppl.)* 2:306-327.
Sperry, R. W., 1970a, Cerebral dominance in perception, *in:* "Early Experience and Visual Information Processing in Perceptual and Reading Disorders," F.A. Young and D.B. Lindsley, eds., Natl. Acad. Sci., Washington, D.C.
Sperry, R. W., 1970b, Perception in the absence of the neocortical commissures, *in:* "Perception and its Disorders," Vol. 48, Res. Publ. Assoc. Nerv. Ment. Dis., Williams and Wilkens, Baltimore.
Spreen, O. and Strauss, E., 1991, "A Compendium of Neuropsychological Tests," Oxford University Press, New York.
SPSS, 1982, "SPSS: Discriminant," SPSS Inc., Chicago.
Temple C.M., Jeeves, M.A., and Vilarroya, O.O., 1989, Ten pen men: rhyming skills in two children with callosal agenesis, *Brain Lang.* 37:548-564.
Temple C.M., Jeeves, M.A., and Vilarroya, O.O., 1990, Reading in callosal agenesis, *Brain Lang.* 39:325-253.
Trevarthen, C., 1974, Cerebral embryology and the split-brain, *in:* "Hemispheric Disconnection and Cerebral Function," M. Kinsbourne and W.L. Smith, eds., Charles C. Thomas, Springfield, IL.
Wechsler, D., 1974, "Wechsler Intelligence Scale for Children - Revised," The Psychological Corporation, New York.
Wechsler, D., 1981, "Wechsler Adult Intelligence Scale - Revised," The Psychological Corporation, New York.
Winer, B.J., 1962, "Statistical Principles in Experimental Design," McGraw-Hill, New York.

THE BEHAVIORAL AND DEVELOPMENTAL CONSEQUENCES OF CORPUS CALLOSAL AGENESIS AND AICARDI SYNDROME

Gregory O'Brien

Department of Psychiatry
Morpeth, Northumberland
NE61 3BP
U.K.

INTRODUCTION

From the standpoint of developmental psychiatry, studies of the behavioral consequences of agenesis of the corpus callosum (ACC) have to date fallen broadly into two types. First of all, most "behavioral" studies of ACC are accounts of research carried out in controlled laboratory settings, and these include studies involving motor tasks and measures of interhemispheric transfer time, and work in the area of psycholinguistics. These studies, which have been reviewed extensively elsewhere (e.g. Milner and Jeeves, 1979; Lassonde, 1986; Ramaekers, 1991) and must surely be the most informative body of research in the investigation of the psychological concomitants of callosal agenesis, are well represented in the present volume. Secondly, there is a body of research concerning the naturally-occurring behavioral problems of acallosal people (as distinct from behavioral observations from laboratory work). There are two notable examples of this work. Firstly, there is a long history of clinical reporting and research into the purported association between ACC and major mental illness, particularly schizophrenia (Gulman et al., 1982; Hatta et al., 1984). The findings in this area of research are at best inconclusive: it is as yet unclear whether the occurrence of schizophrenia in acallosal individuals is related to the callosal lesion, or to other factors (David, 1989; Njiokiktjien, 1991). The ongoing work into Andermann's Syndrome (Andermann et al., 1972), where periods of hallucinatory experience - particularly early in adult life - are reported to occur, may shed further light on the specificity of this association. Secondly, in learning-disabled acallosal children, a high frequency of behavior disturbance has been reported, but as yet neither the nature of such behavioral problems nor the extent to which they are related to callosal dysgenesis has been clarified (Roeltgen and Roeltgen, 1989; Njiokiktjien and Ramaekers, 1991).

Against this general background, the consideration of a wide range of research findings - taken in the light of the factors of known etiological importance in the psychiatric disorders of developmentally disabled children (reviewed in Fraser and

Callosal Agenesis, Edited by M. Lassonde and
M.A. Jeeves, Plenum Press, New York, 1994

Murti Rao, 1991) - suggests certain research questions which might determine the most fruitful lines of inquiry into the behavioral phenotype of ACC.

First of all, much of the behavior disturbance in learning disabled acallosal children reported to date has not been discussed under any diagnostic or other classification system, but has been referred to as "behavior disorder". Such terminology often in fact refers to certain "challenging" behaviors which are common in mentally handicapped children and adults in general, such as self-injurious, aggressive or destructive behaviors (Emerson et al., 1989). Given a suitably large sample it should be possible to clarify whether problems of this type are indeed common in acallosal children. If such problems were found, then one likely diagnosis - in this population in which a major brain anomaly is present - would be hyperactivity. It is therefore important to clarify whether brief attention span and motor restlessness are common in acallosal children, and also to explore the concept of irritability (Taylor, 1986). Secondly, given the reports of linguistic anomalies and deficiencies in acallosal people, including those of normal intelligence (Chiarello,1980; Jeeves and Temple, 1987), it is of great theoretical interest to determine whether persistent developmental disorders of the autistic type are common in acallosal children. Thirdly, the recent reports of EEG anomalies, particularly the work on coherence and REM sleep (Nielsen et al., 1993, in press; Kuks et al., 1987) suggest that in their daily lives acallosal individuals may show behavioral difficulties in relation to sleep. In line with conventional classification and clinical experience (Thompson, 1988), these might include: difficulty in settling down to sleep; nighttime wakening; certain behavioral difficulties (parasomnias) in association with nighttime wakening; early morning wakening; difficulty in attaining and maintaining a regular 24-hour sleep rhythm. Fourthly, a wealth of research in the area of depression and other mood disorders has recently clarified that such disorders do occur in learning disabled people, and are probably more common in this population than in the general population (Day, 1990). It would therefore be of value to clarify whether depressive and related mood disorders are common in acallosal children.

Much of our present knowledge of the developmental consequences of ACC comes from the same body of research as cited above. In particular, the laboratory-based neuropsychology studies (Milner and Jeeves, 1979; Lassonde, 1986; Jeeves, 1986) are clearly informative regarding the developmental level attained on specific tasks. Also, there is a body of research in which intelligence quotients and other psychometric characteristics have been reported. However, given the heterogeneity of the population in question, the difficulties of interpreting and generalizing the results of these studies has been long-recognized (Ferris and Dorsen, 1975). Additional evidence comes from the research concerning the longitudinal development of acallosal children, including the prospective data from the large Dutch series (Njiokiktjien and Ramaekers, 1991a) and from the successive reports of the carefully documented subjects of Lassonde and Jeeves. The consensus emerging from all this work is that there may not be such thing as the "asymptomatic acallosal child". Indeed, the questions now being posed are more concerned with the plasticity and the pluripotentiality of alternative circuits and pathways (Jeeves, 1990), and also issues related to whether the extent of the acallosal lesion or deficiency itself is of developmental importance (Njiokiktjien and Ramaekers, 1991b).

This issue is quite central, for on first principles it would seem that if callosal agenesis *per se* is of any developmental consequence, then it may also be possible to demonstrate some association between the degree of agenesis and the severity of concomitant developmental disability. In other words, if ACC does result in some degree of developmental disability, then the degree of disability might show some correlation with the extent of the callosal deficiency or lesion. Now, to date, the evidence is conflicting regarding this issue. On the one hand a number of writers have stated that the degree of agenesis itself is of no importance, and this opinion is so widespread that it is expressed explicitly in certain standard text books (e.g. Lindsay, 1989). On the other hand, more recent controlled comparative prospective work from Holland has suggested that learning disabled children with partial agenesis may be less developmentally disadvantaged than learning disabled children with total agenesis (Njiokiktjien and Ramaekers, 1991b). It would now be desirable, therefore, to conduct a large series comparison of partial and total agenesis children. At the same time, it must, of course, be borne in mind that a host of other factors which may operate in the learning disabled acallosal child may contribute to developmental disability (Ferriss and Dorsen, 1975).

One contribution to the resolution of this debate may be made by the inclusion of three sub-groups in a study cohort: total callosal agenesis; partial callosal agenesis (hypogenesis); Aicardi syndrome. This allows the comparison between ACC and callosal agenesis compounded by an additional set of disabilities, in addition to comparison between partial and total agenesis.

CORPAL

The recent formation of a support group for the families of ACC subjects offers a unique large cohort for investigation. Before moving on to more detailed consideration of a project carried out with their kind collaboration, the following general points are important to note.

There is already an extensive literature concerning the experience of working with the support societies which have been formed to help the families of mentally handicapped children, particularly research into the behavioral aspects of such syndromes. It is the experience of the Society for the Study of Behavioural Phenotypes (O'Brien, 1992a) that:

1. The families who approach such syndrome support societies tend to be those who have more "difficult" children, i.e. those who are more learning or physically disabled, and/or those who are more behaviorally disturbed - generally both.
2. Despite the preselected and biased nature of such samples, it is possible to obtain reliable information from such families even by postal studies, and a specific instrument has therefore been developed for this purpose (O'Brien, 1992b).
3. The families who make contact with these societies tend to be very motivated toward research inquiry - for very understandable reasons - and therefore a high degree of commitment and co-operation is the rule (Clarke et al., 1989; Udwin, 1990).
4. Clearly from the above, this is one research strategy by which useful data on large samples can be obtained comparatively rapidly. This latter point is particularly important with reference to ACC, where the relative rarity of the condition is the main reason for the lack of extended samples in the literature to date.

THE STUDY

The present study set out to investigate the frequency of a range of behavioral problems and developmental attainments among total callosal agenesis, partial callosal agenesis (hypogenesis) and Aicardi Syndrome subjects. Also, in order to add to the general extended database of information concerning dysmorphology and associated other physical and medical problems, other background and health data were recorded.

Instrument

Data were collected using the Society for the Study of Behavioural Phenotypes Postal Questionnaire. This is a highly structured questionnaire for parental completion, and consists of 98 questions, which are in two parts. The first part concerns developmental attainments, and is arranged in eight domains, namely: sensory; general mobility; dexterity; feeding; continence; dressing; hygiene; communication. Section two focuses on behavioral problems, and consists of questions divided into the following behavioral domains: feeding; sleep; social behavior; difficulties with language; movement activities; unusual movements/interests; self-injurious behavior and aggression; anxiety and mood; others. It should be noted that the present study is part of the wider research initiative of the Society for the Study of Behavioural Phenotypes, in which a common methodology is used to accumulate data on the behavioral features of a wide variety of mentally handicapping syndromes, which are then stored on one common database, in order to facilitate the further investigation of behavioral phenotypes.

Subjects

The subjects were the 47 parent members of "Corpal" who agreed to participate in the study. These comprised 31 cases of total callosal agenesis (ACC), 8 cases of partial callosal agenesis (ACC/p), and 8 of the Aicardi Syndrome (AS). All cases were radiologically diagnosed by CT brain scan (only a very small minority have also received MRI scan), and all cases were clinically diagnosed by a pediatrician with expertise in this area. In line with previous reports, the complete and partial agenesis subjects featured both male and female subjects (ACC/p: 6 male and 2 female; ACC: 17 male and 14 female) while all 8 Aicardi Syndrome subjects were female. Across the whole sample, the ages ranged from 2 to 16 years (mean 6.5, standard deviation 3.9). In the three groups the corresponding ages were, in years: ACC/p - mean 7, s.d. 4.8 ; ACC - mean 5.7, s.d. 3.2; AS - mean 9.2, s.d 5.0. In two cases (4%) there was a first or second degree family history of an affected member.

RESULTS

Dysmorphology, sensory disability and epilepsy

Regarding the facial dysmorphology of the children, the sample was characterized by the rarity rather than the occurrence of the previously reported features of the syndrome. First of all, wide-set eyes were rare, occurring in only 1(12.5%) ACC/p child, 3(9.5%) ACC children, and 1(12.5%) AS child. The typical chorioretinal lacunae of Aicardi syndrome were present in all of that group, and also in 2 of the ACC (6.5%) children, but in none of the ACC/p children. Cleft lip was a problem of only 1 child, in the AS group. Cleft palate occurred in 1 ACC/p child, 3 ACC children and 1 AS child. The findings on skull size and dysmorphicity were of more interest. First of all, a small skull was reported in 1(12.5%) ACC/p child, 9(29%) ACC children and 3(37.5%) AS children, while a large skull was reported in none of the ACC/p or AS cases, but in 5(16.1%) ACC children. Also a normal sized, but very obviously peculiarly shaped, skull was a feature of 5(16.1%) ACC and 2(25%) AS children, but not of the ACC/p cases. Putting these skull findings together, while only 1 (12.5%) ACC/p child had a dysmorphic skull, 19 (61.5%) ACC children had such a problem, and 5 (62.5%) AS children similarly had a dysmorphic skull.

In line with the findings on general dysmorphology, the major sensory disabilities were also infrequent in this cohort. Firstly, blindness: no ACC/p children were blind, 6 (19.5%) ACC children were blind, and no AS children were blind. Deafness was even less common. None of the AS children were deaf, only 1 of the ACC/p children (12.5%) was deaf and 3 (9.5%) of the ACC children were deaf.

In addition to the developmental issues elaborated below, certain key questions in the area of epilepsy were explored. From these data it emerged that of the 47 subjects, 28 are currently diagnosed as epileptic, 33 had a previous history of epilepsy, while interestingly only 36 (77%) of the whole sample had ever at any time had an EEG. Of those now diagnosed as having epilepsy, 23 are currently on anti-convulsants, and only 3 have the major hypsarrhythmia, or the classical "Salaam" attacks, said to be associated with Aicardi Syndrome. Two of these were indeed Aicardi Syndrome cases, 1 an ACC case. Examining the epilepsy data further, a significant trend across the 3

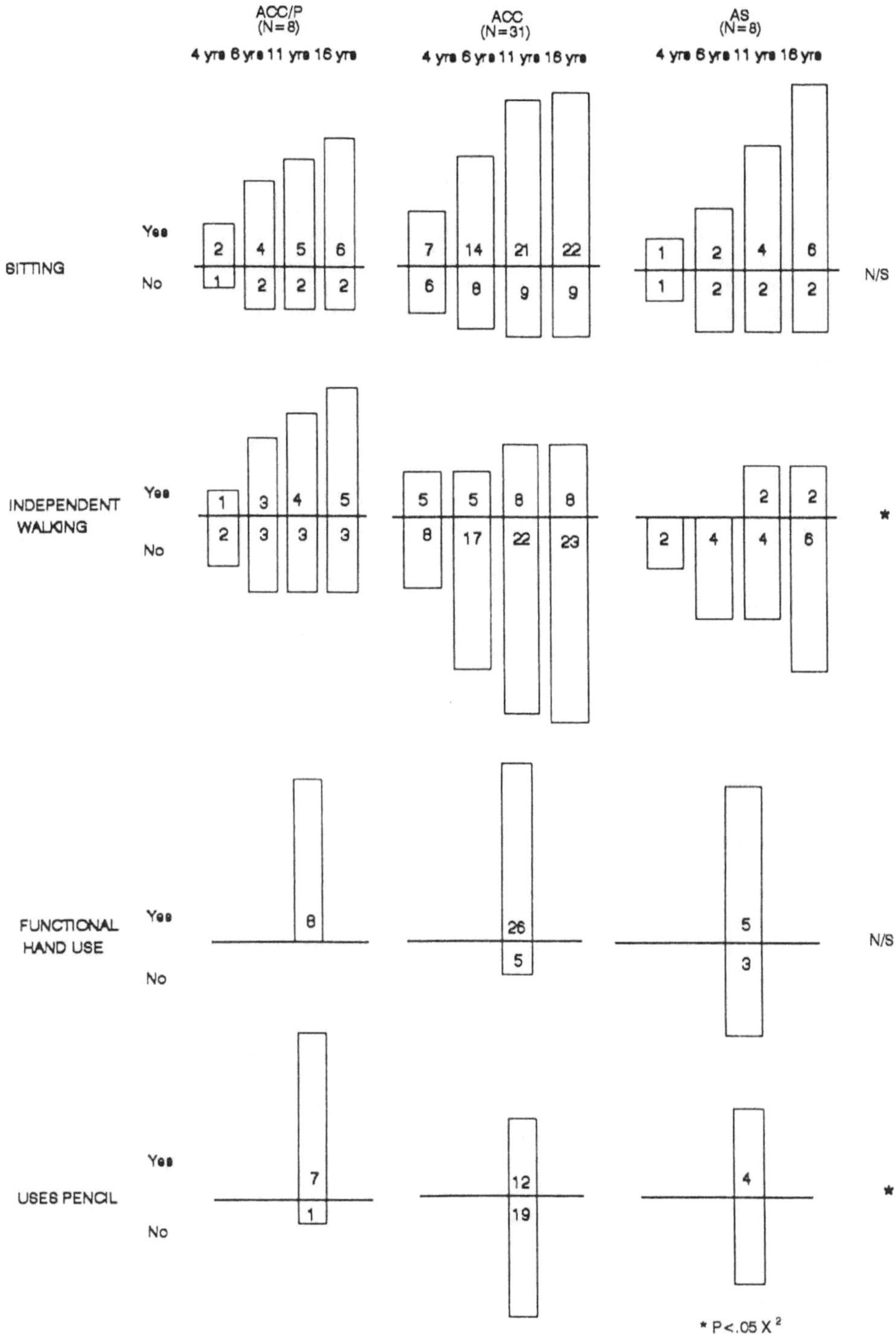

Figure 1. Developmental attainments. The data on sitting and independent walking are presented in cumulative fashion. In the ACC/ P group, therefore, 2 of the three children aged under 5 years are able to sit up, and these are included in the 4 aged under 6 years who have this ability.

groups was found. For, while only 2 of the ACC/p group are currently epileptic (25%) 18 of the ACC group (58%) and all 8 (100%) of the AS subjects have epilepsy ($p<.01$, x^2).

Developmental Domains

First of all, expressive language; of the 8 ACC/p children, 4 (50%) had no expressive language. This compared with 24 (77.5%) of the ACC group and 6 (75%) of the AS children. Similarly, 2 of the 8 ACC/p children (25%) had minimal delay in their expressive language, while only 4 (13%) of the ACC children, and none of the AS group, were found to be functioning at this higher level of expressive language. The other children show "marked delay": ACC/p 2 (25%) ACC 3 (10%) AS 2 (25%).

Figure 1 summarizes the data on the major developmental attainments of sitting independently, walking independently, having some functional use of hands and the ability to use a pencil to mark paper. It can be seen that the 3 groups are quite similar with respect to sitting ability, being characterized by a proportion showing some degree of marked delay, but this proportion being similar across the 3 groups. However, in considering the more demanding task of independent walking, substantial differences emerge. For, by 4 years, while none of the AS children are walking, 1 of the 3 ACC/p children is walking; by 6 years, none of the AS children are walking, while half of the ACC/p group are walking, and only just over a quarter of the ACC group. This extends across the range of the figure, so that in the oldest group, more than half of the ACC/p children are walking independently, but only a quarter of the ACC and AS groups are walking. Similarly, with respect to the ability of some independent meaningful hand use, all of the ACC/p children have this ability, while 6(19%) of the ACC children do not, and 3(3.75%) of the AS children do not. Also, it is in consideration of the more demanding manual task of using a pencil - compared to using hands functionally - that significant differences between these 3 groups become apparent within the domain of manual dexterity. Seven (87.5%) of the ACC/p children possess this skill, compared to less than half (12 - 39%) of the ACC children, and only half of the AS children.

Behavioral Domains

The findings on overall activity are shown in Table 1. These reveal that none of the AS children, less than 10% of the ACC children and 2 (25%) of ACC/p show any hyperactive tendencies. On the other hand, lethargy, the finding of a degree of inactivity beyond that which might be expected given the degree of physical disability, is rare in the ACC/p group (1:12.5%), fairly common in the ACC group (13:42%), and an almost constant feature in the AS group (7:87.5%), showing a striking trend of increasingly problematic lethargy as we move from ACC/p through ACC to AS. Next, a full account is given of the findings on self-injurious behavior (s.i.b.) and aggression: these problem behaviors are mostly uncommon, with the exception of s.i.b. in the AS group.

Interestingly, the most "difficult to handle" behaviors - attacking children or adults, suddenly hitting out for no apparent reason, destroying objects, at times being so violent that parents cannot handle their child without help - do not occur in the ACC/p group. The findings on anxiety and mood are also displayed in Table 1. Here, dysphoric mood, being a screening item which would suggest the likelihood of depression, is neither common nor uncommon across the sample, and there are no significant differences between the 3 groups. Similar considerations apply to the question of whether the acallosal child is often fearful, in other words shows diffuse anxiety (see Table 1), and the results on temper tantrums. Irritability was uncommon, in line with the findings on hyperactivity.

As detailed in the Introduction, it is of great theoretical interest to determine whether autism might be common in this group. First of all, and most importantly, the phenomenon of social indifference is explored. "Social indifference" refers to the kind of lack of empathic interest in the other person which is at the core of autism. As Table 2 demonstrates, this was present in only 16% of the ACC group, only. Next, there is the

Table 1. Behavioral domains.

	ACC/P (8)	ACC (31)	AS (8)
Activity			
Hyperactive	2 (25%)	3 (9.5%)	0
Lethargic	1 (12.5%)	13 (42%)	7 (87.5%)*
Self-injurious Behavior (s.i.b.) and Aggression			
S.i.b.	2 (25%)	8 (26%)	5 (62.5%)
Attacks children	0	3 (9%)	2 (25%)
Attacks adults	0	4 (13%)	1 (12.5%)
Hits out	0	7 (22.5%)	2 (25%)
Destroys objects	0	5 (16%)	2 (25%)
Need help to handle	0	4 (13%)	1 (12.5%)
Anxiety and Mood			
Dysphoric	2 (25%)	6 (19.5%)	3 (37.5%)
Often fearful	3 (37.5%	4 (13%)	3 (37.5%)
Irritable	2 (25%)	6 (19.5%)	3 (37.5%)
Temper tantrums	1 (12.5%)	8 (26%)	2 (25%)

* $P<.05$, X^2

issue of emotional non-communicativeness: the extent to which parents find it impossible to know at any one time what their child is feeling. This was common throughout the whole sample. Moreover, it was increasingly common from the ACC/p through to the AS group, although not significantly so in statistical terms. Moving on to linguistic anomalies in the autistic domain we find that, as expected, echolalia was fairly common among those who did have some expressive language, but even more common

Table 2. Social behavior and language.

	ACC/P (8)	ACC (31)	AS (8)
Indifference	0	5 (16%)	0
Emotional non-comm.	4 (50%)	19 (61.5%)	7 (87.5%)
Some useful exp. lang	4 (50%)	7 (22.5%)	2 (25%)
Echolalia	1 (12.5%/25%)	6 (19.5%/86%)	1 (12.5%/50%)
"Meaningless" or "Out of place"	2 (25%/50%)	10 (32%/>100%)	1 (12.5%/50%)
Repetitive movements	3 (37.5%)	12 (38.5%)	5 (62.5%)
Routine obsessions	1 (12.5%)	(22.5%)	3 (37.5%)
Pica	1 (12.5%)	7 (22.5%)	1 (12.5%)

was "meaningless "or "out of place" conversation, being a pattern of habitually irrelevant or difficult to understand conversation, occurring particularly in someone who does have some useful expressive language. Three other behavioral constellations in this domain were considered. Repetitive movements were common only in the AS group, obsession for sameness was uncommon throughout the three groups, as was pica.

Data on sleep are presented on Table 3. "Rhythm" refers to the possession of a regular 24 hour sleep cycle: the table details the numbers who did not have such a cycle. "Disturbed behavior" refers to other behavioral problems occurring within the sleep period - common examples are head-banging and screaming. "Settling difficulty", "Frequent wakening" and "Day hypersomnolence" are self-explanatory. Frequent night-time wakening was the most common of these, and shows a non-significant increasing trend across the Table, from ACC/p through ACC to AS children.

Table 3. Sleep and associated behavior.

	ACC/P (8)	ACC (31)	AS (8)
Rhythm	2 (25%)	9 (29%)	1 (12.5%)
Settling difficulty	3 (37.5%)	7 (22.5%)	1 (12.5%)
Frequent waking	2 (25%)	11 (35.5%)	4 (50%)
Day hypersomnolence	2 (25%)	5 (16%)	2 (25%)
Disturbed behavior	1 (12.5%)	9 (29%)	3 (37.5%)

DISCUSSION

This study is unusual in both the extent and size of the sample, and the wide remit of the data base. Consideration of the importance of the findings demands that account be taken of certain general issues.

First of all, concerning the sample, to what extent can it be seen as representative of the 3 groups, partial callosal agenesis, total callosal agenesis and the Aicardi Syndrome? On an even more basic level, how confident can we be in the diagnoses? As described in the methodology section, the sample was collected in collaboration with the family support society Corpal, and this process generally results in a sample which is more severely disabled and behaviorally disturbed, but also well-investigated and carefully documented (see O'Brien, 1992). It would seem sensible to conclude that the children in this study will be similar in many respects to the populations from which they are drawn, but that they will indeed be more disabled, and probably more behaviorally disturbed. For this reason, they are a useful group for such study, providing we can be confident in their diagnoses. In this latter connection, all subjects have received at least CT scan, and all subjects were diagnosed by a pediatrician with expertise in this field, as outlined above. Also, many of the general findings - including those of chorioretinal lacunae in the Aicardi Syndrome group (Aicardi et al. 1965), and unusual linguistic anomalies in the complete agenesis group (Jeeves and Temple, 1987) - are consistent with the findings of other workers on these groups. Other important corroborative evidence includes the data on family history and genetic predisposition. In conclusion, we can be confident that these children have been appropriately diagnosed. In discussion of the results, it must be remembered that we are dealing with a more disabled group than the population of callosal agenesis from which they were drawn, but also that this was a uniform sample bias, present throughout the 3 subgroups.

Regarding the data base, certain key points must be borne in mind. Firstly, this is unusually wide: given the complex inter-play of factors which are known to be important in the etiology and maintenance and natural history of behavioral problems in handicapped children, this is necessary for any psychiatric study in behavioral phenotypy (Fraser and Murti Rao, 1991). Also, data were collected over a 1 month period. This gives a unique perspective on the period prevalence of behavioral problems in these children. Thirdly, although any questionnaire study will have limitations - and will clearly have to be followed up by more detailed investigation - not only is this a quick and reliable method of completing a wide database, it is one which is particularly applicable to the initial and innovatory stages of research (Clarke et al., 1989; Udwin, 1990). The reliability of the instrument used in the present study has been established (O'Brien, 1992b).

A sample of this size and nature, with data collected in a reliable uniform fashion, may shed light on the extent of the various stigmata of dysmorphology which have previously been reported as common concomitants of ACC. In this connection, it must be borne in mind that there are, of course, myriad etiological factors which lie behind callosal agenesis itself. However, the present study takes callosal agenesis *per se* as a focus of inquiry, and may therefore be informative in this respect. First of all, cleft lip and cleft palate have not emerged as common, but only sporadic in appearance. This is also true of the major sensory disabilities, blindness and deafness. Given the comparatively large size and more disabled nature of the present sample, these findings may well more accurately reflect the true extent of such problems in acallosal children than the previous smaller series. More central to the line of inquiry of the present study, however, is the issue of any difference across the 3 groups. Only in respect of the composite measure "skull dysmorphology" did any emerge in this section. The developmental significance of this finding is uncertain, but it is consistent with the developmental and behavioral findings outlined above and discussed below, namely, an increasing trend of behavioral and developmental consequence from partial through total agenesis, to the Aicardi Syndrome.

Moving on to the neurological data, epilepsy is common throughout the whole sample, as expected. Given this finding, and the recent EEG evidence (Nielsen et al., 1993, in press; Kuks et al., 1987), it is perhaps surprising to find that not all ACC cases are routinely tested for epilepsy. From the point of view of the present study, the findings of differences across the 3 groups from partial through total Agenesis to the Aicardi Syndrome in respect of the presence of epilepsy itself is important. Such a clear step-wise increasing prevalence of epilepsy, a current problem over the time period of the study, is striking. A difference between the Aicardi Syndrome and other agenesis subjects is to be expected, for epilepsy has been a virtually constant feature of the latter syndrome (Aicardi et al., 1987), while epilepsy is a common but not an invariable concomitant of ACC (Lilienthal and Tarlau 1969). Given that pathognomenic features other than acallosality are present in the Aicardi Syndrome, this is consistent to some extent with the widespread belief that the neurological complications of Aicardi Syndrome and callosal agenesis are due to factors other than the callosal agenesis itself (Lindsay 1989). However, the present study demonstrates an association between the presence of epilepsy and the extent of the callosal dysgenesis. This suggestion, that the extent of the callosal lesion may be of etiological importance, is consistent with the writings of others (e.g. Lassonde 1986, Jeeves 1990) who have stressed the crucial role of the callosal lesion in neurological developmental and behavioral status of affected children, and with the recent evidence that learning disabled children with partial agenesis may be less developmentally-disadvantaged than those with total agenesis (Njiokiktjien and Ramaekers, 1991b).

This hypothesis is further supported by the results in the developmental domains. First, we will consider the data on the ability to sit up and to walk independently. Many of the children do not have the former ability, even the older children. However, there are no significant differences across the 3 groups in this respect. It is only in consideration of the more demanding task of independent walking that differences emerge, in the same direction as elsewhere in this study, namely that the partial agenesis subjects are least affected and the Aicardi Syndrome subjects most affected. Also, in order to clarify that the cross-group differences are not age-related, the results have been considered in cumulative fashion by age. Secondly, the data on manual dexterity. Here, the sample as a whole is characterized by a degree of disability, but no clear differences have emerged between the 3 groups. Once again however, the more demanding task - of using a pencil to make a mark on paper - yields a significant difference across the 3 groups in the same direction as before.

In consideration of the extent and nature of the behavioral concomitants of callosal agenesis, the results in the area of social behavior and language are crucial. Here, deficits and disorders of an autistic type are uncommon, despite the severity of other developmental disabilities found in these children: against such a background, social and linguistic developmental findings on the autistic spectrum could well be anticipated (Gillberg, 1990). Moreover, the present sample demonstrates the same linguistic anomaly in the complete callosal agenesis subjects - i.e. meaningless, repetitive or out-of-place talk - as has been extensively documented elsewhere (Jeeves and Temple, 1987). Furthermore, there is the finding of emotional non-communicativeness in these children. Now, this is the curious finding that the parents of these children have typically reported that they find it difficult or well-nigh impossible to know at any one time what their child is feeling or experiencing emotionally. It should be noted that this is not commonly reported by the parents of mentally handicapped children, and certainly not by the kind of motivated caring parents with whom we deal when contacting a parent and family support society such as Corpal (O'Brien 1992). This is therefore an important finding, and is in keeping with the report of alexithymia in callosal agenesis individuals of normal intelligence (Zeitlin et al., 1989).

In summary, the findings on social behavior and language describe a profile of: linguistic anomaly which is in keeping with previous reports; a striking rarity of deficits and disorders in the autistic domain; and a curious variant of empathic disorder, which may well be an equivalent of alexithymia. For whereas acallosal individuals of normal intelligence find it difficult to verbalize their emotional experience, the severely developmentally-disabled acallosal children in the present study seem to demonstrate their lack of emotional communicativeness in non-verbal fashion according to their parents. Clearly, this is speculative, but in the light of corroborative findings elsewhere, it may well be important in determining the social functioning and development of acallosal individuals, and therefore merits further study.

As detailed in the Introduction, previous research indicated the need for a systematic study of a wide variety of behavioral problems in acallosal individuals, in order to clarify the nature of the behavioral phenotype. First, hyperactivity was found to be rare in the present study. Given that we are dealing with callosal agenesis with a major brain lesion, and one which may well result in a failure of inhibition in certain social cognitive domains, (Ramaekers, 1991) the relative rarity of hyperactivity is of interest, and merits further study. Secondly, lethargy, the unwillingness to initiate or engage in activity to an extent not accountable by the level of physical and intellectual disability, is common. The issue of lethargy merits further study, and, given the putative role of the corpus callosum in conative processing (Ferriss and Dorsen, 1975), may well be important in the ongoing social development and habilitation difficulties of acallosal individuals.

The next set of behavioral results which merit discussion are those on self-injurious behavior and aggression. It is in line with the rarity of hyperactivity that such problems are not particularly common in the study group - the only exception being self-injurious behavior in Aicardi Syndrome. This suggests that the earlier studies which had reported "behavior disorders" to be common in acallosal children (e.g. Njiokiktjien and Ramaekers, 1991b) were referring to problems of different kinds, such as possibly lethargy, linguistic anomalies or even sleep disturbance, all of which the present study found to be more characteristic of callosal agenesis than such "challenging behaviors". It is of interest that these problems are even less common among partial agenesis subjects, but the numbers here are so small that no firm conclusions can be drawn. The results on anxiety and mood - that just over one fifth of a severely learning disabled group of people have dysphoric mood, and a similar proportion are diffusely anxious - are consistent with other research in the mental handicap population (Day, 1990). The data on irritability and temper tantrums are in line with the findings on hyperactivity in the present study.

The final set of results for consideration are those on sleep. In general, sleep-related problems have been found to be not uncommon among the present study group, but no clear pattern has emerged concerning the nature of disturbance. However, the findings concerning frequent night-time wakening are important, not only in terms of their implications for families, but also in theoretical terms, where a parallel may be drawn with the findings on REM sleep from other studies (Kuks et al. 1987; Nielsen et al. 1993, in press).

CONCLUSION

In conclusion, a behavioral phenotype of emotional non-communicativeness and a linguistic anomaly in association with lethargy, but in the absence of autism, is suggested. The developmental and neurological findings suggest that the extent of the callosal lesion itself is of etiological importance.

REFERENCES

Aicardi, J., Chevrie, J.J., and Baraton, J., 1987, Agenesis of the corpus callosum, *in:* "Handbook of Clinical Neurology: Malformations," Vol. 6, N.C. Myrianthopoulos, ed., Elsevier, North Holland Amsterdam.

Aicardi, J., Lefebvre, J., and Lerique-Koechlin, A.,1965, A new syndrome: Spasms in flexion, callosal agenesis, ocular abnormalities, *Electroencephalogr. Clin. Neurophysiol.* 19:609-610.

Andermann, F., Andermann. E., Joubert, M., Karpati, G., Carpenter, D., and Melanson, D., 1972, Familial agenesis of the corpus callosum with anterior horn cell disease. A syndrome of mental retardation, areflexia and paraplegia, *Trans. Am. Neurol. Ass.* 97:242-244

Chiarello, C., 1980, A house divided? Cognitive functioning with callosal agenesis, *Brain Lang.* 11:128-158

Clarke, D.J., Waters, J., and Corbett, J., 1989, Adults with Prader-Wili Syndrome, *J. R.. Soc. Med.* 82:21-24.

David, A., 1989, Reading about the split-brain syndrome, *Br. J. Psychiatry* 154:422-425.

Day, K., 1990, Depression in moderately and mildly mentally handicapped people, *in:* "Depression in Mentally Retarded Children and Adults," A. Dosen and F. Menolascino, eds., Logan Publications, Leiden.

Emerson, E., Cummings, R., Hughes, H., Toogood, A., McCool, C., and Barrett, S, 1989, Challenging behavior and community services, 6. Evaluation and overview, *Ment. Handicaps* 15:166-168.

Ferriss, G.S. and Dorsen, M.M., 1975, Agenesis of the Corpus Callosum I. Neuropsychological studies, *Cortex* 1:95-122.

Fraser, W.I. and Murti Rao, J., 1991, Recent studies of mentally handicapped young people's behavior, *J. Child Psychol. Psychiatry* 32:79-108.

Gillberg, C., 1990, Autism and pervasive developmental disorders, *J. Child Psychol. Psychiatry* 31:99-120.

Hatta, T., Yamamoto, M., and Kawabata, Y., 1984, Functional hemispheric differences in schizophrenia: Interhemispheric transfer deficit or selective hemisphere dysfunction? *Biol. Psychiatry* 19:1027-1036.

Jeeves, M.A., 1986, Functional and neuronal developmental adaptations, *in:* "Two Hemispheres - One Brain: Functions of the Corpus Callosum," F. Lepore, M. Ptito, and H.H. Jasper, eds., Alan R. Liss, New York.

Jeeves, M.A., 1990, Agenesis of the corpus callosum, *in:* "Handbook of Neuropsychology, Vol. 4," F. Boller and J. Grafman, eds., Elsevier Science, Amsterdam.

Jeeves, M.A. and Temple, C., 1987, A further study of language function in callosal agenesis, *Brain Lang.* 32:325-335.

Kuks, J.B.M., Vos, J.E., and O'Brien, M.J., 1987, Coherence patterns of the infant sleep EEG in absence of the corpus callosum, *Electroencephalogr. Clin. Neurophysiol.* 66:8-14.

Lassonde, M., 1986, The facilitatory influence of the corpus callosum on interhemispheric processing, *in:* "Two Hemispheres - One Brain: Functions of the Corpus Callosum," F. Lepore, M. Ptito, and H.H. Jasper, eds., Alan R. Liss, New York.

Lilienthal, F. and Tarlau, M., 1969, The EEG in congenital absence of the corpus callosum, *Electroencephalogr. Clin. Neurophysiol.* 26:635.
Lindsay, M.P., 1989, "Dictionary of Mental Handicap," Routledge, London, U.K.
Milner, D. and Jeeves, M.A., 1979, A review of behavioral studies of agenesis of the corpus callosum, *in:* "Structure and Function of the Cerebral Commissures," I.S. Russell, M.W. van Hof, and G. Berlucchi, eds., Macmillan, London
Nielsen, T., Montplaisir, J., and Lassonde, M., 1993, Sleep architecture in agenesis of the corpus callosum: assessment of four cases, *Eur. Neurol.* 33, 173-176.
Nielsen, T., Montplaisir, J., and Lassonde, M., in press, Decreased interhemispheric EEG coherence during sleep in agenesis of the corpus callosum, *J. Sleep Res.*
Njiokiktjien, C., 1991, A historical perspective, *in:* "The Child's Corpus Callosum," G. Ramaekers and C. Njiokiktjien, eds., Suyi, Amsterdam.
Njiokiktjien, C. and Ramaekers, G., 1991a, Developmental Interhemispheric disconnection and learning disabilities, *in:* "The Child's Corpus Callosum," G. Ramaekers and C.Njiokiktjien, eds., Suyi, Amsterdam.
Njiokiktjien, C. and Ramaekers, G., 1991b, Developmental interhemispheric disconnection in learning and motor disabilities, *in:* "The Child's Corpus Callosum," G. Ramaekers and C. Njiokiktjien, eds., Suyi, Amsterdam.
O'Brien, G., 1992a, Behavioral phenotypy in developmental psychiatry - a guide to the available measurement schedules, *Eur. Child Adolesc. Psychiatry (Suppl.)* 1:1-61.
O'Brien, G., 1992b, Behavioral phenotypes and their measurement *Dev. Med. Child Neurol.* 34:379-381.
Ramaekers, G., 1991, Functions of the corpus callosum in adults, *in:* "The Child's Corpus Callosum," G. Ramaekers and C. Njiokiktjien, eds., Suyi, Amsterdam.
Taylor, E., 1986, Overactivity, hyperactivity and hyperkinesis: problems and prevalence, *in:* "Clinics in Developmental Medicine, Nr. 79: The Overactive Child. Spastics International Medical Publications" E. Taylor, ed., Blackwell, Oxford.
Thompson, P., 1988, Methods and problems in the assessment of behavior disorders in epileptic patients, *in:* "Epilepsy, Behavior and Cognitive Function," M. R. Trimble and E. H. Reynolds, eds., Wiley, Chichester, U.K.
Udwin, O, 1990, A survey of adults with Williams syndrome and idiopathic infantile hypercalcaemia, *Dev. Med. Child Neurol.* 32:129-141.
Zeitlin, S.B., Lane, R.D., O'Leary, D.S., and Schrift, M.D., 1989, Interhemispheric transfer deficit and alexithymia, *Am. J. Psychiatry* 146:1434-9.

AUDITORY VERBAL LEARNING AND MEMORY IN CASES OF CALLOSAL AGENESIS

Gina M. Geffen,[1] Glenys M. Forrester,[2]
Dean L. Jones,[1] and Donald A. Simpson[3]

[1]Cognitive Psychophysiology Laboratory
Psychology Department and Clinical Studies Unit
University of Queensland and
Queensland Health Department
Brisbane, Australia, 4072
[2]Research Unit, Julia Farr Centre, Fullarton and
Flinders University of South Australia
Adelaide, Australia, 5063
[3]NH and MRC Road Accident Research Unit
University of Adelaide and Department of Neurosurgery
Adelaide Children's Hospital, Adelaide, Australia, 5001

INTRODUCTION

The importance of the corpus callosum in memory is not well established. Memory appears to function primarily through the consolidating action of subcortical structures like the hippocampus and amygdala on the cerebral cortex (see reviews by Mishkin and Appenzeller, 1987; Squire, 1987). The corpus callosum may play significant secondary roles, however, by transferring stimuli to the opposite cortex, inhibiting ipsilateral input, and/or facilitating processing within each hemisphere (see Cook, 1986; Jeeves, 1990; Lassonde, 1986).

Studies of both commissurotomized patients and cases of callosal agenesis provide equivocal evidence for the participation of the corpus callosum in memory. Studies of commissurotomized patients tend to suggest that memory deficits, particularly in acquisition processes, can follow callosal surgery (Geffen et al., 1980; see reviews by Clark and Geffen, 1989; Zaidel, 1990). The conclusions of such studies are confused, however, by two points. Firstly, reviewers disagree as to the extent of post-operative decline in memory. On the one hand, Clark and Geffen (1989) suggested that where memory was tested before and after surgery, only one case showed a small decrease in performance. Zaidel (1990), in contrast, quoted four studies that suggest post-operative memory dysfunction. A second problem with commissurotomy studies is that many patients also undergo sections of the hippocampal commissure and/or fornix. Since these structures interconnect areas well implicated in memory (Mishkin and Appenzeller, 1987; Squire, 1987), damage to these regions, rather than the corpus callosum, may produce the impairment (Clark and Geffen, 1989; Zaidel, 1990).

Studies of callosal agenesis are beset with similar problems. Firstly, the condition is most frequently only a part of a more extensive brain malformation (see reviews by Aicardi et al., 1987; Jeeves, 1990), making it difficult to dissociate the effects of callosal agenesis from other brain abnormalities. This criticism particularly applies to studies by Ferriss and Dorsen (1975) and Bigler et al. (1988). Although these studies showed deficits in verbal learning, their cases had Full Scale IQs from 54 to 77, which could be associated with confounding disorders of attention and language comprehension. Ferriss and Dorsen (1975), moreover, did not compare their data with a control group. A better approach for reducing the possible effects of other impairments is to use cases with relatively normal general intelligence.

A second problem with studies of callosal agenesis (and many of those with commissurotomized patients) is that they frequently use tasks/tests which measure memory function too broadly. Pirozzolo et al. (1979), for instance, used the Wechsler Memory Scale, while Dennis (1981) used a simple auditory recall and recognition test, both of which allow only limited inference about components of memory, such as acquisition, retrieval, and metamemory skills. Both Pirozzollo et al. (1979) and Dennis (1981) reported normal memory performance in their respective single cases. Such results, however, may be an artifact of using tests which provide only global memory scores.

The present study therefore sought to re-examine memory in cases of callosal pathology. Four cases of callosal agenesis were verified by magnetic resonance imaging (MRI) and examined with the Auditory Verbal Learning Test (AVLT; Lezak, 1976; Rey, 1941, 1964). Of the four cases, two had relatively normal intelligence and two had below average (performance scale) intelligence. Unlike many previous assessments of memory in acallosals, the AVLT measures several different aspects of memory performance and has been shown to be sensitive to various memory problems in different patient populations (Butters, 1985; Delis et al., 1987; Geffen et al., 1990; Mungas, 1983; see review by Peaker and Stewart, 1989). In particular, the AVLT allows assessment of acquisition, the effect of pro- and retroactive interference, the effect of serial position of presentation, delayed recall, recognition, and metamemory. Such sub-processes of memory may be differentially affected by callosal pathology.

METHOD

Subjects

Four subjects with agenesis of the corpus callosum verified by MRI were tested. The results of the acallosal subjects were compared with normative data for adults (Geffen et al., 1990) and children (Forrester and Geffen, 1991), matched for age and sex. The case histories for the acallosal subjects are described below and their radiological evidence, intelligence test results, handedness and age at testing are shown in Table 1.

PH. PH (born 7/5/54) has been described previously (Jeeves, 1978; Geffen et al., 1980). As shown in Figure 1, MRI at age 36 showed complete agenesis of the corpus callosum.

The anterior commissure was reported as present, seen as a signal intense structure in the region of the lamina terminalis. A hippocampal commissure could not be visualized. A small amount of lipomatous material was also present in the midline between the frontal lobes antero-superiorly in relation to the falx. There were also abnormalities of the ventricular system, and at the brain stem level. The lateral ventricles were enlarged, the 3rd ventricle was poorly formed, and a Chiari type I malformation could be seen at brain stem level. Furthermore, the A1 and A2 segments of the anterior cerebral arteries were displaced posteriorly and were vertical in position. A small, fluid filled cleft in the right frontal lobe could also be visualized, presumably being a consequence of ventriculography at age 6.

Table 1. Chief clinical characteristics for acallosal cases.

Case	Age at Testing	Extent of Agenesis	VIQ	PIQ	Handedness
PH[1]	37	Complete	92	69	Right
AS[2]	22	Partial	83	77	Left
MR[3]	12,14	Complete	94	82	Left
DC[3]	8,10	Complete	112	108	Right

[1] Tested at age 27 on WAIS
[2] Tested on WAIS-R
[3] Tested on WISC-R

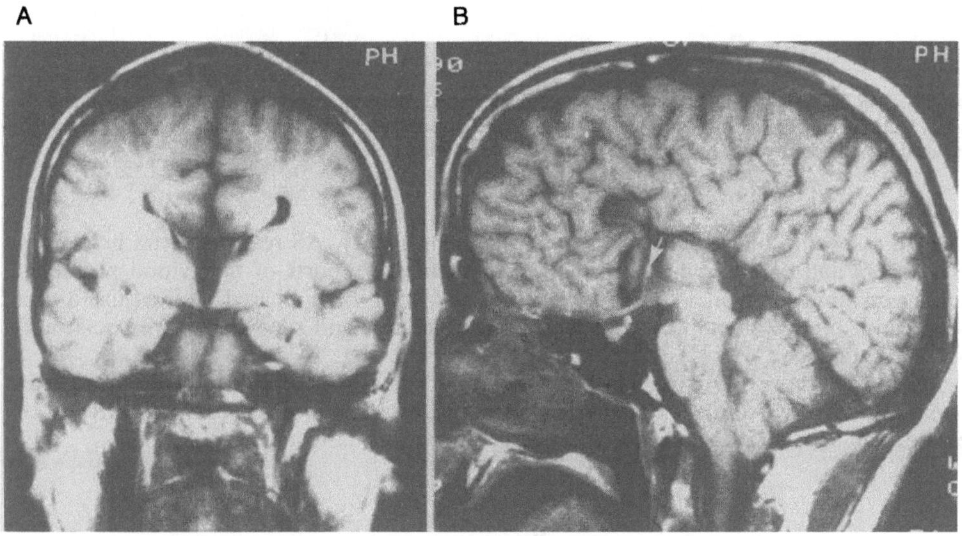

Figure 1. MR T_1 weighted images of PH: **A.** coronal section indicating absence of corpus callosum, and **B.** sagittal section (the white arrow indicates the anterior commissure).

AS. AS (born 20/8/68) was referred at age 12 years, as a result of learning difficulties and associated behavioral problems. An MR scan at age 22 indicated *partial* absence of the corpus callosum.

The genu or anterior corpus callosum was visualized but the body and splenium were absent. The hippocampal commissures also appeared to be deficient. The anterior commissure, however, might be represented by the thickened lamina terminalis. There were also a number of ventricular abnormalities: (a) the lateral ventricles showed wide separation of the anterior horns (with Probst bundles following the medial walls) and dilation of the occipital horns, and (b) the third ventricle was "high riding" at the site of the deficient body of corpus callosum. With respect to the cerebral hemispheres, the gyri along the medial aspect were radially arranged and oriented more vertically than normal. Furthermore, the left cerebral hemisphere appeared smaller than the right, and the left frontal region showed an abnormal gyral pattern and heterotopic grey matter. The poorly formed grey matter suggested a disorder of neuronal migration to this region. Finally, at brain stem level, a Chiari type I malformation could be visualized.

A B

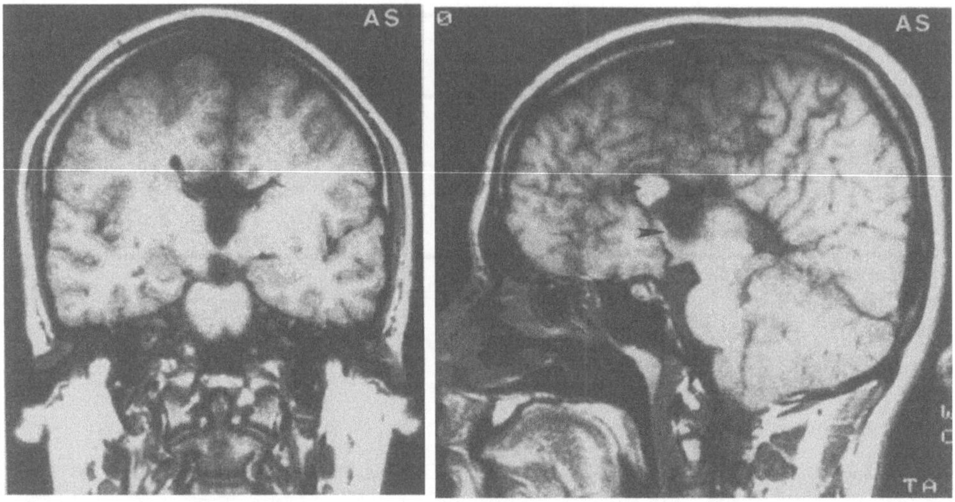

Figure 2. MR T_1 weighted images of AS: **A.** coronal section showing absence of corpus callosum, and **B.** sagittal section (the black arrow indicates the genu of the corpus callosum and the small arrow indicates the thickened lamina terminalis which could be the anterior commissure).

With respect to his educational and employment history, AS required extra help from his teachers and parents while at school, and there are reports of behavioral problems such as irritability. Nonetheless, at the time of testing he appeared well adjusted and co-operative. He was employed as a tire manufacturer, and reported active social and sporting interests.

MR. MR (born 10/6/76) has been monitored since birth, at which his head was abnormally large (37 cm). Computerized tomography (CT) at nine years and MRI at 13 years of age showed the complete absence of the corpus callosum (Figure 3).

A B

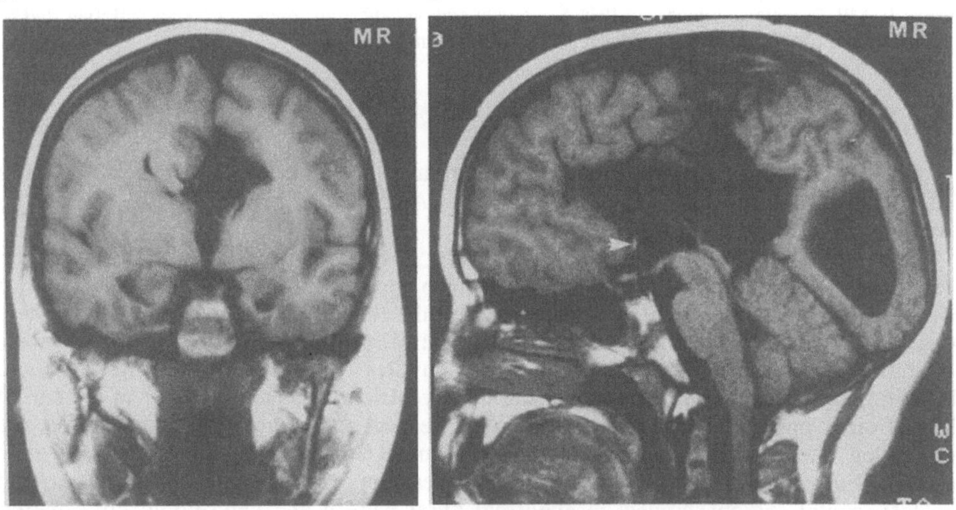

Figure 3. MR T_1 weighted images of MR: **A.** coronal section showing absence of corpus callosum and **B.** sagittal section (the white arrow indicates the anterior commissure).

The anterior commissure could be visualized as a signal intense structure in the region of the lamina terminalis. In contrast, a hippocampal commissure could not be imaged. The ventricular system was notably abnormal, with upturned lateral angles and severe atrial dilation of the lateral ventricles with thinning of the occipital cortex and extensive cystic upward extension of the 3rd ventricle, particularly on the left side of the falx to the level of the cingular sulcus. Probst's bundles were also probably demonstrated.

Earlier investigations at various ages also suggested posterior cerebral disorders. CT at nine months showed considerable thinning of the left occipital cortex. Electroencephalograms (EEGs) at two months and at eight years respectively indicated an excess of delta waves in the posterior regions, and an excess of sharp wave activity in the right post frontal to central regions.

At 8 years of age this boy was coping well with school work, with excellent reading, factual memory and calculation skills but with signs of diminished ability for abstract thought and a reluctance to attempt creative thinking. Bimanual skills were weak and eye hand coordination lagging in what was otherwise normal development. He writes with his left hand. MR is now 14 years old. He has some problems interacting with his parents who report behaviors that are not uncharacteristic of adolescent males such as untidiness and unwillingness to do chores.

DC. DC (born 27/2/81) had his first radiological examination (ultrasound) as a neonate, following an observation of large head size. Ultrasound revealed agenesis of the corpus callosum that appeared complete. At birth, he had a convergent squint, associated with bilateral VIth nerve weakness. This was subsequently treated by surgery. He walked and began to say a few words at 13 months. He was communicating in sentences by two years and knew his colors by four years. His speech was fluent at five years, though he displayed some dyslalia. His capacity to draw constructively was poor for his age.

At the age of ten years, MRI confirmed complete absence of the corpus callosum (Figure 4).

A somewhat shrunken anterior commissure was identified as a signal-intense structure seen in the T_1 sagittal sections, although the anterior commissure could not be visualized in the horizontal sections. The lateral ventricles showed widely

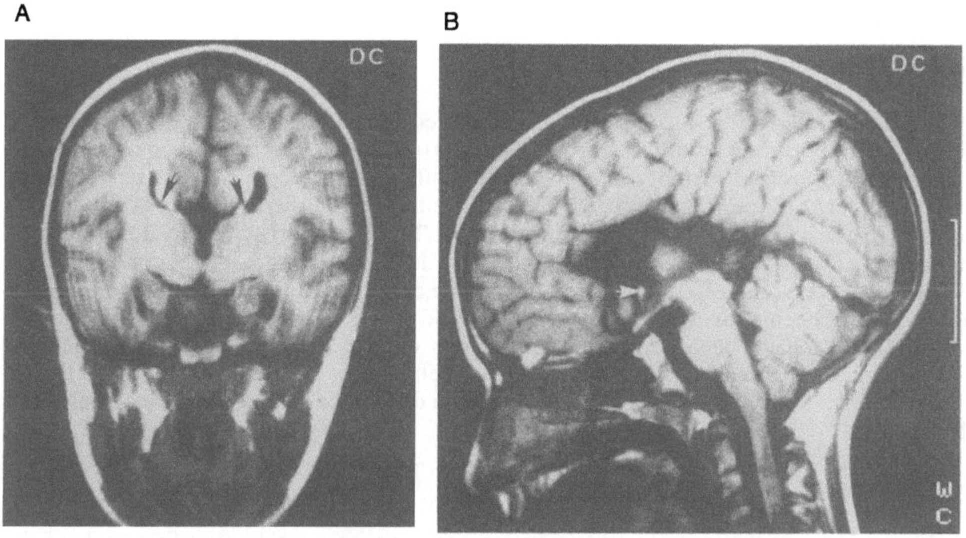

Figure 4. MR T_1 weighted images of DC: **A.** coronal section showing absence of corpus callosum (the black arrows indicate Probst's bundles and the fornices and **B.** sagittal section (the white arrow indicates the anterior commissure).

separated anterior horns and markedly dilated occipital horns, the 3rd ventricle demonstrated superior prolongation and the temporal horns were mildly prominent; the hippocampal formations were present. Probst's bundles were seen along the medial walls of the anterior horns; it seems likely tat these were associated with the fornices. The cerebral cortex also showed some abnormalities, including bilateral occipital thinning, more marked on the left, and radial and vertical arrangements of the sulci in the parieto-occipital region. Nonetheless, there was no visible neuronal migration disorder, i.e. pachygyria or heterotopic grey matter.

Procedure

The AVLT was administered according to the Lezak (1983) format and the results were scored using a specially written program. The Lezak (1983) format, including a twenty minute delay between Trials 7 and 8, was also used for the establishment of normative data for adults (Geffen et al., 1990) and children (Forrester and Geffen, 1991). PH and AS were tested once at age 37 and 22 respectively. The other acallosal subjects, however, were tested twice over a period of two years: MR at ages 12 and 14 and DC at ages 8 and 10. For MR and DC, AVLT Form 1 was used on the first occasion, while the alternate Form 3 was used at the second session. Forms 1 and 3 have been shown to be equivalent (Geffen et al., unpublished manuscript).

Handedness was determined by the six primary items of the Annett (1970) questionnaire. Three subjects were administered a general intelligence test, either the Wechsler Intelligence Scale for Children-Revised (WISC-R, Wechsler, 1974) or the Wechsler Adult Intelligence Scale-Revised (WAIS-R, Wechsler, 1981). PH refused an intelligence test and his results at age 27 on the WAIS are reported (Table 1).

RESULTS

Scoring of the AVLT provided measures for each trial and a number of derived indices for verbal memory processes (see Geffen et al., 1990). The results of the acallosals were compared against normative data and reported as outside normal limits if they varied from the norms by more than one standard deviation (SD). Before considering the memory findings, however, reference to the intelligence assessment is necessary.

General Intelligence

Verbal and performance IQ scores are presented for all acallosal subjects in Table 1. AS is the only subject with a Verbal IQ (VIQ) below the normal range. Thus, in examining AS' verbal memory data (below), it should be realized that his performance may be affected by more generalized verbal deficits. (It will be seen, however, that AS actually showed minimal difficulties on the AVLT). By contrast with VIQ scores, all subjects except DC had Performance IQs (PIQs) in the low range. Although deficits in PIQ might have some effect on AVLT performance, this effect is probably small since only VIQ has been shown to correlate with AVLT scores (Geffen et al., 1990). Indeed, results for MR have been compared against two control groups: a group matched for chronological age and a group matched for mental age (as determined by WISC-R performance). Such a comparison made very little difference to the significance values of the findings.

Acquisition

Figure 5 shows the learning or acquisition performance over the first five trials of the AVLT for the acallosal subjects and their respective control groups.

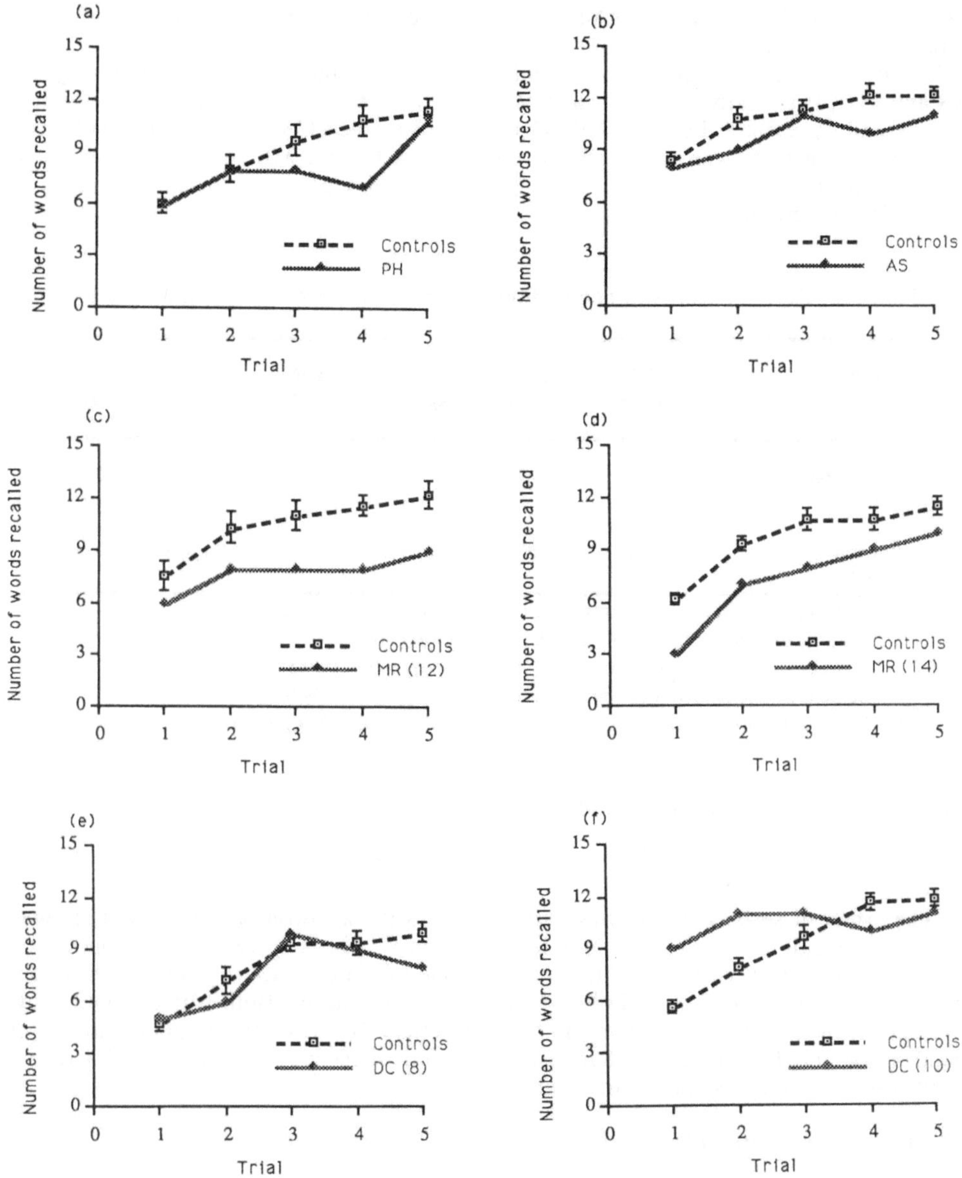

Figure 5. Number of words recalled in the five learning trials for the acallosal subjects and normative groups.

Both PH and AS showed relatively normal performance in recalling the words presented in the first trial and in total recall over the first five trials (recall was within one SD of the normative group for both measures). MR and DC, however, showed a poor level of initial recall (excepting a first trial score of 6 for MR at age 12), and a deficient level of total recall over trials 1-5. Table 2 extends these observations by showing two derived indices: (1) the rate of acquisition over trials 1-5, and (2) the ratio of the words recalled in trial 1 against the Digit Span forward score (from the Wechsler IQ sub-test). The latter measure indexes the effect of overloading memory with 15 items of information (as in the AVLT word list), compared with the small number of items in recalling digits.

The overall rate of acquisition was within normal limits for the two adult subjects (AS and PH), but abnormal for DC and MR. The effect of overloading memory with a long verbally presented list (the AVLT) compared to the effect of a smaller such list (the Digit Span test) was also variable: DC at age 10 and MR at age 14 showed a deficit in information overload, while AS in fact showed superior performance compared to his reference group.

Table 2. Derived scores of acquisition performance for all acallosal subjects.

Index	DC Age 8	DC Age 10	MR Age 12	MR Age 14	AS Age 22	PH Age 37
Acquisition Rate (Trials 1-5)[1]	0.90*	0.30***	0.60*	1.60*	0.70	0.90
Information Overload[2]	0.71	1.29*	0.75	0.38**	1.60+	0.86

1 Slope of the least squares regression line of best fit for words recalled in trials 1-5.
2 Number of words recalled in trial 8/Digit Span forward.
*/+ Score is inferior/superior to normative group mean by 1.00 - 1.99 SD.
**/++ Score is inferior/superior to normative group mean by 2.00 - 2.99 SD.
***/+++ Score is inferior/superior to normative group mean by 3.00 SD or greater.

Effect of Serial Position on Recall

Figure 6 shows recall over trials 1-5 as a function of serial position in the presentation list.

The effect of serial position was variable across the acallosal subjects. PH showed very good recall of the earlier and later presented words (effects of primacy and recency), with very poor recall of the middle group of words. AS, in contrast, showed a recency effect more than one SD *below* his normative group. Both MR and DC show reversals in their performance over the two year intervals. MR showed poor primacy of recall at age 12, but at age 14 demonstrated deficient recency of recall. DC's recent recall was poor at age 8 but enhanced at age 10.

Effect of Interference of Recall

Table 3 shows raw and derived scores for the effect of presenting the new (distractor) list of words at trial 6.

As with the effect of serial position, interference effects varied considerably across the acallosal cases. Only AS showed normal recall after previous learning (proactive interference) and normal recall of the previous list following new learning (retroactive interference). MR and DC showed relatively normal performance, except that MR was sensitive to retroactive interference at age 12, and DC was sensitive to proactive interference at age 10. (MR and DC actually showed *facilitation,* or less interference, at two points. MR showed superior proactive recall at age 14 and DC showed enhanced retroactive recall at age 8).

PH was the most sensitive to interference effects. Both his pro- and retroactive interference indices were more than one SD below the normative group.

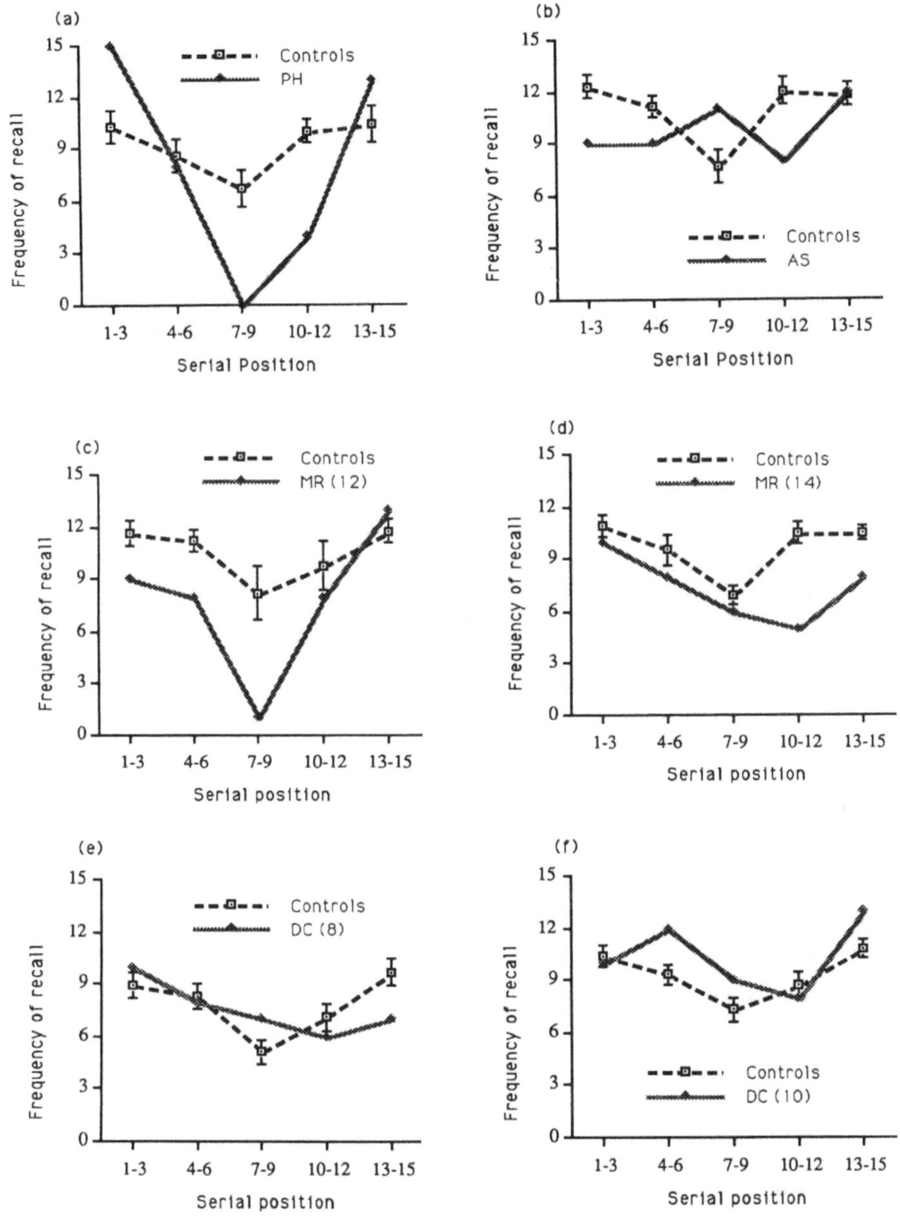

Figure 6. Frequency of word recall over trials 1-5 according to serial position for the acallosal subjects and normative groups.

Delayed Recall and Recognition

Table 4 shows measures of recall and recognition after the 20 minute delay for the acallosal subjects. Three derived scores were obtained: (a) a Forgetting index (a ratio of the words recalled before and after the delay), (b) Signal Detection Measures of Recognition (recognition scores corrected according to the number of false positive

Table 3. Raw and derived scores on interference trials for all acallosal subjects.

Score	DC Age 8	DC Age 10	MR Age 12	MR Age 14	AS Age 22	PH Age 37
<u>Words Correct</u> (Raw Scores)						
Trial 6 (Distractor list)	4	3*	6	7	5	3*
Trial 7 (Retention list)	10+	9	3**	9	11	6*
<u>**Derived Scores**</u>						
Proactive Interference (Trial 6/Trial 1)	0.80	0.33**	1.00	2.33++	0.62	0.50*
Retroactive Interference (Trial 7/Trial 8)	1.25++	0.82	0.33**	0.90	1.00	0.55***

*/+ Score is inferior/superior to normative group mean by 1.00 - 1.99 SD.
**/++ Score is inferior/superior to normative group mean by 2.00 - 2.99 SD.
***/+++ Score is inferior/superior to normative group mean by 3.00 SD or greater.

Table 4. Raw and derived scores on delayed recall and recognition trials for all acallosal subjects.

Score	DC Age 8	DC Age 10	MR Age 12	MR Age 14	AS Age 22	PH Age 37
<u>Words Correct</u> <u>(Raw Scores)</u>						
Trial 8 (Delayed recall)	4**	8*	4**	8	10	6*
Trial 9 (Recognition)						
Original list	11***	14	14	15+	15+	8***
Distractor list	6	6	5*	8	9	3
Number of false positives	1	2	5*	5	2	4*
<u>**Derived Scores**</u>						
Forgetting (Trial 8/Trial 7)	0.40***	0.89*	1.33	0.89*	0.91	1
Signal Detection Measures of Recognition[1]						
p(A) Original list	0.84***	0.92	0.87*	0.95	0.95+	0.69***
p(A) Distractor list	0.70	0.70	0.64*	0.69	0.80	0.57***
Retrieval Efficiency[2]	0.32*	0.58*	0.31**	0.56*	0.70	0.58*

1 p(A) = 0.5 (1 + hit rate - false positive rate) (see Forrester and Geffen, 1990).
2 Proportion words correct trial 8 / p(A).
*/+ Score is inferior/superior to normative group mean by 1.00 - 1.99 SD.
**/++ Score is inferior/superior to normative group mean by 2.00 - 2.99 SD.
***/+++ Score is inferior/superior to normative group mean by 3.00 SD or greater.

identifications: see Green and Swets, 1961), and (c) an index of Retrieval Efficiency (the ratio of words recalled after the delay to corrected recognition).

All subjects (except AS) recalled relatively fewer words after the delay than their control groups. Moreover, DC and MR showed some forgetting of words over the delay period. In the case of PH, however, no forgetting was seen over the break, since he had sub-normal recall *before the delay* period. Three subjects *recognized* a high proportion of the original words after the delay. PH was the exception. With relatively poor delayed recall and high recognition, all subjects (excluding AS) demonstrated a sub-normal Retrieval Efficiency. Such a result suggests that the information was registered (as evidenced by the good level of recognition), but that the information was ineffectively retrieved from memory.

Metamemory

Scores reflecting the ability to self-monitor recall and recognition are shown in Table 5.

The monitoring of recall appeared to be generally within normal limits with normal levels of repeated and intruded words. The exceptions here were high levels of extra-list intrusions by DC (age 8), MR (age 14) and AS. By contrast, the monitoring of recognition processes was relatively deficient in at least two subjects. Both PH and MR (at ages 10 and 12) showed relatively large numbers of false positive identifications (see Table 4) and misassignments of words between lists. AS and DC (at ages 8 and 10), on the other hand, were within normal limits in making false identifications and misassigning words during recognition.

Table 5. Metamemory measures for all acallosal subjets.

Score	DC Age 8	DC Age 10	MR Age 12	MR Age 14	AS Age 22	PH Age 37
<u>Recall</u>						
Repeated Words (Trials 1-5)	6	4	0	0+	0+	2
Extra-list Intrusions (Trials 1-5)	3*	1	5	2**	5***	0
Between-list Intrusions						
A -> B[1]	0	0	0	0	0	2**
A -> B[1]	0	0	0	0	0	0
<u>Recognition</u>						
Total Misassignments	2	0	4***	3*	2	6***

1 Between list intrusions
*/+ Score is inferior/superior to normative group mean by 1.00 - 1.99 SD.
**/++ Score is inferior/superior to normative group mean by 2.00 - 2.99 SD.
***/+++ Score is inferior/superior to normative group mean by 3.00 SD or greater.

DISCUSSION

Auditory memory was examined in four cases of callosal agenesis using the AVLT. PH, an adult with complete agenesis, showed high sensitivity to pro- and retroactive interference, poor self-monitoring of recognition processes and an apparent deficit in retrieval (i.e. poor delayed recall with high recognition). AS, another adult but with partial callosal agenesis, was normal in most respects, except for some difficulty in recalling recently presented words. AS' normal memory function is particularly striking in view of the fact that he has the lowest VIQ. MR and DC, two children with complete agenesis, consistently demonstrated (a) a poor level of acquisition (but normal rate of learning), (b) poor monitoring of recognition, and (c) inefficient memory retrieval. Deficits in PIQ seem unlikely to have significant effects on AVLT performance. This view is supported by the similarities in AVLT findings for MR (who has a low PIQ) and DC (who has a normal PIQ). In sum, the only consistent deficit in subjects with *complete* callosal agenesis was poor delayed recall combined with relatively high recognition (i.e. poor retrieval).

The above findings may extend previous observations of poor memory in subjects lacking a corpus callosum (see reviews by Clark and Geffen, 1989; Zaidel, 1990), since they might specify the deficit as one of deficient retrieval. A problem in retrieval may arise in one of two ways; through an impairment in retrieval processes themselves, or through weak processes of acquisition. The second account does not suggest that memories are not formed; it suggests that memory traces are poorly consolidated. These weaker memory traces would make retrieval inefficient. Indeed, the traces may be adequate for recognition but insufficient for the more demanding process of free recall.

At a neural level, these problems might arise in four main ways. Firstly, without the corpus callosum to transfer stimuli efficiently between the hemispheres, the memory trace may be consolidated over a restricted and largely unilateral area of cerebral cortex. Secondly, the corpus callosum has been said to *inhibit* sensory information from the ipsilateral hemisphere reaching the contralateral hemisphere (see review by Cook, 1986), and also to *facilitate* processing within each hemisphere (Lassonde, 1986). If these effects do occur, then an absent corpus callosum may result in "noisy" processing within each hemisphere. Poor intrahemispheric processing might produce poor consolidation or faulty retrieval. A third explanation would emphasize the apparent absence of the commissure joining the hippocampi in all acallosal subjects. The hippocampi are well implicated in memory consolidation (Mishkin and Appenzeller, 1987; Squire, 1987), and their joining commissure may assist in establishing a stronger bihemispheric memory trace. A final account of the present data may be in terms of extra-commissural dysfunction. All subjects have ventricular abnormalities (which may cause re-alignment of cerebral fibers), and structural deficiencies in other parts of the cerebrum (particularly in MR). Such abnormalities might produce generalized impairments that in turn cause problems in retrieval (e.g. deficits in sustaining attention.)

Of the above explanations, the arguments that refer to extra-commissural dysfunction and the absent hippocampal commissure can be downplayed. Subject AS showed almost no memory deficits (excepting an absence of a recency effect in recall), and yet his MRI showed a number of other abnormalities (i.e. ventricular enlargement, abnormal gyri and heterotopic grey matter in the left frontal region, a Chiari malformation, and no hippocampal commissure). If such pathology disrupted memory dramatically, AS might be expected to show more pervasive deficits. The lack of major verbal memory impairments in AS may be due to the fact that the pathology has been present since birth, which in turn has allowed the developing brain to make compensatory adjustments.

AS' data may also be used to *support* an explanation in terms of callosal function. AS differs from the other subjects in that he has a partial corpus callosum (the genu is intact): the lack of pervasive problems on the AVLT may suggest that problems arise when agenesis is complete, as in the other cases. Thus, the explanations above that refer to callosal function may be the most applicable accounts of the data. Indeed, the genu might be critical for auditory verbal memory processes (consolidation and/or retrieval). The latter argument must be qualified, however. Fibers from other parts of the corpus callosum (which may be as critical as those in the genu) may have been rerouted through the anterior commissure and/or genu during the development of AS' brain.

CONCLUSION

Thus, there is some basis on which to argue that the corpus callosum plays an indirect but important role in verbal memory processes, particularly those necessary for efficient retrieval. However, further investigation is needed, using two sets of cases: (a) those with minimal extra pathology and with the hippocampal commissure intact (see Zaidel, 1990), and (b) those with congenital abnormalities that do not involve the corpus callosum. Such cases may allow double dissociation of any callosal memory functions. Further, an examination of visual-spatial memory, with specific comparison of visual recall and recognition, may extend the finding of retrieval deficits to other types of memory. It is also desirable that the reliability of MRI diagnosis should be validated eventually by autopsy and full neuroanatomical examination. MRI far surpasses earlier methods of in vivo diagnosis of callosal agenesis, but cannot exclude re-routing of commissural fibers, e.g. through the anterior commissure, nor can it visualize histological abnormalities in apparently intact structures such as the hippocampus and mammillary bodies.

In conclusion, three cases of complete agenesis of the corpus callosum demonstrated poor recall but good recognition of auditorily presented words (i.e. poor memory retrieval). A case with incomplete agenesis showed minimal problems in verbal memory processes, as assessed by the AVLT. The memory problems in complete agenesis might therefore be related to roles of the corpus callosum in transferring, inhibiting or facilitating information flow between or within the hemispheres; in particular, that information used for forming or retrieving memories.

Acknowledgments

Grants from the Australian Research Council (no. A788 3211) and from the Julia Farr Centre supported this project.

We are grateful to Drs Lloyd Morris and Gary Le Quesne for their reports on the MRIs of the subjects, and to Dr. J. Nilsson for facilitating access to her clients.

REFERENCES

Aicardi, J., Chevrie, J.J., and Baraton, 1987, Agenesis of the corpus callosum, *in:* "Handbook of Clinical Neurology. Malformations," Vol. 6(5), N. C. Myrianthopoulos, ed, Elsevier, North Holland, Amsterdam.

Annett, M., 1970, A classification of hand preference by association analysis, *Br. J. Psychol.* 61:303-321.

Bigler, E. D., Rosenstein, E. D., Roman, M., and Nussbaum, N. L., 1988, The clinical significance of congenital agenesis of the corpus callosum, *Arch. Clin. Neuropsychol.* 3:189-200.

Butters, N., 1985, Alcoholic Korsakoff's syndrome, *J. Exp. Clin. Neuropsychol.* 7:181-210.

Clark, C. R. and Geffen, G. M., 1989, Corpus callosum section and recent memory, *Brain* 112:165-175.

Cook, N. D., 1986, "The Brain Code: Mechanisms of Information Transfer and the Role of the Corpus Callosum," Methuen, London.

Delis, D. C., Kramer, J. H., Kaplan, E., and Ober, B. A., 1987, "California Verbal Learning Test: Adult Version. Manual, " Psychological Corporation and Harcourt Brace Jovanovich, Toronto.

Dennis, M., 1981, Language in a congenitally acallosal brain, *Brain Lang.* 12:33-53.

Ferriss, G. S. and Dorsen, M. M., 1975, Agenesis of the corpus callosum: 1. Neuropsychological studies, *Cortex* 11:95-122.

Forrester, G. and Geffen, G. M., 1991, Performance measures of 7- to 15-year-old children on the Auditory Verbal Learning Test, *Clin. Neuropsychologist* 5:345-359.

Geffen, G. M., Forrester, G., and Butterworth, P., unpublished manuscript, The persistence of verbal memory deficits after closed head injury: Validity of the Auditory Verbal Learning Test and a new form test-retest reliability and alternate form equivalence.

Geffen, G. M., Moar, K. J., O'Hanlon, A. P., Clark, C. R., and Geffen, L. B., 1990, Performance measures of 16- to 86-year-old males and females on the Auditory Verbal Learning Test, *Clin. Neuropsychologist* 4(1): 45-63.

Geffen, G., Walsh, A., Simpson, D., and Jeeves, M., 1980, Comparison of the effects of transcortical and transcallosal removal of intraventricular tumors, *Brain* 103:773-788.

Green, D. M. and Swets, T. A., 1966, "Signal Detection Theory and Psychophysics, " Wiley, New York.

Jeeves, M. A., 1978, Some limits to interhemispheric integration in cases of callosal agenesis and partial commissurotomy, *in:* "Structure and Function of the Cerebral Commissures," I. S. Russell, M. van Hof, and G. Berlucchi, eds, Macmillan, New York.

Jeeves, M. A., 1990, Agenesis of the corpus callosum, *in:* "Handbook of Neuropsychology," Vol. 4, F. Boller and J. Grafman, eds, Elsevier, North Holland, Amsterdam.

Lassonde, M., 1986, The facilitatory influence of the corpus callosum on intrahemispheric processing, *in:* "Two Hemispheres - one Brain: Functions of the Corpus Callosum," F. Lepore, M. Ptito, and H. H. Jasper, eds., Alan R. Liss, New York.

Lezak, M. D., 1976, "Neuropsychological Assessment," Oxford University Press, New York.

Lezak, M. D., 1983, "Neuropsychological Assessment," 2nd Edition, Oxford University Press, New York.

Mishkin, M. and Appenzeller, T., 1987, The anatomy of memory, *Sci. Amer.* 256:62-71.

Mungas, D., 1983, Differential clinical sensitivity of specific parameters of the Rey Auditory Verbal Learning Test, *J. Consult. Clin. Psychol.* 51:848-855.

Peaker, A., and Stewart, L. E., 1989, Rey's Auditory Verbal Learning Test - A review, *in:* "Developments in Clinical and Experimental Neuropsychology," J. R. Crawford and D. M. Parker, eds., Plenum, New York.

Pirozzolo, F. J., Pirozzolo, P. H, and Ziman, R. B., 1979, Neuropsychological assessment of callosal agenesis: Report of a case with normal intelligence and absence of the disconnexion syndrome, *Clin. Neuropsychol.* 13-16.

Rey, A., 1941, L'examen psychologique dans les cas d'encéphalopathie traumatique, *Arch. Psychol.* 28:286-340.

Rey, A., 1964, "L'Examen Clinique en Psychologie," Presses Universitaires de France, Paris.

Squire, L.R., 1987, "Memory and Brain," Oxford University Press, New York.

Wechsler, D., 1974, "Wechsler Intelligence Scale for Children-Revised Manual," Psychological Corporation and Harcourt Brace Jovanovich, New York.

Wechsler, D., 1981, "Wechsler Adult Intelligence Scale - Revised Manual," Psychological Corporation, New York.

Zaidel, D.W., 1990, Memory and spatial cognition following commissurotomy, *in:* "Handbook of Neuropsychology," Vol. 4, F. Boller and J. Grafman, eds., Elsevier, North Holland, Amsterdam.

SOUNDS AND SHAPES: LANGUAGE AND SPATIAL COGNITION IN CALLOSAL AGENESIS

Christine Temple and Joanne Ilsley

Developmental Neuropsychology Unit
Department of Psychology
University of Essex
Wivenhoe Park
Colchester C04 3SQ, U.K.

INTRODUCTION

Our current understanding of the functions of the corpus callosum in man derives in part from the studies of the commissurotomy patients, in whom the corpus callosum had been sectioned in adulthood as a treatment for intractable epilepsy.

Whilst commissurotomy patients provide evidence about the role of the corpus callosum during active cognitive processing, children with congenital absence of the corpus callosum provide evidence about its role in the development of cognitive processes. In particular, there is interest in the potential limits to the functional plasticity and compensation which accompany callosal absence. Consistent deficits across acallosal children may relate to functions for which the corpus callosum is critical and for which other brain systems cannot substitute. This issue will be discussed here in relation to the development of general cognitive functions rather than the development of inter-hemispheric transfer.

PREVIOUS STUDIES

Intelligence

The studies discussed here look at children with callosal agenesis who are of normal intelligence and investigate specific impairments which are out of line with the children's other intellectual abilities. Whilst the early neurological literature is scattered with cases of callosal agenesis combined with severe learning difficulties and mental retardation, it is now known that some children with callosal agenesis may have normal intellectual profiles on intelligence tests. This can be illustrated by looking at performance on the Wechsler Intelligence Scale which is composed of several sub-tests: verbal and non-verbal. Each sub-test has an average score of 10, with a standard deviation of 3; scores between 7 and 13 are entirely normal. The sub-test scores for one of our acallosal children S.B. are as follows:

Callosal Agenesis, Edited by M. Lassonde and
M.A. Jeeves, Plenum Press, New York, 1994

Verbal Scale		Performance Scale	
Information	12	Picture Completion	11
Similarities	12	Picture Arrangement	9
Arithmetic	10	Block Design	11
Vocabulary	9	Object Assembly	11

S.B. has a normal profile with little variation across sub-tests. The subtest scores correspond to an overall Verbal I.Q. of 105 and a Performance I.Q. of 104. A second acallosal child, K.W., displays a similar type of profile:

Verbal Scale		Performance Scale	
Information	12	Picture completion	14
Similarities	11	Picture Arrangements	8
Arithmetic	9	Block Design	12
Vocabulary	10	Object Assembly	11
Verbal I.Q.	103	Performance I.Q.	106

The neurological literature indicates that callosal agenesis often co-occurs with other neurological abnormalities and in such cases of callosal agenesis, severe and widespread learning difficulties may result from the associated abnormalities rather than from callosal agenesis *per se*. The intellectual profiles of S.B. and K.W. indicate that callosal agenesis is compatible with normal intelligence. The issue then becomes whether, even in those of normal intelligence, there are other more specific cognitive impairments which may emerge from fine grain analysis of cognitive development.

Language

Recent research has explored the possibility of specific language deficits in children and adults with callosal agenesis. Dennis (1981) suggested that there might be a specific deficit in syntactic function in callosal agenesis following her investigation of a case study. She concluded that there was "failure to use or understand the syntactic-pragmatic component of language." A reanalysis of some of the data from that study led to the conclusion that in addition to the syntactic problems in Dennis' patient there was also evidence of difficulty in certain phonological tasks that required explicit phonological analysis and the organization of speech sounds (Jeeves and Temple, 1987). An investigation of two further acallosal adults confirmed deficits in phonological language skills (Jeeves and Temple, 1987) but did not show any consistent syntactic impairment.

A subsequent exploration of similar skills on a broader range of tests targeted at the phonological deficit was conducted in children with callosal agenesis. These children also displayed specific impairments on tasks of explicit phonological processing and sound organization despite having normal levels of intelligence (Temple, Jeeves and Vilarroya, 1989).

There was no gross phonological impairment in our acallosal subjects. Their articulation of speech is clear and well formed. Their gross language skills are normal. Yet, for example, on rhyming tasks which require more explicit decomposition, analysis and synthesis of speech-related material, there are significant deficits in performance. These impairments in explicit phonological processing were in some ways unexpected and raised a series of questions. One reason that the results were unexpected is because phonological functions are normally considered to be subserved by the left hemisphere of the brain, and to be strongly lateralized. Interhemispheric cooperation is not typically considered to be crucial in the activation of lateralized functions. It is also known that in callosal agenesis there is a normal direction to lateralization of function, and the phonological difficulties cannot simply be attributed to an artifact of bilateralization of language. However, whilst the corpus callosum may not be necessary for the general development of lateralization, as had once been hypothesized (e.g.

Trevarthen, 1974; Selnes, 1974), it may nevertheless be involved in fine tuning the extremes of lateralization (Temple et al., 1989; Jeeves, this volume).

A further reason for interest in the observed phonological deficits related to the suggestion in the dyslexia literature that impairment in phonological processing could be causal in the onset of developmental dyslexia (e.g. Bradley and Bryant, 1978). This raised a question about the nature of literacy development in acallosal children. Specifically, if impairments in phonological processing are causal in developmental dyslexia, then since our acallosal children had phonological processing deficits they should also be dyslexic.

Another question that arose from our results was whether the phonological difficulties, in the children with callosal agenesis, were specifically related to the complex nature of the rhyming tasks involved, or whether there would also be impairments on phonological processing tasks employing simple speech sounds. This issue addresses the cognitive level at which the deficit is manifest.

A final question that arose from our results was whether very specific impairments such as the phonological processing deficit would only be found in the language domain or whether there would also be very specific impairments in visuo-spatial domains.

In summary, three questions will be addressed in this chapter. Do the deficits in phonological processing have an impact upon the development of reading and literacy? Are deficits in phonological processing also evident on tasks assessing the discrimination of speech sounds? Are specific cognitive impairments in callosal agenesis restricted to language or are they also seen in areas of spatial skill?

CURRENT STUDIES

Reading

The first of these questions was addressed in a study which investigated reading performance in acallosal subjects. It was found that despite the severe impairment on phonological language skills in the acallosals, their overall reading levels were normal. Reading ages were on a par with chronological ages (see Table 1). Since the overall word recognition skills were normal, the acallosal children could not be classified as dyslexic.

Table 1. Reading ages.

Subject	Chronological Age	Reading Age
S.B	7.1	7.1
S.B	8.1	8.5
S.B.	9.4	9.1
K.W.	13.2	13.4

However, when the nature of the reading performance was looked at in more detail, the pattern of reading was abnormal (Temple et al., 1990). There was a significant impairment in the ability to pronounce aloud non-words. These were letter strings with no meaning, which conformed to the orthographic rules of language, e.g. *gip, sutter*. There was also persistence of a lexicality effect with age. Normal young children are better at reading words than non-words. In the early stages of reading they have not mastered all the letter-sound rules which are used in reading non-words or unfamiliar words. As normal children get older, they master the letter-sound rules more effectively; the discrepancy between the ability to read words and non-words, the so-called lexicality effect, declines and disappears. However, as acallosal children get older the lexicality effect persists with continuing difficulty in reading non-words aloud.

Despite the non-word reading difficulty, there is no overall impairment in the reading levels in terms of word recognition. This is true even for the recognition of irregular words, i.e. words that do not conform to standard spelling-to-sound rules, e.g. *yacht,* which when read according to logical rules sounds like "yatched". For irregular words, it is essential to recognize word-specific information to identify the word as an individual. The acallosal children could read irregular words like *yacht* but had difficulty with non-word stimuli like *gip*.

In terms of current psycholinguistic jargon, the acallosals' phonological reading route is impaired in its development but the lexical reading route is normal. The difficulties with explicit phonological processing do not cause dyslexia but could nevertheless be causal in the impaired development of the phonological reading route. Spelling is similarly impaired and there is particular difficulty in spelling unfamiliar items where phonological rules would normally be of benefit. This seems to have a greater impact in the classroom and many of the children require remedial spelling help.

Phonemic discrimination

The second question, raised earlier, related to whether phonological problems would also be evident on tasks involving simple speech sounds. A recent investigation has explored the discrimination of speech sounds in callosal agenesis (Temple and Ilsley, in press). One task looked at simple phonemic discrimination. Pairs of phonemes were presented and the subject was asked to say whether or not they were the same sound or whether the sounds were different. They were required to make a simple judgment about whether or not the two phonemes were identical. When the phonemes differed, they did so by only one significant feature. The second task was Wepman's auditory discrimination task which is a standardized clinical test where pairs of words are spoken aloud to the subject. Some of the words are identical e.g. *lack* - *lack* and other words differ by one sound, e.g. *tub* - *tug*. The subject has to judge whether or not the pairs are identical. A further auditory discrimination task was used in which one of the sounds in each pair of words from the Wepman task was altered to produce a set of non-word pairs. These pairs match the Wepman pairs but as they are not words, they have no meaning e.g. *thumble* - *sumble*. The same response was required with the subject asked to make an identity judgment. A final task was developed by Snowling et al. (1986). Words (e.g. *eskimo*) and non-words (e.g. *istibo*) are spoken aloud and the subject is asked to repeat them.

The control data to which the acallosal data was compared were either normative data from the standard tests or control data collected from normal children. At least ten normal children matched for sex and age were tested for comparison with each acallosal subject. They were also tested for reading age, so that no child control had a reading impairment or a reading superiority. All had a reading age within six months of the acallosal subjects. Table 2 summarizes the tests for which deficits appear.

For the acallosal K.W., there was an impairment on all four tests, and for the acallosal S.B., there was an impairment on three of the tests. The third subject C.K. is an acallosal boy who is of lower intellectual level, with a Verbal I.Q. of 68 and a Performance I.Q. of 75. C.K.'s results illustrate that it is not necessarily the case that the acallosal subjects who are more limited in intellectual ability have the most severe deficits. C.K. was slightly less impaired on these tasks than the acallosal children of normal intelligence. Nevertheless, he was impaired and there were consistent deficits on two of the tasks.

It can be concluded from these studies of language function that there is a consistent deficit in phonological processing in children with callosal agenesis. More recent testing of further subjects confirms this picture. In terms of education, the impairment does not necessarily affect the overall development of reading in the sense of word recognition but may nevertheless have an impact on literacy development. The difficulty in phonological aspects of non-word reading and in spelling is not simply a feature of the complexity of the task since the phonological problems extend to the discrimination of simple speech sounds.

Table 2. Phonemic discrimination.

Tasks	S.B.	K.W.	C.K.
1. Phonemic Discrimination	–	X	–
2. Wepman's Auditory Discrimination	X	X	X
3. Non-word Discrimination	X	X	–
4. Snowling's Repetition	X	X	X

X=Significant impairment ($p<0.05$)

Spatial cognition

The final question raised earlier, related to whether selective deficits were restricted to language or could also be found in non-verbal domains. If the phonological difficulties reflect a particular problem with extremely lateralized tasks it might imply that extremely lateralized spatial tasks would also be affected. More generally, it is of theoretical interest to establish whether language is unusual in the fractionation (Foder, 1983) of its development with, for example, certain reading skills, such as word recognition, developing normally and others, such as phonological reading skills, developing abnormally.

Do visuo-spatial skills also fractionate in development? A range of tests of non-verbal functions were either selected from the neuropsychological literature or were novel modifications (see Table 3). The first tests which were selected cover two broad areas of visuo-spatial and visuo-perceptual functions. A conceptual distinction between visuo-spatial and visuo-perceptual functions is found in the cognitive psychological literature and also received substantiation from neuropsychological studies conducted with patients who have sustained focal missile injuries to the brain (Mehta et al., 1987). These indicate a distinct anatomical basis for the two different groups of tasks.

The two visuo-spatial tasks selected here were line orientation and mental rotation. In Benton's line orientation task (Benton et al., 1978) the subject is required to look at two lines of different angles or orientations and match them to lines from an array. The mental rotation task (Vandenburg and Kuse, 1978) presents two-dimensional pictures of three dimensional arrays of cubes. The subject is then presented with a set of further pictures and has to decide which pictures depict the same spatial array viewed from a different angle and which depict a completely different spatial array. In principle it requires the mental rotation of the abstract stimulus.

The visuo-perceptual tasks selected here included the Mooney faces (Landsell, 1968). In this task, the subject is presented with drawings of human faces which are partially obscured by shadows and highlights. They are told that the item is a face. The subject is then requested to judge whether it is the face of a boy or a girl, a young man or an old man, or a young woman or an old woman. This is a test of closure and it requires the integration of the patterns of black and white to form a gestalt of the face, in order that the appropriate judgments regarding age and sex can then be made. The second visuo-perceptual task employed was the Street (Street, 1931). This task also requires visual integrative skills, to identify the series of objects or animals which are represented by patterns of black and white. The third visuo-perceptual task was Benton's face recognition task (Benton et al., 1983). In this task the subject has to match faces that are unfamiliar. The pictures of the faces are partially obscured with

Table 3. Spatial tasks results.

Tasks	S.B.	M.J.	K.W.
Visuo-spatial			
1. Line Orientation	–	–	–
2. Mental Rotation	–	–	X
Visuo-perceptual			
3. Mooney Faces	–	X	–
4. The Street	–	X	–
5. Benton Faces	–	–	–
Visuo-spatio-motor			
6. Cornish Jigsaw	X	X	X
Tactile			
7. Tactile Line Orientation	–	–	–
8. Mazes	–	–	–
9. Formboard	X	X	X
Graphic			
10. Drawing	X	X	x

X=Impaired performance; -- =Normal performance; x= Normal performance but defect at earlier age

shadows and highlights. In the first part of the task, the subjects are shown a single front-view photograph of a face and must make a forced choice selection of the same face from a set of six. In later items the subject must match a target face to three other photographs of the same person viewed from different angles. This is not a memory task. When the subject is matching the faces they continue to have the initial face in view.

A further visuo-spatial task was also included: the Cornish Jigsaw. Here the subject had to carry out a visuo-spatial operation but there was a further motor constructional component. In this task, a set of black and white pieces is scattered in front of the subject, who is requested to put them together in order to construct an abstract shape. A picture of the target shape is also given. Although this is a constructional task there is no semantic content, in that the shape constructed is meaningless. It does not relate to the child's knowledge of the world.

Three tactile tasks were performed out of vision. The tactile line orientation task was constructed in the laboratory and was a direct parallel to Benton's visual line orientation task (Benton et al., 1978). The child feels raised lines made by pieces of wire and has to match them to those lines in an array below, which are of the same orientation. The tactile line orientation task has exactly the same principles as the visual line orientation task but is given out of vision, by the use of an arm screen or blindfold. A maze task was also conducted out of vision. The final tactile task was a formboard, in which wooden pieces had to be placed into comparable holes in a board. The shapes were constructed as far as possible to be abstract though some were more easily given verbal labels than others.

The last task in the spatial battery was drawing. The subject was requested to draw both a picture of a man and a picture of a bicycle.

This battery of spatial tasks has formed part of another study which has been addressing individual differences in normal performance (Temple and Cornish, under revision; Cornish and Temple, under revision). The first subject whose results will be presented was compared to this normal control data. S.B. was given the spatial battery at the age of ten (Temple and Ilsley, in preparation-a). She was compared to sixteen subjects, aged nine to eleven, who were matched to her for sex, handedness and familial sinistrality. In interpreting the acallosal results, our criteria for impaired performance were either a z-score associated with a $p<0.05$ or an acallosal score in comparison to which only one control subject performed more poorly, thereby placing the acallosal in the bottom percentile group.

On many of the tasks in our study of individual differences, significant sex differences are found in children of this age. Normal females perform significantly more poorly than normal males on the visuo-spatial, visuo-spatio-motor and tactile tasks (Temple and Cornish, under revision; Cornish and Temple, under revision). S.B. is compared to normal girls. Had she been compared to a sample of boys or a control sample mixed for sex, even more significant differences would have emerged. However the interest here is not in sex differences *per se* but in the degree to which S.B. differs from normal girls. S.B. performed entirely normally on the visuo-spatial tasks and the visuo-perceptual tasks but she was significantly impaired on three tasks: the Cornish Jigsaw task which is the spatial task with the constructional component; the formboard; and drawing.

The mean control score for completion of the Cornish Jigsaw was 259 seconds, with a standard deviation of 138. S.B. took 495 seconds to complete the task. So whereas controls took just over 4 minutes, S.B. took twice as long, taking 8 minutes to complete the task.

On the formboard task, S.B. used the right hand. After 17 minutes, she was unable to place any piece correctly. In comparison, the controls had placed all the pieces correctly after an average of 9 minutes and 33 seconds.

In drawing, the construction of the bicycle was poor with only 6 features indicated. The control average was 8.9 features (S.D.=1.9). S.B.'s score is at a borderline level of significance, with a probability of 0.06. However, only one of the sixteen control subjects performed more poorly than S.B., so she was in the lowest percentile grouping, and was therefore categorized as displaying impaired performance by our second criterion.

The next two acallosal children were tested at the age of nearly fourteen and just turned sixteen (Temple and Ilsley, in preparation-a). They were compared with control subjects in that age range. Twelve control boys were tested. The acallosal subjects were matched to the controls for sex, age and type of school. The result from the second acallosal, M.J., are similar to those of S.B. but in addition there is a visuo-perceptual impairment. M.J. had even more problems than S.B. on the Cornish Jigsaw. The controls at this age took an average of 96 seconds (approximately one and half minutes) to complete the Jigsaw. M.J. took 962 seconds (approximately 16 minutes). On the formboard only one of the twelve controls was slower than M.J., so he was in the bottom percentile group. Drawing was significantly impaired for both the man and the bicycle.

At the time M.J. was tested he was fourteen years old and Figure 1 depicts his drawing of a man at this age. In addition to the impairments on the Jigsaw, the formboard and drawing which M.J. shares with S.B., M.J. was also impaired on the closure tasks: the Street and the Mooney faces.

The third acallosal subject differed from the previous two acallosal subjects in that he was left-handed. This may have influenced the results. K.W. did not have visuo-perceptual problems like M.J. but was significantly impaired on the mental rotation task with only one control subject performing more poorly. The visuo-perceptual problems of M.J. and the mental rotation difficulties of K.W. illustrate the potential variation across the acallosal subjects. They can have a selective focal impairment within the spatial system which the other acallosal subjects do not necessarily share. These findings support a view of modularity in the development of spatial skills.

However, K.W., like the other acallosals, also has difficulty on both the Cornish Jigsaw and the formboard task. On the Jigsaw task K.W. is significantly impaired, taking 270 seconds in comparison to a control mean of 96 seconds. Initially when K.W. was tested on the formboard, the acallosal children were using the preferred hand to accomplish the task. For the two previous right-handed acallosal children the task was

Figure 1. Figure of a man drawn by M.J. (age 14).

carried out with the right hand. K.W. is left-handed and he, in contrast, was using the left hand. When K.W. performed with the left hand, he did poorly but he was not outside the range of normal children. However, when we asked him to perform the task with the right hand he was extremely impaired. So the acallosal subjects show consistent impairment on right hand attempts at the formboard.

It then seemed of interest to retest the previous acallosal subjects using the left hand. It was found that the left hand, even in these right-handed subjects was markedly better than the right hand at this task. This appears to be evidence of a disconnection effect. There is also significant failure of transfer of learning on this task.

Unlike the other acallosals, K.W. did not have deficient drawing (tested with the preferred left hand), at the time the spatial battery was given. However, when he had been seen several years prior to the current investigation, he did have impaired drawing, and this had been one reason for our original interest in drawing performance in acallosals. The nature of the impairment at the earlier age was of a specific sort. K.W. elongated and distorted the spatial dimensions of his drawings, so that everything was drawn with an extension either on the vertical or on the horizontal dimension. K.W. learned to recognize his difficulty and for several years worked systematically at correcting his errors. By the time the spatial battery was administered he had learned specifically to modify his drawings to more appropriate proportions.

In summary, all of the acallosals were deficient in certain sorts of visuo-constructional skills where it is required that the child coordinates a series of different movements in order to perform the spatial task. These are tasks that require complex execution of motor movements and the construction of spatial material.

Visual memory

Consistent with the results from the spatial battery there is also difficulty in copying tasks. Figure 2 illustrates S.B.'s attempt, at age seven, to copy the Figure of Rey (Osterrieth, 1952). The extreme disorganization is evident. Figure 2 also illustrates her recall of the pattern after 30 minutes. Whilst it is unsurprising that visual recall of the information is poor, since it had been very poorly analyzed and processed at the time of copying, the deficit is nevertheless very marked. Across the acallosal subjects a consistent impairment is found in the recall of complex visual information (Temple and Ilsley, in preparation-b). This can be illustrated with another acallosal subject. Figure 3 illustrates a copy of the same figure which is of a much better standard and was done by M.J. at the age of eleven. It is an impaired copy for a child of this age in comparison to

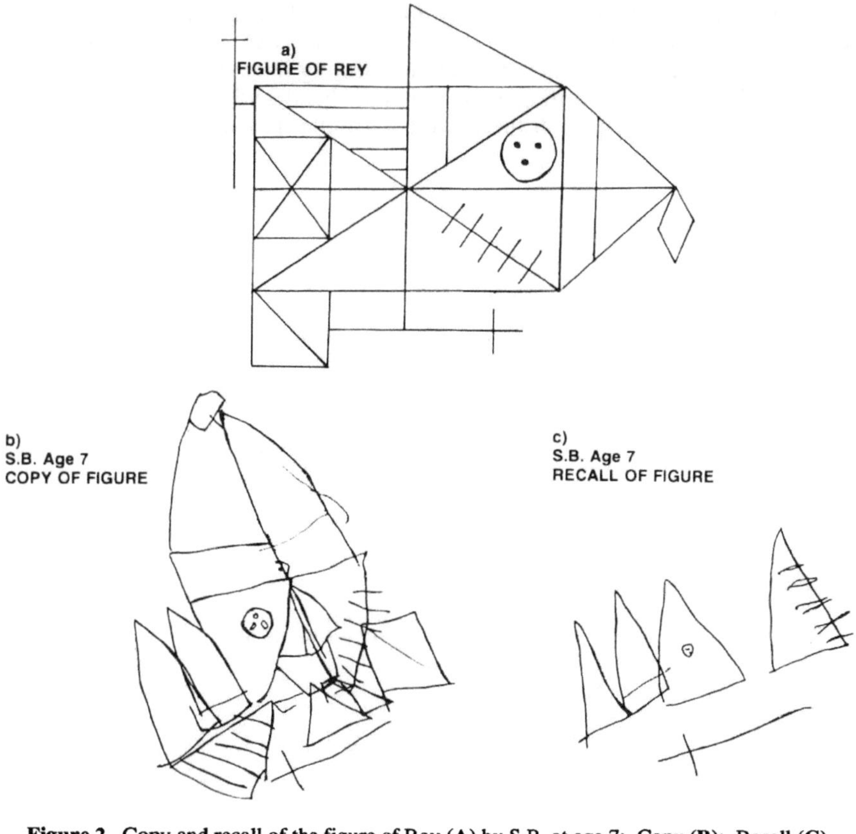

Figure 2. Copy and recall of the figure of Rey (**A**) by S.B. at age 7: Copy (**B**); Recall (**C**).

normative data but it is much less severely impaired than seen in the younger child S.B.. However, after a delay of only three minutes, only minimal information could be recalled (see Figure 3). Whilst there is consistent memory impairments with the recall of complex visual information, if the visual information is very simple, the deficits may be less severe (Temple and Ilsley, in preparation-b). With complex visual information there is also an organizational component involved. The deficit may not reflect a simple impairment in visual memory but may particularly concern the recall of information whose organization at the time of encoding is important.

Visual neglect

The final set of tasks, that will be mentioned briefly, relates to the orientation of visual attention. One of the tasks that is used in the investigation of visual neglect and attention is line bisection. Lines of different length are presented and subjects are requested to mark the mid-point. This is an easy task to conduct with both children of different ages and people of different levels of intellectual ability. When normal subjects carry out this task they make a slight error, tending to dissect the line very slightly to the left hand side of the center, though there is some individual difference in this pattern. One interpretation is that the right hemisphere is more involved in control and orientation of attention and therefore slightly greater significance is attached to the left hand side of the world. This causes slight underestimation of the right hand side, so that in the line bisection task there is very mild "pseudo-neglect". For the acallosal study, control data was taken from normal fourteen year old children. Figure 4 represents the errors made by normal children in bisecting lines of different lengths from 2cm to 20cm. They are performing with the right hand and the lines are presented horizontally across midline vision. If there were a similar proportion of error on each

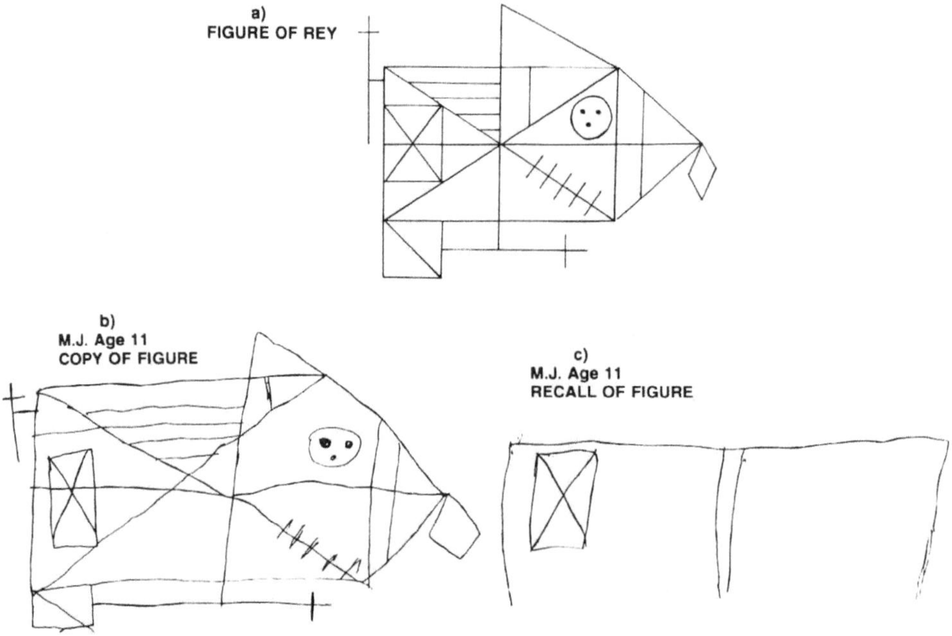

Figure 3. Copy and recall of the figure of Rey (**A**) by M.J. at age 11: Copy (**B**); Recall (**C**).

line, then the connecting line through the responses would be straight. Data will also be presented from one acallosal subject, M.J., who has been the focus of a case study of neglect (Temple and Ilsley, in preparation-c). For M.J. a different pattern of response is evident when bisecting in the midline (see Figure 5). Instead of there being right pseudo-neglect there is evidence of left neglect. The lines are bisected to the right of the midline rather than the left and the magnitude of the error is greater than for controls. One speculative interpretation was that because the right hand was being employed and being controlled by the left hemisphere, there might not be good integration with attentional control mechanisms from the right hemisphere. If this is a viable hypothesis then by presenting the stimulus lines out into right hemispace the neglect might be made worse. In contrast, if the lines were presented in left hemispace it might make the neglect better, since the lines would then be in the space more directly controlled by the right hemisphere. Figure 6 illustrates performance with the right hand in right hemispace. The neglect becomes more exaggerated, which is consistent with the prediction. In left space (see Figure 7), with lines up to 10cm, the neglect reduces to the normal direction and size of pseudo-neglect and is comparable to control performance in this hemispace. This is also consistent with the hypothesis. However, with longer lines in left hemisphere, a more conspicuous right neglect becomes manifest, which lies outside the control range. The results partially substantiate a disconnection in attentional control mechanisms and certainly indicate that the attentional mechanisms are abnormal.

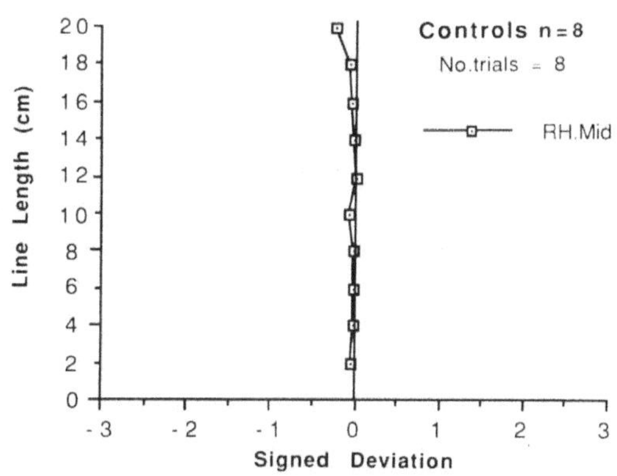

Figure 4. Deviations from the midpoint in line bisections by control children: Midline presentation and right hand response.

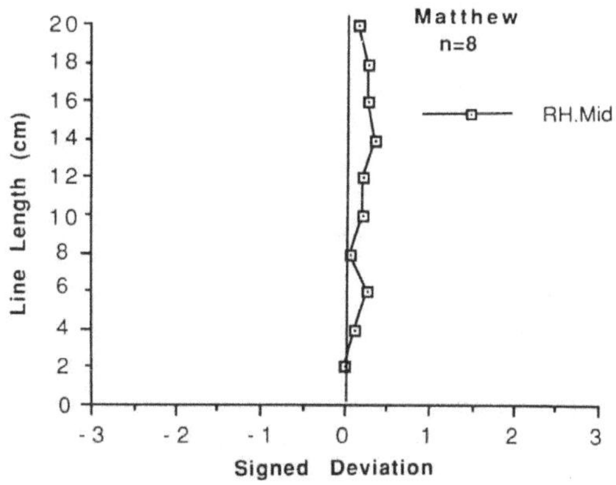

Figure 5. Deviations from the midpoint in line bisections by M.J.: Midline presentations and right hand response.

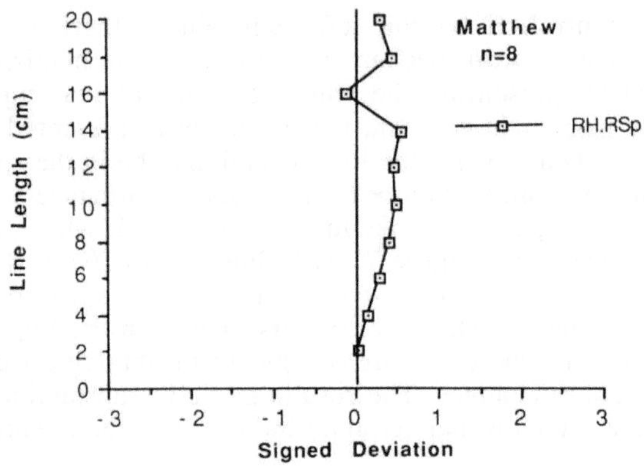

Figure 6. Deviations from the midpoint in line bisection by M.J.: Right hemispace presentation and right hand response.

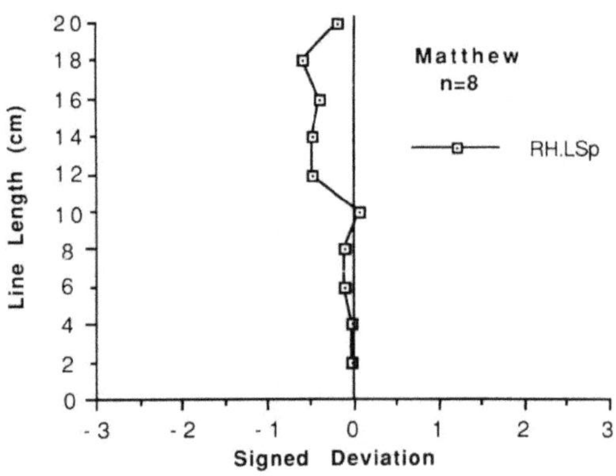

Figure 7. Deviations from the midpoint in line bisection by M.J.: Left hemispace presentation and right hand response.

CONCLUSION

Our investigations of verbal skills and literacy in callosal agenesis indicate several highly specific impairments in the explicit phonological processing required for rhyming tasks and in phonemic discrimination. Reading may be at a normal level in relation to word recognition but there are problems in the development of the phonological reading route. Spatial skills indicate normal development in certain areas but consistent impairments on Jigsaw, formboard and drawing tasks all of which include constructional components. Copying difficulties are common, as are problems with visual memory for complex spatial material. There is also evidence of visual neglect in a case study of an acallosal. The results therefore implicate a number of selective cognitive domains in which the corpus callosum may play a critical role in normal performance. The results also indicate that acallosal subjects provide useful evidence about the nature of the modular subcomponents of cognitive skill which may develop in relative independence of each other.

Acknowledgments

This research was supported in part by grants from the Wellcome Trust and the Wolfson Foundation.

REFERENCES

Benton, A.L., Hamsher, K. deS., Varney, N.R., and Spreen, O., 1983, "Facial Recognition," Oxford University Press, New York.

Benton, A.L., Varney, N.R., and Hamsher, K. deS., 1978, Visuo-spatial judgement: a clinical test, *Arch. Neurol.* 35:364-367.

Bradley, L. and Bryant, P., 1978, Difficulties in auditory organization as a possible cause of reading backwardness, *Nature* 271:746-747.

Cornish, K. and Temple, C.M., under revision, Spatial cognition in young adults: Spatial factors and the influence of biological variables, *Quart. J. Exp. Psychol.*

Dennis, M., 1981, Language in a congenitally acallosal brain, *Brain Lang.* 12:33-53.

Foder, J., 1983, "The Modularity of Mind," MIT Press, Cambridge, Mass.

Jeeves, M. and Temple, C., 1987, A further study of language function in callosal agenesis, *Brain Lang.* 32:325-335.

Landsell, H. , 1968, Effect of extent of temporal lobe ablations on two lateralized deficits, *Physiol. Behav.* 3:271-273.

Mehta, Z., Newcombe, F., and Damasio, H., 1987, A left hemisphere contribution to visuo-spatial processing, *Cortex* 23:447-467.

Osterrieth, P.A., 1944, Le test de la copie d'une figure complexe, *Arch. Psychol. (Chicago)* 30:206.

Selnes, O.A. , 1974, The corpus callosum: Some anatomical and functional considerations with special reference to language, *Brain Lang.* 1: 111-139.

Snowling, M., Stackhouse, J., and Rack, J., 1986, Phonological dyslexia and dysgraphia: A developmental analysis, *Cog. Neuropsychol.* 3:309-340

Street, R., 1931, "A Gestalt Completion Test," Teachers College, Columbia University, New York.

Temple, C.M. and Cornish, K., under revision, Spatial cognition in school children, *Br. J. Psychol.*

Temple, C.M. and Ilsley, J., in press, Phonemic discrimination in callosal agenesis, *Cortex.*

Temple, C.M. and Ilsley, J., in preparation-a, Spatial Cognition in callosal agenesis.

Temple, C.M. and Ilsley, J., in preparation-b, Memory in callosal agenesis.

Temple, C.M. and Ilsley, J., in preparation-c, Attention and neglect: A case study in callosal agenesis.

Temple, C.M., Jeeves, M.A., and Vilarroya, O.O., 1989, Ten pen men: Explicit phonological processing in two children with callosal agenesis, *Brain Lang.* 37:548-564.

Temple, C.M., Jeeves, M.A., and Vilarroya, O.O., 1990, Reading in callosal agenesis, *Brain Lang.* 39:235-253.

Trevarthen, C., 1974, Cerebral embryology and the split-brain *in:* "Hemispheric Disconnection and Cerebral Function," M. Kinsbourne and W. Smith, eds., Thomas, Springfield, Il.

Vandenburg, S.G. and Kuse, A.R. ,1978, Mental rotation, a group test of three dimensional spatial visualization, *Percept. Mot. Skills* 47:599-604.

DISCONNECTION SYNDROME IN CALLOSAL AGENESIS

Maryse Lassonde

Groupe de Recherche en Neuropsychologie Expérimentale
Département de Psychologie
Université de Montréal
C.P. 6128, Succ. A
Montréal, Qué., H3C 3J7, Canada

INTRODUCTION

The previous chapters have confirmed the notion that callosal agenesis may be considered as a natural model of hemispheric disconnection. In fact, this observation initially prompted several investigators to evaluate acallosal subjects in a number of tasks requiring interhemispheric communication (for reviews, see Milner and Jeeves, 1979; Chiarello, 1980; Jeeves, this volume; Milner, this volume). Most of the early work, however, emphasized the fact that callosal agenesis did not produce the disconnection syndrome observed after surgical section of the callosal commissure. Typically, acallosal subjects were reported to have no difficulties cross-matching objects, shapes or other stimuli placed in each hand out of view and they showed no impairments in tachistoscopic tasks requiring half-field comparisons. A similar pattern of results has been reported for tasks involving recognition of material presented unilaterally. With a few exceptions (e.g. Ettlinger et al., 1972; Donoso and Santander, 1982), acallosals, unlike callosotomized patients, did not show hemialexia for letters or words tachistoscopically projected to the left visual hemifield nor did they display anomia for objects palpated by the left hand (e.g. Milner and Jeeves, 1979; Sauerwein et al., 1981; Sauerwein and Lassonde, 1983).

More recently, however, a series of studies have indicated that acallosal patients may display impairments in interhemispheric communication. For instance, Geffen et al. (this volume) have reported deficits in cross-localization of serial finger stimulation in acallosals. Similarly, interhemispheric transmission times have repeatedly been found to be abnormally long in acallosal subjects (e.g. Jeeves, 1969; Milner et al., 1985; Milner, this volume). Impairments have also been observed in tasks requiring asynchronous distal movements, suggesting poor interhemispheric integration of bimanual activities (Jeeves, 1986; Jeeves et al., 1988; Silver and Jeeves, this volume). Finally, Temple and Ilsley (this volume) have indicated that visuo-spatial and visuo-perceptual skills may occasionally be affected by callosal absence.

Considering the richness of the connections normally assumed by the corpus callosum (e.g. Myers, 1955; Innocenti, 1986; Pandya and Seltzer, 1986) and the

Callosal Agenesis, Edited by M. Lassonde and
M.A. Jeeves, Plenum Press, New York, 1994

anatomical defects that may accompany callosal agenesis (Wahlsten and Ozaki, this volume; Olavarria et al., this volume; Schmidt, this volume), it is surprising that acallosal subjects should display such a limited amount of deficits and so little of the disconnection signs observed in callosotomized patients. One explanation of this phenomenon may be that most of the tasks in which no interhemispheric deficits were observed required either visual identification and discrimination or tactile identification and matching of objects. These two functions could, in fact, be assumed by compensatory pathways. Indeed, visual information could be conveyed through subcortical commissures (Milner, this volume) or through the anterior commissure, which is almost always present in cases of callosal agenesis (Rauch and Jenkins, this volume) and tactile discriminations could be achieved using bilateral somesthetic pathways (Dennis, 1976; see also Jeeves, this volume).

On the other hand, the tasks that are most often failed by acallosal subjects involve some form of motor control or spatial skills (e.g. Silver and Jeeves, this volume; Temple and Ilsley, this volume). It may well be that the acallosal brain has developed no compensatory means to assume interhemispheric communication between the regions involved in these activities, namely the associative frontal and posterior parietal areas. The present chapter presents evidence that, in tasks addressing the cognitive abilities of these two cortical areas, acallosal subjects do, in effect, show deficits of the same magnitude as those observed in split-brain patients.

"FRONTAL" TASKS

Interhemispheric Transmission Times

In reaction time (RT) experiments requiring a manual response to a lateralized visual stimulus, the speed of the response depends on the anatomical and spatial relationships between the hemifield in which the stimulus appears and the hand used to respond. In the ipsilateral stimulus-response (S-R) condition, the response is integrated intrahemispherically because the hemiretinae receiving the stimulus project to the hemisphere that controls the responding hand whereas in the contralateral S-R condition, an interhemispheric integration is necessary to coordinate visual input and motor output. The difference between ipsilateral and contralateral RTs may then be regarded as an indirect measure of interhemispheric transmission time (ITT). Some years ago, Jeeves (1969; see also Milner, this volume) had demonstrated that ITTs were much longer in acallosals than in normals. This finding has later been corroborated by several researchers (e.g. Milner, 1982; Lines, 1984; Clarke and Zaidel, 1989).

These studies and those conducted on split-brain patients (e.g. Clarke and Zaidel, 1989) clearly indicated that interhemispheric integration of simple visuomotor responses was mediated by the callosal pathway. However, little was known about the specific callosal region involved in this intermodal integration. Early work by Arutiunova and Blinkov (1960) and Berlucchi et al. (1977) suggested that the callosal connections between visual cortices were not crucial for bilateral interhemispheric integration. Along with MariRosa Di Stefano and Hannelore Sauerwein (Di Stefano et al., 1992), we undertook the testing of this assumption by comparing the ITTs of two acallosal adults to that of a callosotomized patient with an intact splenium.

We used a traditional paradigm (e.g. Berlucchi, 1978) in which the subjects had to press a key with either their left or their right hand whenever a light-emitting diode (LED), situated 10° to the left or the right of a central fixation point, was turned on. ITTs were computed using the RT difference between the contralateral (left hand/right visual field; right hand/left visual field) and ipsilateral (left hand/left visual field; right hand/right visual field) conditions.

It can be seen from Table 1 that our estimate of interhemispheric transmission time in the normal control group (3 msec) is comparable to that reported by other authors (e.g. Berlucchi et al., 1977). This table further shows that both the acallosals and the

Table 1. Mean reaction times (msec) and interhemispheric transmission times (ITT) obtained on the simple RT task.

	S-R CONDITION		
Groups	**Ipsilateral**	**Contralateral**	**ITT**
Controls Ss	235.9	238.9	3.0
Acallosal Ss	486.1	537.4	51.3
Partially-callosotomized Ss	511.3	569.6	58.3

partially-callosotomized patient have ITTs that greatly exceed those found in the control group. Interestingly, the ITT obtained from the callosotomized patient with sparing of the splenium was comparable to that measured in the two acallosal subjects. It is worth noting that the residual splenial fibers connecting the visual areas of each hemisphere were found to be functional in visual transfer studies carried out in this patient (Lassonde et al., 1988). If results from this single patient reflect a general mechanism of visuomotor processing[1], they imply that S-R interaction is normally carried out intrahemispherically and that the response is conveyed to the other hemisphere through the motor components of the callosal pathway. More importantly, if this assumption is correct, it also implies that there is no compensation for the disconnection of the frontal motor areas in acallosal subjects. The next experiment was undertaken to further test this hypothesis.

Interhemispheric Transfer of a Unilateral Motor Skill

The study of interhemispheric transfer of motor information has generally been restricted to tasks based on bimanual motor control rather than on cross-manual motor learning. Typically, these studies have revealed that both acallosal and split-brain patients are deficient with respect to speed of bimanual coordination (e.g. Jeeves, 1965; Sauerwein et al., 1981) and they show impairments in tasks requiring bilateral interdependent motor control (Preilowski, 1972; Zaidel and Sperry, 1977; Jeeves, 1986).

Another means of assessing the functional communication between homologous motor areas consists of testing the interhemispheric transfer of a motor learning skill that has been mastered intrahemispherically. In effect, it is well known that transfer of tactuo-motor learning from the dominant to the non-dominant hand is abolished after transection of the corpus callosum (Goldstein and Joynt, 1975). Deficits in tactuo-motor transfer of formboard learning and kinesthetic maze learning have also been observed in acallosals (Ferris and Dorsen, 1975; Gott and Saul, 1978; Jeeves, 1979) but results obtained with those tests are controversial (Reynolds and Jeeves, 1977; Sauerwein et al., 1981; see also Sauerwein et al., this volume). Furthermore, although these tasks involved the cross-manual transfer of an acquired performance, they did not require a systematic learning taking place over a considerable number of trials. In conjunction with Anne-Josée Chicoine and Luc Proteau (Chicoine et al., under revision), we therefore undertook the task of studying the interhemispheric transfer of such a motor learning.

Four acallosal subjects and four age- and IQ-matched controls were required to make a hundred aiming movements from a starting position towards either the middle or the side target ipsilateral to the hand used, while maintaining central fixation. Visual feedback was given to the subjects after each trial in order to guide them in their subsequent trials. Prior to training, a pre-test was performed with the hand contralateral to the one used during acquisition. Once the acquisition phase was completed, the subjects were submitted to a post-test with the untrained hand.

The spatial accuracy of the movement was measured to the nearest millimeter on its frontal plane (the X axis) and its sagittal plane (the Y axis). Root mean square errors (square root of $(x_i^2 + y_i^2)$) were used as an overall measure of accuracy. All subjects (acallosals and controls) learned the unilateral aiming task as demonstrated by a reduction in spatial errors with an increasing amount of practice trials. However, as illustrated in Figure 1, these two groups differed on their performance on the transfer task. Whereas the untrained hand of the IQ-matched controls benefited from the learning process, the acallosal subjects were deficient on the interhemispheric transfer of the visuo-motor learning.

These findings question the efficiency of the ipsilateral pathways as an alternative route for the integration and transfer of motor functions. More importantly, this study, along with the previous one, suggests that compensation for congenital absence of the corpus callosum is extremely limited when interhemispheric transfer of motor performance, and hence of "frontal" functions, is required.

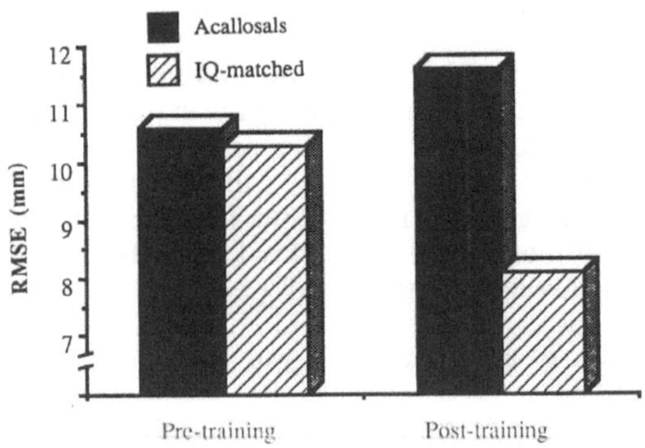

Figure 1. Pre- and post-training performances of acallosal and IQ-matched control subjects. Spatial accuracy of movement is expressed using the Root Mean Square Error (RSME) of aiming performance.

"PARIETAL" TASKS

Visuo-Spatial Localization as Assessed by a Pointing Response

Difficulties in effecting visually-directed hand movements, often referred to as "optic ataxia", are generally produced by parietal lobe lesions (Perenin and Vighetto, 1983; Jakobson et al, 1991). Such imprecision in visually-guided gestures has occasionally been described in acallosal (Rohmer et al., 1960) and split-brain patients (Gazzaniga et al., 1965; see also Trevarthen, 1991). In these initial descriptions, the specific disorder in visuomotor coordination was reported only for the hand ipsilateral to the seeing hemisphere and has been attributed to a disconnection between the visual areas of the stimulated hemisphere and the motor areas of the responding hemisphere (Rondot et al., 1977). The deficit was interpreted as being part of the disconnection syndrome and was thus labeled "bilateral crossed optic ataxia".

There are other instances, however, in which functional loss of the callosum seems to produce deficits in pointing that differ from the conventional crossed optic ataxia. For example, accurate pointing with the left hand to stimuli in the right visual field, but errors in indicating with the right hand the location of stimuli in the contralateral visual field were observed in a patient with posterior section of the corpus callosum (Levine and Calvanio, 1980) and in three acallosal subjects (Rohmer et al., 1960). In view of these discrepancies, Micheline Benoit and I decided to compare the precision of two acallosal subjects to that of a callosotomized patient in assessing the location of a visual stimulation. Three subjects with average intelligence and three IQ-matched controls were also submitted to the task.

The task consisted of pointing with either hand in the direction of a luminous square (2°) that appeared for 150 msec on one of two monitor screens. These monitors, located one in the left, the other in the right visual field, could be rotated along a giant perimeter with a half-diameter of 1.14 m. On each trial, the monitors were positioned at different eccentricities (± 5°, 20°, 35°, 65°, 80°). The subjects were seated in a completely darkened room in front of a centrally located illuminated diode. Once the subjects had pointed to the luminous source, a curtain was closed to prevent them from seeing their responding hand, thus making the experimental situation an open-loop condition (e.g. Perenin and Vighetto, 1983).

The results are depicted in Figure 2. The shadow area represents the control subjects' medians of errors, with a confidence interval of 95%. Any performance falling outside this interval may thus be regarded as abnormal. The solid line in Figure 2 corresponds to the mean performance of the acallosal and callosotomized subjects. It can be seen that these patients were quite inaccurate in their left visual field, regardless of the hand used for the pointing response. They also showed some imprecision of gestures in the right visual field when they used their left hand (crossed condition) although their performance in that condition fell within the distribution of the controls' responses. In summary, although deficits were seen in the crossed conditions (left hand/right visual field; right hand/left visual field), impairments were also observed in the ipsilateral left-hand/left visual field condition. In fact, the "callosal" group showed a normal performance in only condition, the one in which the responses were made by the right hand to visual presentation in the homolateral visual field.

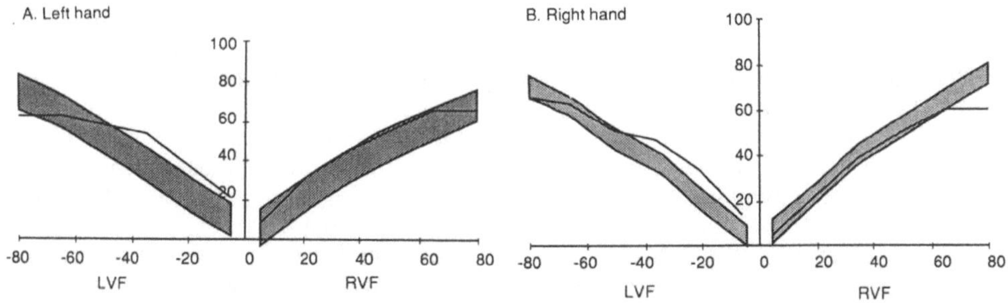

Figure 2. Response accuracy of acallosal subjects (solid lines) in the pointing task. The shadow area represents the control subjects' median errors with a confidence interval of 95%.

This pattern of results suggests that deficits in pointing occur whenever the response and/or the visual stimulation are being processed within the right hemisphere. It may be that the left hemisphere is more proficient in this type of task, as is often the case for tasks requiring some form of motor control (e.g. Kimura and Archibald, 1974). Whether or not our findings indicate a left hemisphere superiority in visually-guided motor activities, the fact remains that our acallosal subjects behaved in the same manner as the totally callosotomized patient.

The results of this study, and those of the previous ones, indicate the presence of disconnection signs that are not compensated for by the early absence of the callosal commissure. These studies also have in common the fact that all of them required a manual response to a visual stimulation. In order to see whether the performance observed in acallosal subjects was not just specific to the visuo-motor domain but really reflected a general disconnection, we designed another experiment aimed at tackling "parietal" functions without comprising any visuo-motor elements.

Interhemispheric Transfer of Tactuo-Spatial Analysis

By contrast to commissurotomized patients, acallosal subjects generally show no difficulties in identifying or recognizing verbal material presented to either hemisphere (e.g. Sauerwein and Lassonde, 1983). Indeed, their good performance in these verbal tasks has been taken as evidence of a lack of disconnection syndrome following congenital absence of the corpus callosum. A different pattern of results, however, has been obtained for unilateral processing of visuo- or tactuo-spatial cues. For instance, right visual field impairments for judgment of stimulus location and word orientation were reported in a case of callosal agenesis (Martin, 1985; but see Jeeves and Milner, 1987, for different findings). Similarly, right-hand deficits in spatial processing are sometimes, but not always, observed in the so-called De Renzi test (De Renzi et al., 1971), a task that requires, through blind palpation, the reproduction of the orientation of a model rod on a copy (Meerwaldt, 1983; Jeeves and Silver, 1988). These left-hemisphere deficits in spatial analysis have also been found in callosotomized patients. Split-brain subjects usually show difficulties in discriminating geometrical shapes with their right hand (Franco and Sperry, 1977), their impairment being especially marked when short-term memory is involved (Milner and Taylor, 1972).

The results reviewed above suggest that the acallosal brain may not be able to compensate for the disconnection between the cortical regions -presumably the parietal areas- responsible for spatial analysis. In collaboration with Simone Benoit-Dubrocard (Benoit-Dubrocard and Lassonde, submitted), we undertook to test this assumption by comparing the ability of acallosal subjects and callosotomized patients to perform tactuo-spatial discriminations.

The sample consisted of two acallosal subjects and two callosotomized patients as well as four IQ-matched control subjects. Only one of the acallosal subjects underwent MRI examination, which confirmed the presence of the anterior commissure. MRI also revealed an intact splenium in one of the callosotomized patients (Lassonde et al., 1988). The task involved the comparison of tactual stimuli designed in such manner that they could be processed on the basis of their spatial characteristics only. The shapes selected were a letter (P), a familiar symbol (star) and a meaningless shape faintly resembling a fish. Following an adaptation of Franco and Sperry's paradigm (1977), the shapes were presented under three geometrical transformations: 1) Euclidean, the "usual" presentation; 2) projective, in which the Euclidean representation of the stimulus was projected at a 45° angle; and 3) topological in which the Euclidean projection of the object was represented with a broken contour. The stimuli were put on a console in an open box designed to conceal them from the subject's view. Subjects were required to make simultaneous bimanual discriminations of the geometrical representations of each stimulus (e.g. "topological" and "Euclidean" stars). Response accuracy and response times were computed with the aid of an IBM micro-computer.

The percentages of correct responses obtained by the two "callosal" groups and their IQ-matched controls are depicted in Figure 3. Whereas the controls succeeded the task at a 75% accuracy level, both acallosal and callosotomized subjects showed difficulties in this bimanual task, with a respective performance of 60 and 54% . Thus, the two "callosal" groups performed at, or close to, chance level. Their response times were also longer than those of the control groups (RT IQ-matched controls: 1.32; RT acallosals: 1.68; RT callosotomized: 2.11).

Whereas the possibility of extracallosal damage may be invoked to account for the observed deficit, the fact that the two experimental groups performed in a similar fashion despite the different etiology of their "hemispheric disconnection" rather indicates that their tactuo-spatial impairments were directly related to the absence of the callosal

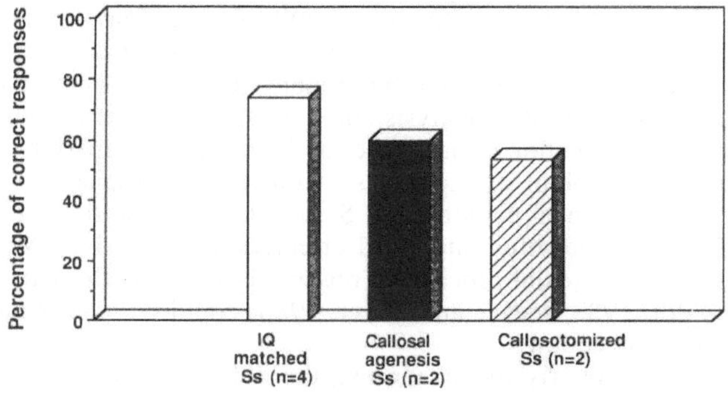

Figure 3. Percent correct responses obtained by acallosal, callosotomized and control subjects in the task requiring a bimanual comparison of the geometrical transformations of shapes.

commissure. In accord with the findings by Martin (1985), our results indicate that the anterior commissure -intact in the callosotomized patients and present in at least one of our acallosal subjects- is insufficient to assume interhemispheric integration of spatial properties. Interestingly, both the callosal agenesis subjects and the callosotomized patient with an intact splenium showed the same deficits as the completely callosotomized patient. The latter result suggests that the trunk of the corpus callosum -relating, among others, the parietal areas- may be crucial in mediating interhemispheric transfer of tactuo-spatial analysis.

CONCLUSION

We have reported a series of experiments confirming the presence of disconnection signs in callosal agenesis. We have further shown that these disconnection signs may be of the same amplitude as those observed in split-brain patients. These observations are not isolated: as reported in the Introduction, an increasing amount of reports point to the presence of a disconnection syndrome in callosal agenesis.

Several of the deficits observed in acallosals involve a motor component. Thus, acallosals are commonly reported to be clumsy and slow in bimanual operations (Jeeves, 1965; Chiarello, 1980; Sauerwein et al., 1981). More recent studies indicate that acallosal subjects perform in a manner similar to anterior commissurotomy patients in a bimanual tracking task requiring asymmetrical input from the two hands (Jeeves et al., 1988). Deficits in intermanual transfer of formboard or maze learning have also been reported (Ferris and Dorsen, 1975; Gott and Saul, 1978; Jeeves, 1979) and interhemispheric transmission times estimated from our visuo-motor RT paradigms were as abnormal in the acallosal as in the partially split-brain subject. Finally, our task involving a unilateralized visuo-motor learning unequivocally points to a lack of interhemispheric transfer of motor skills in acallosal subjects.

Acallosals also display other types of "motor" deficits that, at first glance, may not appear to be related to a lack of interhemispheric communication. These include defects in grip formation as well as in reaching toward a visual stimulus (e.g. Silver and Jeeves, this volume). However, many of these results could also be interpreted within the general framework of the disconnection syndrome, assuming that one hemisphere is incapable of carrying out an analysis in the absence of communication with the more proficient one. In fact, our own results on the precision of pointing in acallosals may be imputed to a lack of communication between the left hemisphere, specialized in this motor activity, and the less proficient right hemisphere. Conversely, the impaired performance of the right hand in the right hemispace in Temple and Ilsley's visual neglect task (this volume) could be attributable to the dissociation of the left hemisphere from the right one, the latter being more involved in attentional mechanisms. Left hemisphere deficits in spatial analysis could also account for the results obtained in spatial tasks. Indeed, a functional disconnection between the superior right hemisphere and the left one may explain our subjects' inability to compare the shapes' spatial characteristics during bimanual exploration. Similarly, the right visual field impairments for judgment of stimulus location and word orientation reported by Martin (1985) and the right hand deficits in spatial processing observed in the tactile part of the De Renzi Rod Test (Meerwaldt, 1983; Jeeves and Silver, 1988) could be other examples of such a interhemispheric disconnection.

At the risk of being overly localizationist, the deficits enumerated above appear to be mostly related to frontal (bimanual activities, transfer of motor learning, etc.) or parietal (visual neglect, spatial analysis, etc.) activities. This is somewhat expected since the callosal connections, apart from relating the primary sensory areas involved in midline analysis, are predominantly relating associative cortical areas. Lepore et al., (this volume) and Jeeves (1991) have shown that acallosals do, indeed, show sensory impairments when midline functions are involved. The present review also emphasizes the dysfunctions related to the dissociation between homologous frontal and parietal associative areas. Surprisingly, few disconnection signs involving the temporal areas have been reported. For example, one could expect impairments in visual discrimination, especially when the discriminanda are presented in different hemifields, but in general, no such deficits are observed. It should be recalled, however, that the temporal areas are also related by the anterior commissure which is almost always present in cases of callosal agenesis (Rauch and Jenkins, this volume). In fact, a recent report (Fischer et al., 1992) indicates that whilst visual interhemispheric transfer is normal in an acallosal subject having an anterior commissure, it is impaired in an acallosal patient whose anterior commissure is missing. Therefore, it appears that the pattern of disconnection deficits observed in callosal agenesis reflects, grosso modo, the disruption of anatomical connections between the two hemispheres.

However, acallosal subjects display other problems that may be difficult to interpret within the framework of interhemispheric disconnection. The visuo-constructive deficits seen in some of Temple and Ilsley's patients (this volume) as well as their memory impairments for complex material are typical examples. Although it may be suggested that these deficits may somehow be related to the motor component involved in those tasks, there are other instances in which impairments are observed in tasks devoid of any motor aspect. For example, Temple et al. (1990) have shown that the phonological reading route is deficient in acallosals. Similarly, Temple and Isley (this volume) have reported visuo-perceptual problems and deficits in phonological processing in some acallosal subjects. We have also found acallosal subjects to be as impaired as split-brain patients in a task that required a verbal judgment of the relative distance between objects (Lassonde, 1986). In the latter task, the estimation of distance was equally deficient whether the judgment had to be made upon stimuli presented intra- or interhemispherically. This observation, as well as others, prompted us to postulate a facilitatory or modulating influence of the corpus callosum on the activity of both hemispheres (Lassonde, 1986). In fact, we have hypothesized that the absence of transcallosal enhancement of cortical activity could explain the appearance of uni- or bi-lateral symptoms following agenesis or section of the corpus callosum. This hypothesis may perhaps account for the specific cognitive impairments reported above, such as the visuo-constructive defects.

In summary, a series of disconnection signs may thus be observed in callosal agenesis. These manifestations generally follow the pattern of callosal connectivity. It should be stressed, however, that acallosal individuals have access to a series of compensatory pathways or mechanisms (see Jeeves, this volume) that may alleviate or obscure these symptoms, depending on the strength of their utilization. Some of these mechanisms may be privileged over others in certain individuals, thus producing different manifestations of the disconnection deficits. Individual differences would then arise and indeed, there are numerous examples of such interindividual variability. Therefore, in order to determine the real extent of compensation in individuals who spent an entire life without a corpus callosum, extensive case studies may be the most fruitful avenue of future investigations

Footnote

[1]It should be noted that the lack of involvement of splenial fibers in the production of ITTs is still a matter of debate, other researchers (Aglioti et al., 1991) having reported normal ITTs in callosotomized patients with a splenium intact.

Acknowledgments

This work was supported by grants from the National Science Research Council of Canada (NSRCC) and from the Québec Formation de Chercheurs et Aide à la Recherche (FCAR).

REFERENCES

Aglioti, S., Berlucchi, G., Pallini, R., Rossi, G., and Tassinari, G., 1991, Interhemispheric transfer in patients with partial and total callosotomy, *Abstracts of 3rd IBRO World Congress*, Montreal., Canada.

Arutiunova, A.S. and Blinkov, S.M., 1962, Latent period of motor reactions with hemianopia, *J. Higher Nerv. Activity* 12:432-436.

Benoit-Dubrocard, S. and Lassonde, M., submitted, Callosal involvement in tactuo-spatial analysis.

Berlucchi, G., 1978, Interhemispheric integration of simple visuomotor responses, *in:* "Verbal Correlates of Conscious Experience," P.A. Buser and A. Rougeul-Buser, eds., Elsevier, North Holland, Amsterdam.

Berlucchi, G., Crea, F., Di Stefano, M., and Tassinari, G., 1977, Influence of spatial stimulus-response compatibility on reaction time of ipsilateral and contralateral stimuli, *J. Exp. Psychol. Hum. Perc. Perf.* 3:505-517.

Chiarello, C., 1980, A house divided? Cognitive functioning with callosal agenesis, *Brain Lang.* 11:128-158.

Chicoine, A.J., Proteau, L., and Lassonde, M., under revision, Unilateralization of a motor engram: evidence from cases of congenital and functional absence of the corpus callosum.

Clarke, J.M. and Zaidel, E., 1989, Simple reaction times to lateralized light flashes: Varieties of interhemispheric communication routes, *Brain* 112:849-870.

Dennis, M., 1976, Impaired sensory and motor differentiation with corpus callosum agenesis: a lack of callosal inhibition during ontogeny?, *Neuropsychologia* 14:455-469.

De Renzi, E., Faglioni, P., and Scotti, G.G., 1971, Judgment of spatial orientation in patients with focal brain damage, *J. Neurol. Neurosurg. Psychiatry.* 34:489-495.

Di Stefano, M., Sauerwein, H.C., and Lassonde, M., 1992, Influence of anatomical factors and spatial compatibility on the stimulus-response relationship in the absence of the corpus callosum, *Neuropsychologia* 30:177-185.

Donoso, A.D. and Santander, M., 1982, Sindrome de desconexion en agenesia del cuerpo calloso, *Nervol. Colombia* 6:177-180.

Ettlinger, G., Blakemore, C. B., Milner, A. D., and Wilson, J., 1972, Agenesis of the corpus callosum: A behavioural investigation, *Brain* 95:327-346.

Ferris, G. S. and Dorsen, M. M., 1975, Agenesis of the corpus callosum: 1. Neuropsychological studies, *Cortex* 11:95-122.

Fischer, M., Ryan, S.B., and Dobyns, W.B., 1992, Mechanisms of interhemispheric transfer and patterns of cognitive function in acallosal patients of normal intelligence, *Arch. Neurol.* 49:271-277.

Franco, L. and Sperry, R. W., 1977, Hemisphere lateralization for cognitive processing of geometry, *Neuropsychologia* 15:107-114.

Gazzaniga, M.S., Bogen, J. E., and Sperry, R.W., 1965, Observation on visual perception after disconnexion of the cerebral hemispheres in man, *Brain* 88(2):221-235.

Goldstein, M.N., Joynt, R.J., and Hartley, R.B., 1975, Long-term follow-up of a callosal-sectioned patient: report of a second case, *Arch. Neurol.* 32:52-53.

Gott, P.S. and Saul, R.E., 1978, Agenesis of the corpus callosum: limits of functional compensation, *Neurology* 28:1272-1279.

Innocenti, G.M., 1986, General organization of callosal connections in the cerebral cortex, *in:* "Cerebral Cortex," vol. 5, E.G. Jones and A. Peters eds., Plenum Publishing Corporation, New York.

Jakobson, L.S., Archibald, Y.M., Carey, D.P., and Goodale, M.A.,1991, A kinematic analysis of reaching and grasping movements in a patient recovering from optic ataxia, *Neuropsychologia* 29:803-809.

Jeeves, M.A., 1965, Psychological studies of three cases of congenital agenesis of the corpus callosum, *in*: "Functions of the Corpus Callosum," E.G. Ettlinger, ed., Churchill, London.
Jeeves, M.A., 1969, A comparison of interhemispheric transmission times in acallosals and normals, *Psychon. Sci.* 16:245-246.
Jeeves, M.A., 1979, Some limits to interhemispheric integration in cases of callosal agenesis and partial commissurotomy, *in:* "Structure and Function of the Cerebral Commissures," I.S. Russell, M.W. van Hof, and G. Berlucchi, eds., Macmillan, London.
Jeeves, M.A., 1986, Callosal agenesis: neuronal and developmental adaptations, *in:* "Two Hemispheres-One Brain: Functions of the Corpus Callosum," F. Lepore, M. Ptito, and H.H. Jasper, eds., Alan R. Liss, New York.
Jeeves, M.A., 1991, Stereoperception in callosal agenesis and partial callosotomy, *Neuropsychologia* 29:19-34.
Jeeves, M.A. and Milner, A.D., 1987, Specificity and plasticity in interhemispheric integration: evidence from callosal agenesis, *in*: "Duality and Unity of the Brain - Unified Functioning and Specialization of the Hemispheres", D. Ottoson, ed., Macmillan:, London.
Jeeves, M.A. and Silver, P.H., 1988, Interhemispheric transfer of spatial tactile information in callosal agenesis and partial commissurotomy, *Cortex* 24:601-604.
Jeeves, M.A., Silver, P.H., and Jacobson, I., 1988, Bimanual co-ordination in callosal agenesis and partial commissurotomy, *Neuropsychologia* 6:833-850.
Kimura, D. and Archibald, Y., 1974, Motor functions of the left hemisphere, *Brain* 97:337-350.
Lassonde, M.,1986, The facilitatory influence of the corpus callosum on intrahemispheric processing, *in:* "Two Hemispheres-One Brain: Functions of the Corpus Callosum," F. Lepore, M. Ptito, and H.H. Jasper, eds., Alan R. Liss, New York.
Lassonde, M, Sauerwein, H C., McCabe, N., Laurencelle, L., and Geoffroy, G., 1988, Extent and limits of cerebral adjustment to early section or congenital absence of the corpus callosum, *Behav. Brain Res.* 30:165-181.
Levine, D.N. and Calvanio, R., 1980, Visual discrimination after lesion of the posterior corpus callosum, *Neurology* 30:21-30.
Lines, C.R., 1984, Nasotemporal overlap investigated in a case of agenesis of the corpus callosum, *Neuropsychologia* 22:85-90.
Martin, A., 1985, A qualitative limitation on visual transfer via the anterior commissure, *Brain* 108:43-63.
Meerwaldt, J.D., 1983, Disturbance of spatial perception in a patient with agenesis of the corpus callosum, *Neuropsychologia* 21:161-165.
Milner, A.D., 1982, Simple reaction times to lateralized visual stimuli in a case of callosal agenesis, *Neuropsychologia* 20:411-419.
Milner, A.D. and Jeeves, M.A., 1979, A review of behavioural studies of agenesis of the corpus callosum, *in*: "Structure and Function of the Cerebral Commissures," I.S. Russell, M.W. van Hof, and G. Berlucchi, eds., Macmillan, London.
Milner, A.D., Jeeves, M.A., Silver, P.H., Lines, C.R., and Wilson, J., 1985, Reaction times to lateralized visual stimuli in callosal agenesis: Stimulus and response factors, *Neuropsychologia* 23:323-331.
Milner, B. and Taylor, L., 1972, Right-hemisphere superiority in tactile pattern-recognition after cerebral commissurotomy: Evidence for nonverbal memory, *Neuropsychologia* 10:1-15.
Myers, R.E., 1955, Interocular transfer of pattern discrimination in cats following section of crossed optic fibers, *J. Comp. Physiol. Psychol.* 48:470-496.
Pandya, D. P. and Seltzer, B., 1986, The topography of commissural fibers, *in*: "Two Hemispheres-One Brain: Functions of the Corpus Callosum," F. Lepore, M. Ptito, and H. H. Jasper, eds., Alan R. Liss, New York.
Perenin, M.T. and Vighetto, A., 1983, Optic ataxia: A specific disorder in visuo motor coordination, *in:* "Spatially-Oriented Behaviour," A. Hein and M. Jeannerod, eds., Springer, New York.
Preilowski, B.F.B., 1972, Possible contribution of the anterior forebrain commissures to bilateral motor coordination, *Neuropsychologia* 10:267-277.
Reynolds, D. and Jeeves, M. A., 1977, Further studies of tactile perception and motor co-ordination in agenesis of the corpus callosum, *Cortex* 13:257-272.
Rohmer, F, Wackenheim, A., and Vrousos, G., 1960, Les agénésies du corps calleux, *in*: "Comptes Rendus du Congrès de Psychiatrie et de Neurologie de la Langue Française," LVIIe session, P. Cossa, ed., Masson., Tours.
Rondot, P., De Recondo, J., and Ribadeau-Dumas, J.L., 1977, Visuomotor ataxia, *Brain.* 100:355-376.
Sauerwein, H.C. and Lassonde, M., 1983, Intra- and interhemispheric processing of visual information in callosal agenesis, *Neuropsychologia* 21:167-171.
Sauerwein, H.C., Lassonde, M., Cardu, B., and Geoffroy, G., 1981, Interhemispheric integration of sensory and motor functions in agenesis of the corpus callosum, *Neuropsychologia* 19:445-454.62.
Temple, C.M., Jeeves, M.A., and Vilarroya, O.O., 1990, Reading in callosal agenesis, Brain Lang. 39:235-253.
Trevarthen, C., 1990, Integrative functions of the cerebral commissures, *in* : "Handbook of Neurology," Vol. IV, Section 7, R.O. Nebes, ed., Elsevier, North Holland, Amsterdam.
Zaidel, D. and Sperry, R.W., 1977, Some long-term motor effects of cerebral commissurotomy in man, *Neuropsychologia* 10:103-110.

CALLOSAL AGENESIS - A NATURAL SPLIT-BRAIN

OVERVIEW

Malcolm A. Jeeves

Psychological Laboratory
University of St. Andrews
St.Andrews, Fife
Scotland, KY16 9JU

INTRODUCTION

Though diverse, the contributions to this book have all focused on a common topic. Such diversity, whilst potentially enriching the depth of understanding afforded, at the same time makes the task of pulling together the various threads into a coherent pattern, unusually challenging. This final chapter attempts, first, to highlight the salient points contained within each of the main sections of this volume. Inevitably this requires radical selection and necessarily omitting some of the important points covered in detail in earlier chapters. This selection in turn affords the opportunity to look for recurring themes which may be running through the earlier specialist sections.

Despite the diversity of perspectives of the participants it emerges that there are indeed recurring themes. These, taken together, enable us to sketch out a composite neuropsychological and phenotypical profile of callosal agenesis patients which raises questions about possible underlying compensatory mechanisms, neural and/or behavioral. Any discussion of such compensatory mechanisms, of necessity, makes certain prior assumptions about what are the normal functions of the forebrain commissures. A brief section of this chapter therefore reassesses the putative normal functions of the forebrain commissures in the light of the earlier discussions and in particular of the possible compensatory mechanisms. Finally, an attempt is made to

Callosal Agenesis, Edited by M. Lassonde and
M.A. Jeeves, Plenum Press, New York, 1994

look ahead and to identify some of the future trends of research in this area which are likely to be most productive.

SOME SALIENT POINTS FROM EARLIER SECTIONS

Common Factors and Diverse Etiologies

More than fifty disorders have been reported which are associated with agenesis of the corpus callosum (Wisniewski and Jeret, this volume; Geoffroy, this volume). In most cases, as Geoffroy makes clear, the condition is probably a chance occurrence. There are, however, five syndromes (see Aicardi and Chevrie, this volume; Geoffroy, this volume) in which this association is almost constant: the Aicardi, Andermann, Acrocallosal, Shapiro and Menkes syndromes. There are others in which it is frequent, for example, neurofibromatosis, Apert syndrome and medial facial cleft syndrome. The potential widespread occurrence of callosal agenesis underlined by Wisniewski and Jeret is a salutary reminder that no simplistic view of callosal agenesis is tenable in the light of the extant evidence concerning its clinicopathological heterogeneity both in presentation and etiology. That, however, must not be over emphasized to such a degree that it minimizes the importance of the accumulating and rapidly increasing evidence of clinically "quasi-asymptomatic" cases - a point emphasized by Sauerwein, Nolin and Lassonde (this volume). The diversity of origins, presentation and associated malformations has to be set firmly in the context, on the one hand, of those who are grossly disabled and intellectually retarded and, on the other hand, of those who are apparently normal.

Aicardi and Chevrie's review of the Aicardi syndrome (this volume), as well as being a timely and valuable summary of the syndrome in its own right, also serves as a forceful reminder that in some cases of callosal dysgenesis any conclusions claiming to link the absence of the callosum with cognitive and behavioral changes, should be scrutinized very carefully. Thus the effects of the frequent and diverse associated abnormalities with callosal agenesis must never be forgotten and always given due weight.

The detailed examination of the brains of two patients with Andermann's syndrome (Carpenter, this volume) amplifies further how the presence or absence of the corpus callosum in some cases is a secondary outcome of a more basic abnormality of the nerve roots of the cells of origins of callosal fibers in which there is a swelling of axons accompanied by a relative decrease in the packing density of neurofilaments. Andermann offers the opinion that it is likely that those fibers that normally begin to cross from hemisphere to hemisphere to form the corpus callosum are exposed to some unidentified factor in the CSF environment which damages them, resulting in agenesis of the corpus callosum.

Neuroradiological Characteristics

The rapid expansion in the availability of new imaging techniques has opened up fresh avenues for careful research into the cognitive and behavioral consequences of callosal agenesis and how this relates to identifiable structural changes in the brain. In particular, the examples of postnatal sonographic findings can form the starting point for systematic longitudinal studies of developing acallosal children. The report by Cioni et al. (this volume) documents the, perhaps, surprisingly high incidence of 0.81% cases of partial or complete callosal agenesis in a population of 1,359 infants studied. The infants included in their Ultrascan study were either born prematurely or at full term but were included because they had presented with risk conditions for the CNS (i.e. perinatal asphyxia, low birth weight, malformations etc.). Thus there may well have been yet other infants affected with callosal agenesis that escaped observation. It needs to be remembered, however, that this was probably a high-risk sample for agenesis, thus inflating the apparent incidence rate.

The widening availability of MRI has certainly added a significant new dimension to the diagnosis of malformations of the forebrain commissures and to the possibility of carrying out correlative studies linking extent of dysgenesis with the results of neuropsychological investigations. In particular, Rauch and Jinkins (this volume) on the basis of MRI studies, make the important point that we must seek to distinguish between agenesis, hypogenesis and hypoplasia of the corpus callosum. By hypoplasia they refer to the thinning of the corpus callosum which may result from an injury which damages the developing cerebral cortex, perhaps at the site of origin of the axons which will potentially cross in the corpus callosum. This condition may be difficult to detect and more widespread than hitherto suspected, adding further support for the likelihood that some forms of loosely defined 'learning difficulties' in children may be linked with callosal thinning (Njiokiktjien, 1991; Ramaekers, 1991).

Rauch's reminder of the possibility of hypoplasia alerts us to the need for care when interpreting observed neuropsychological deficits as being due solely to dysgenesis or agenesis of the corpus callosum and hence ignoring the etiologic importance of cortical abnormalities. Another finding relevant to the issue of compensatory mechanisms is that in 10% of the cases of agenesis of the corpus callosum the anterior commissure is smaller or absent whereas in another 10% it is enlarged. This casts doubt on the currently received wisdom that the anterior commissure, when present, is always enlarged.

Sleep Patterns

It is interesting that two contributors (Kuks and Vos, this volume; Nielsen et al., this volume) picked up the topic of the possibility of sleep abnormalities in callosal agenesis. One of them (Kuks) noted the importance of inhibitory and facilitatory interactions between homologous areas in the brains of normal people. Kuks' work exemplified how evidence gathered using a variety of techniques can shed light on unresolved issues concerning callosal function. For example, the notion that the neonate is a functional split brain requires re-examination in the light of Kuks' study of coherence patterns in infant sleep unless one makes the further (not implausible) assumption that the observed coherence comes from a common sub-cortical generator.

The likelihood that interhemispheric coherence reflects callosal processes finds further support in Nielsen et al.'s report (this volume) that such coherence is lower in all sleep stages for acallosals than in normals. Nielsen's observation that there is such lowered interhemispheric coherence in callosal agenesis in all areas except the occipital lobes itself raises further specific issues. His view is that since the primary visual cortex (area 17) is largely devoid of commissural connections and that interhemispheric coherence is normally subserved by the posterior commissure, that route would be available for compensatory activity in callosal agenesis. We shall take up this point and similar ones later in this review.

Learning from Animal Models

As in other areas of investigation in neuropsychology, animal models, properly used, can shed important light on many issues in human neuropsychology where the precise anatomical status of presumed cerebral abnormalities cannot be known with certainty, even with the help of ultrasound or MRIs. Hence the importance of the contributions to this part of the book.

Innocenti raised the major question of whether the development of cortical neurons is affected by retrograde signals, possibly trophic in nature, originating at the site of termination of their axons. His careful studies of dendritic morphology indicated that there is no evidence to support the notion that such trophic signals from the contralateral target are critical for the maintenance and/or maturation of dendrites of callosally projecting neurons.

Wahlsten and Ozaki, using suitably large numbers of mice, demonstrated that the absence of the corpus callosum does not correlate with an expansion of the sizes of the hippocampal or anterior commissures. The latter is a finding congruent with that of Rauch and Jinkins from their MRI studies. They also believe that when callosal agenesis occurs in mice the problems lie in the substrates of axon guidance at the midsagittal plane and not in the cells of origin or in the axons themselves. Another important finding is that putative callosal axons which fail to cross to the opposite hemisphere, generally do not reroute through the anterior commissure. If this finding applies also in the human condition, then it becomes important when speculating about a possible compensatory role for the anterior commissure in the absence of the corpus callosum.

This latter point is taken up and given added force by the studies reported by Olavarria et al., who concluded that their findings indicate that "gross anomalies do not prevent normal specification of the overall topography of callosal connections during development". And they go on "in addition, our failure to find labeled fibers within the anterior commissure indicates that this commissure does not serve as an alternative route for deviated callosal axons".

If the findings with mice apply also to humans then they make the important point that the overall topography of connections through small callosal remnants may be normal in affected patients. This could be a very significant finding when considering possible compensatory mechanisms in partial callosal agenesis.

Finally in this section, Schmidt demonstrates that early callosal defects do not affect individual asymmetries since these individual asymmetries exist even in the absence of the corpus callosum. Nevertheless, normal patterns of population asymmetries may be disturbed. If this finding can be applied to the human condition, then it underlines yet again the reports that there are cerebral asymmetries in acallosals and that the absence of the callosum does not inevitably lead to bilateral representation as some have maintained (Milner and Jeeves, 1981).

Neuropsychological Description

Is There a Behavioral Phenotype? As O'Brien (this volume) makes clear the criteria for deciding normality can, broadly speaking, be set either on the basis of careful and detailed laboratory studies or on the basis of a more ethological approach which focuses on naturally occurring behavioral problems. There is in addition a third group who may, in the future, receive increasing attention from researchers, namely, those loosely defined as 'learning disabled' children with varying degrees and kinds of callosal pathology. This point has been brought out by other researchers such as Njiokiktjien (1991).

The study of behavioral phenotypes in callosal agenesis naturally widens the interest in the particular issues taken up later in this volume, including sleep disorders, mood disorders and attention span. It is evident from O'Brien's research that there may be "an increasing trend of behavioral and developmental consequences from partial through total agenesis to the Aicardi syndrome". There is, for example, a clear step-wise increasing prevalence of epilepsy going from partial agenesis to total agenesis to the Aicardi syndrome. Aspects so far little studied, but potentially of major importance for management and rehabilitation, include studies of social behavior, e.g. recognizing the lack of empathy evident at times and of subtle language deficits (e.g. Temple et al., 1989, 1990; Temple and Ilsey, this volume). How one best characterizes these and links them with specific putative functions of the forebrain commissures remains a largely open question but there are already hints from, for example, O'Brien's reference to 'failure of inhibition' in certain social and cognitive domains.

Sensory and Sensorimotor Functions. The contributors to this section of the book highlighted how, by devising and using tailor-made experimental techniques, hitherto undetected deficits in sensory and sensorimotor functioning became apparent. The special importance of the findings in this section are the way in which they limit the possible compensatory mechanisms which may be operative where deficits are not present. Equally, where gross deficits are not present, but diminished performance levels when compared with normals are demonstrated, it becomes possible to study how alternative pathways, whether sensory or motor, set limits to the maximum performance achievable in callosal agenesis when no other gross cerebral abnormalities are suspected. These issues are taken up and expanded later in this chapter when we discuss compensatory mechanisms.

Cognitive Functions. As we move on from basic sensory and sensorimotor functions to higher cognitive functions we take an important step in that we begin to raise questions which are of particular relevance to understanding how hitherto unsuspected and undetected deficits may singly or jointly produce some of the handicaps evident amongst acallosals in an educational setting. To identify and quantify such possible handicaps may, in some instances at least, suggest remedial procedures and should in all cases lead to a greater understanding on the part of parents and teachers, as well as of the acallosal individuals themselves.

Sauerwein et al's and Lassonde's papers constitute together the most up to date and comprehensive neuropsychological profile presently available on acallosals. Their work further highlights how in recent years with the use of modern imaging techniques more and more cases are documented in whom there are no gross neurological dysfunctions. Sauerwein's use of stepwise discriminant analysis to analyze her accumulated data has enabled her to study which group of tests would best characterize the three groups of subjects in her studies, the acallosals, the normals and the IQ-matched controls. Taken together, her data indicate that at least some individuals born without the corpus callosum could demonstrate normal cognitive abilities, whilst tending to function at the lower end of the normal range. This point is taken up again in Temple's contribution. The level of functioning may, however, still be further lowered if other important structural abnormalities in the CNS co-exist.

It would be sensible, at this point, to reiterate the wise and timely comment by Sauerwein et al. that 'there is no need to unduly alarm the already distressed parents of newly diagnosed (acallosal) children. Rather, each case should be considered individually with regard to the presence or absence of concomitant CNS pathology, and even then the prognosis should be made prudently in the light of acallosal individuals with multiple malformations detected at autopsy that reportedly lived normal and productive lives'.

The study reported by Geffen, Forrester, Jones and Simpson, though primarily exploratory in nature and therefore interpreted with due caution, is important not least because in an earlier review (Clark and Geffen, 1989) they concluded that the case for the participation of the corpus callosum in memory was equivocal. In their present study they show that three patients with total agenesis demonstrated poor recall but good recognition of auditorily presented words - in short a retrieval problem. A case with partial agenesis reported in the same study showed minimal problems in verbal memory processes.

The findings in this section are especially relevant to discussions and speculations about the role of the forebrain commissures in normal people and thus later in this chapter we shall take up this issue and deal with it in more detail.

SOME RECURRING THEMES IN EARLIER SECTIONS

Cognitive and Behavioral Uniformity Despite Etiologic Diversity

The diversity of associated abnormalities in callosal agenesis serve as a constant warning against facile generalizations or interpretations of the data from neuropsychological studies of such patients. At the same time, this diversity must not be over-stated to such an extent that it ignores a core of recurring deficits which careful research identifies, whether the approach is through detailed neuropsychological testing or the phenotypic approach, exemplified by the work of O'Brien (this volume).

Recognizing and Classifying Degrees of Dysgenesis

It is sensible, for research purposes, to disentangle the varying degrees of callosal dysgenesis. Thus, Aicardi and Chevrie (this volume) suggest a division in terms of partial, total and the Aicardi syndrome. Rauch and Jinkins (this volume) suggest a continuum from agenesis, hypogenesis and hypoplasia. The direction of future research will, to some extent, be guided by the accuracy in detecting anatomical abnormalities using modern imaging techniques. Such techniques will hopefully make possible a clearer, more confident allocation of, for example, the division between partial and total agenesis and in particular distinguishing between those with and without the anterior commissure (see chapters by Milner; Rauch and Jinkins; Wahlsten and Ozaki).

Cortical Abnormalities and Callosal Dysgenesis - a Significant Link?

It is evident, both from human and animal work, that the interpretation of behavioral and neuropsychological abnormalities must be seen not simply against the background of the presence or absence of the corpus callosum but that this in itself may be a secondary outcome of a more basic abnormality. This is evident, for example, in the work of Carpenter (this volume) and his account of the Andermann syndrome. It is likewise evident in the animal work where, for example, Innocenti raises the question of whether the development of cortical neurons is affected by retrograde signals originating at the site of termination of their axons. Or again, Rauch and Jinkins' work provides a timely reminder that in interpreting observed neuropsychological deficits with callosal dysgenesis, account must be taken of cortical abnormalities as well as the callosal agenesis since the two may be intimately interlinked.

New Imaging Techniques - the Gateway to More Precise Neuropsychological Interpretations

The increasingly widespread use of new imaging techniques for detecting and specifying with some precision degrees of callosal dysgenesis is certain to open up new avenues for careful systematic neuropsychological research. This may be of a developmental kind made possible through the use, for example, of sonographic techniques (Cioni et al., this volume) which demonstrated (a) the surprisingly widespread occurrence of dysgenesis (0.81% of the population studied by them) and (b) the possible use of such techniques to embark upon longitudinal studies of callosal dysgenesis. Likewise, work based on MRI studies should now make it possible to distinguish between those patients with agenesis, hypogenesis and hypoplasia of the corpus callosum. This latter group may turn out to be much more extensive than has been suspected in the past and may be associated with what have loosely been categorized as learning disorders in young children in the past.

A Reappraisal of the Functions of the Anterior Commissure

In the past, there has been considerable speculation, based on a small number of cases, that the anterior commissure is enlarged in callosal agenesis. The careful animal studies of Wahlsten and Ozaki (this volume) give no support to this view. The animal studies of Olavarria et al. (this volume) make the further important point that where there are any remnants of the corpus callosum the remaining fibers (albeit very few) maintain the normal overall topography of callosal connections. This is an important finding giving a fresh clue to why, in cases of partial dysgenesis of the callosum, it is extremely difficult to detect any of the so-called typical split-brain signs.

Normal Cerebral Asymmetries

Taken together, the studies, whether animal or human, are consistent with the view that early callosal defects do not, as some have supposed, necessarily affect individual cerebral asymmetries (Milner and Jeeves, 1981). In fact, normal cerebral asymmetries, in kind if not degree, are more likely than bilateral representation.

The Converging Core of a Neuropsychological Profile

The neuropsychological profile emerging from the many studies now carried out on callosal dysgenesis picks up again several of the themes already noted above. We are, for example, reaching consensus on the issue of whether cerebral asymmetries are shown by acallosals, that with partial dysgenesis the overall deficits are reduced significantly, and on the possible involvement of the anterior commissure and of subcortical pathways (e.g. Lassonde; 1986; Milner, this volume). Taken together, the neuropsychological results, well summarized by Sauerwein et al. (this volume), O'Brien (this volume) and Geffen et al. (this volume), underline the recurring theme of the need to distinguish between the differing consequences of the varying degrees of dysgenesis as well as always taking account of the existence of other possible important structural abnormalities in the CNS in the presence of callosal dysgenesis.

COMPENSATORY MECHANISMS IN CALLOSAL AGENESIS

Roger Sperry's work with split brain animals and patients (e.g. Sperry, 1990) convincingly demonstrated the importance of the major forebrain commissures in integrating activity in the two cerebral hemispheres and for transferring information from one hemisphere to the other. Almost all of the early work, whether on split brain patients or on animals, was carried out on adults. There were, however, occasional early attempts to examine the effects of early neonatal callosal sections in animals (e.g. Jeeves and Wilson, 1969). These have since been extended by Ptito and Lepore (1983).

Understandably, the convincing demonstration by Sperry and his co-workers of the vital importance of intact forebrain commissures for many psychological functions raised the question of how it is that people who are born without the major commissure, nevertheless, at times, appear, to common observation, behaviorally normal. Of course even the surgical split-brain patients appear normal in brief social interaction. However, deficits which, in the case of the surgical splits, are very dramatic are only evident in much more subtle form in the so-called 'experiments of nature' - the agenesis and dysgenesis patients - and require highly sophisticated testing to reveal them. These patients, it has been pointed out, have a pre and postnatal lifetime in which to develop appropriate compensatory mechanisms, whether neural or behavioral, for the missing or underdeveloped forebrain commissures.

Sperry's first response on examining an acallosal patient, in whom it so happened that amytal testing could demonstrate bilateral representation of language, was to conclude that the mechanism of bilateral representation was indeed the most likely and plausible overall compensatory mechanism in callosal dysgenesis. Around the same time, Jeeves (1965) and Jeeves and Rajalakshmi (1964) took a somewhat different view and speculated about the possible roles of the anterior commissure and of subcortical pathways. Thereafter, attention tended to focus on the fact that many acallosal patients possess the anterior commissure. The work on the surgical splits by Risse et al. (1978) and others, though conflicting and inconclusive, lent some support for an enhanced role for the anterior commissure, at least in commissurotomy patients. The other area constantly referred to in discussions of compensatory mechanisms was the presence in many, if not all, acallosals of the prominent Probst's bundle, though what this fiber tract did remained unexplored, untested and largely a mystery. However, Rauch and Jinkin's (this volume) observation that Probst bundle does not contain all fibers that could typically have formed the corpus callosum is timely. The remainder, they speculate, disappear following the normal process of axonal diminution which occurs during development of the normal corpus callosum.

The neuropsychological profile in callosal agenesis expounded in earlier contributions to this volume leaves a number of important questions unanswered, such as, if other pathways take over from the callosal pathways then which pathways are these and how do they achieve their compensatory role? More specifically, what are the limits to the functions of the anterior commissure when present? What transfer functions are possible through subcortical pathways? In the latter case, we note in particular the relevance of studies in the past decade by Justine Sergent (1986; 1990) which have suggested the possibility of interhemispheric transfer in the classical split-brain patients of a kind not previously observed. In addition, questions are raised about the possibility of intrahemispheric integration and the extent to which this can replace interhemispheric integration which normally takes place through the corpus callosum.

Some recent speculations about compensatory mechanisms with a brief reassessment and reconsideration of each in the light of the contributions to this volume will be presented in the next section.

Bilateral Representation

Since Sperry's early suggestion that bilateral representation of function in acallosals would obviate the necessity to interrogate the opposite hemisphere through the corpus callosum in testing with unilateral inputs, the evidence in general has failed to support that suggestion. For example, the review by Jeeves and Milner (1987) summarized the available evidence at that time. That review has been updated in the present volume, notably in the contributions by Sauerwein et al. (this volume), as well as in Lassonde's many publications in the intervening years, all of which fail to support the suggestion of bilateral representation of function in acallosals. Instead, the evidence points in favor of normal hemispheric specialization in callosal agenesis. This applies to verbal behavior (Jeeves and Milner, 1987), visuospatial behavior (Meerwaldt, 1983), spatial motor tasks (Meerwaldt, 1983; Jeeves and Silver, 1988b; Lassonde, this volume) and motor dominance (e.g. Jeeves, Silver and Jacobson, 1988; Silver and Jeeves this volume). The evidence, however, does not eliminate the possibility that whilst there is hemispheric specialization in callosal agenesis, it may be of a different degree than in normals. This topic is worthy of further investigation. However, it should be noted that the results of studies on animals summarized in this volume also point to cerebral specialization in callosal agenesis, e.g. the work by Olavarria and the work by Schmidt . Schmidt concludes from his studies on mice that 'there is no evidence of a compensatory mechanism such as absence of cerebral lateralization in BALB/cCF mice with total or partial callosal agenesis '. Nielsen et al. (this volume) interpret their data on EEG coherence in sleep patterns as providing further evidence that 'there is no specialized intrahemispheric compensation for callosal agenesis'.

On balance, the conclusion remains that it is unlikely that bilateral representation of function occurs in acallosal agenesis. However, the possibility of a reduced degree of specialization awaits further investigation.

The Role of the Anterior Commissure and of Subcortical Pathways

Earlier discussions of compensatory mechanisms in callosal agenesis frequently fixed upon the anterior commissure as the most likely candidate to effect interhemispheric transmission and transfer of particular visual information. Some early case reports such as that by Geschwind, reported a significantly larger anterior commissure in a callosal agenesis patient. Others then assumed that this was likely to be the norm. There was, however, little or no evidence for this. The paper by Rauch and Jinkins (this volume) based upon careful MRI studies, makes it clear that in the cases they studied, whilst 10% of the agenesis cases did indeed show an enlarged anterior commissure, equally 10% of the cases showed a smaller one. Consistent with this were the results by Wahlsten and Osaki based on their animal data which showed that the absence of the corpus callosum did not correlate with expansion in the sizes of the hippocampal and anterior commissures. Wahlsten and Ozaki's report gives the further timely reminder that in mice at least, putative callosal axons which fail to cross to the opposite hemisphere generally do not re-route through the anterior commissure. If the same applies in humans it should curb enthusiasm for attributing too great a compensatory role for the anterior commissure in the absence of the corpus callosum.

Milner's (this volume) detailed critical assessment of the relative roles of the anterior commissure and subcortical pathways comes down clearly in favor of the importance of the subcortical pathways rather than the anterior commissure. His work is based entirely on studies of vision. At the same time Lepore's work (this volume) adds the further observation that, in general, compensatory mechanisms are better for those modalities which are more bilaterally organized, for example temperature and audition, than for those which project mainly to contralateral cortical structures, for example touch and vision.

The question remains unanswered whether or not there are instances where "callosal" fibers pass through the anterior commissure in the absence of the corpus callosum and the further question as to whether, if they do, they exhibit normal topography. On balance, the evidence from the papers in this book certainly would support a yes answer to the second of these questions, though to the first question, on the basis of the animal evidence (e.g. Wahlsten and Ozaki) the answer is probably not.

When considering the anterior commissure it is important to maintain a distinction between the transmission of spatial information, which seems unlikely in view of the termination of the fibers passing through the anterior commissure, as compared with the carrying of pattern information which seems plausible in view of the fact that the temporal lobes are cross-connected through the anterior commissure. This latter view is strengthened through the work of Martin (1985) as well as the work by Meerwaldt (1983). Milner (this volume) suggests that it is likely that the intertectal commissure mediates the spatial integration observed in our acallosal subjects (Jeeves and Milner,1987), via the spatial coding known to exist within the superior colliculus.

The conflicting evidence concerning the abilities of acallosals in tasks requiring midline stereopsis can be interpreted as pointing to a compensatory role for the anterior commissure in such tasks. This is a view strengthened by the evidence from animal studies. For example, Cowey (1985) and Hamilton and Vermeire (1986) with monkeys, and Timney et al. (1985, 1988) with cats is relevant here. At the same time, the work of Justine Sergent (e.g. 1986, 1990) would be consistent with Milner's argument that more visual information is transferred subcortically than had previously been suspected.

That some form of interhemispheric communication occurs in the absence of the cerebral commissures now seems established. This includes both low level visual perceptual information (Trevarthen and Sperry 1973) as well as higher level linguistic and non-linguistic information (Cronin-Golomb, 1986; Johnson, 1984; Myers and Sperry,

1985; Sauerwein and Lassonde, 1983; Sergent, 1990; Sperry, Zaidel and Zaidel; 1979). In these studies the stimulus set sizes were small and the stimuli were normally verbalizable. This raises the possibility that the comparisons were done by cross-cueing rather than by subcortical visual transfer. More recent work by Clarke (unpublished Ph.D. thesis 1991) who carried out further studies of some of the Californian split-brain patients, indicated that at least in one of them, N.G., sensory information was being transferred through subcortical visual pathways. It would certainly seem that the human subcortical visual pathways are sensitive to both rudimentary visual characteristics such as movement, stimulus position, and line orientation, and possibly to simple high contrast patterns (Perenin, 1978; Perenin and Jeannerod, 1978; Trevarthen and Sperry, 1973; Weiskrantz, 1986).

At this point, it is worth noting Nielsen et al.'s speculation that their data raises the possibility that there might be some interhemispheric compensatory activity reflected in the occipital coherence index. They comment 'One might question why the posterior commissure, and not the anterior commissure, compensates for the lack of the corpus callosum'. The answer may be, they say, that 'because primary visual cortex (area 17) is devoid of commissural connections, interhemispheric coherence is normally subserved by the posterior commissure and would thus be an available route for compensatory activity'. This suggestion warrants further thought.

The Enhanced Development of Ipsilateral Sensory and Motor Pathways

Jeeves (1979,1986), Jeeves, Silver and Milner (1988) and Lassonde (1986) have emphasized how the enhanced development of ipsilateral somato-sensory and motor pathways would help to explain the success of some acallosal patients on tasks of tactile discrimination whilst they may fail on fine tactile localization or on fine motor control. This possibility was first argued in detail by Dennis (1976). There is certainly evidence which can be interpreted to suggest that acallosals revert to the use of enhanced ipsilateral pathways when more distal processes are insufficient to perform a particular task. Such a development also helps to explain the observation of reaching and grasping behavior in acallosals (Jeeves and Silver, this volume), as well as their behavior on tasks of bimanual coordination (Jeeves, Silver and Jacobson, 1988). The studies by Laget et al. (1976, 1977) and Vanasse et al. (this volume) using evoked potential techniques give support to such enhanced development and elaboration in the ipsilateral pathways. They observed that in normals there was a strong contralateral scalp response to stimulation of the wrist but little or no ipsilateral response. By comparison, in acallosals, there was a strong ipsilateral response presumed to come through the ipsilateral pathways.

Whilst in general such enhanced development clearly confers benefits when carrying out certain kinds of tasks, it needs to be remembered that there may be situations in which it may set limitations to performance. This may be one of the reasons for the observed deficits in some tasks requiring bimanual coordination or the transfer of sensori-motor information. In the absence of presumed transcallosal inhibition (Jeeves, 1991; Preilowski, 1972; Zaidel, 1986) which is normally mediated through the corpus callosum, the enhanced over-developed ipsilateral pathways may compete with the crossed contralateral pathways in such a way as to reduce the overall level of performance when measured carefully in terms of accuracy or speed or both. Such a formulation will be consistent with the finding of a bilateral representation of the hand area of the monkey, Area S2, which is not completely abolished after callosal transection.

It is reasonable to raise the question of why the adult surgically split-brain patients did not benefit from some of the mechanisms of compensation described above. There are a number of obvious reasons. First, the widely accepted belief that cerebral plasticity is closely linked to the degree of functional maturity of the lesioned structure at the time of the injury or insult. Secondly, as Innocenti (1981) and his co-workers have shown, many cortical connections undergo a state of exuberant growth before synaptic stabilization occurs in later development. Thus, when damage occurs during the formative stages, connections which would otherwise have been eliminated during the critical period of axon retraction and synaptic elimination, may persist. Thirdly, it is possible that agenesis or early damage could result in the extension of the normal period of cerebral plasticity. Thus, the decreasing availability of exuberant connections in adulthood would limit the possibilities of neural compensation to a greater extent than when this occurs prenatally, as in callosal agenesis, or when it occurs early postnatally.

There are aspects of the neuropsychological profile presented in earlier chapters which are not easily explained even on the basis of the supposed compensatory mechanisms discussed above. For example, Temple's demonstration that there are some highly specific impairments in the explicit phonological processing required for rhyming tasks and in phonemic discrimination, are not capable of easy explanation. She demonstrated how reading may be at a normal level in relation to word recognition but there would still be problems in the development of the so-called phonological reading route. She also demonstrated the existence of very specific visuospatial and visuoperceptual deficits. In short, her results implicated a number of selective cognitive domains in which the corpus callosum seemed to be playing a critical role in normal performance. Her work also underlined the way in which by hypothesizing a variety of modular sub-components going to make up any typical cognitive skill, such modular sub-components need to be in ready and easy intercommunication for the production of such complex skills, whether linguistic or visuospatial. It may then be the case that the putative compensatory mechanisms discussed above are insufficient to combine these modular subcomponents and result in them developing in relative independence of each other. This would apply particularly in the case of phonological processing where its contribution to reading may require to be integrated by combining the activity in the two cerebral hemispheres in a way not possible in callosal agenesis.

There is also one tantalizing hint in part of Temple's chapter which is worthy of further exploration and investigation. One of her intellectually above-average children had presented elongated and distorted spatial dimensions in his freehand drawings. As a result, everything seemed to be drawn longer either on the vertical or on the horizontal dimension. This was drawn to the attention of the patient in question, who apparently learned to recognize this specific difficulty and as a result worked for several years systematically at correcting the errors which he otherwise would make. The result was that by the time the spatial battery was administered some years after the initial testing, he had learned and succeeded specifically in modifying his drawings so that they approximated more normal and more appropriate proportions. Rehabilitation and relearning is thus, on this slender evidence, an aspect of callosal behavior which should be taken more seriously and investigated accordingly. This same point concerning remediation is made by Sauerwein et al. (this volume) who report that they found that manual dexterity and motor coordination in the young child can be greatly and more quickly improved with the help of physiotherapy and occupational therapy.

One other aspect of possible compensatory mechanisms, which some of the earlier presentations hinted had not received sufficient attention in the past, arose from new light that was shed on the apparently normal topography of callosal fibers even when only a small remnant were present. This may help to explain why it is that in dysgenesis, as compared to agenesis, of the corpus callosum, many of the features of the neuropsychological profile well documented in total agenesis, do not appear. Could this be because the remnant of the callosum which is present contains fibers which represent the normal widespread topographic linking of the two cerebral hemispheres, albeit with drastically reduced numbers of fibers in each case? This, on the face of it, seems an important possibility which has hitherto received insufficient attention.

THE PUTATIVE DIVERSITY OF FUNCTIONS OF THE FOREBRAIN COMMISSURES IN NORMALS RECONSIDERED IN THE LIGHT OF STUDIES OF CALLOSAL AGENESIS

Sperry's work both with animals and with patients, demonstrated dramatically and convincingly, the importance of the major forebrain commissures in integrating activity occurring in the two cerebral hemispheres as well as transferring information from one hemisphere to the other. The concentration of research effort on the integrative function of the forebrain commissures has continued in the tradition established by Sperry and his co-workers. However, in the context of our present discussion of the possible effects of the absence of the callosum from birth on the subsequent development of the brain and in particular of interhemispheric interactions, it is clearly prudent to consider not only the integrative function of the callosum but other functions which have been discussed from time to time. Some of these indeed, it has been argued, (Dennis, 1976) may in this context be more important than the integrative function.

What seems apparent is that in the absence of the callosum from birth there are other pathways which, though in varying degrees less efficient, can nonetheless ensure some interhemispheric transfer of information and a sharing of activity in the two cerebral hemispheres. There may well be, however, exceptions to this, for example in midline processes (Jeeves, 1991; Lassonde, 1986) and in the transfer of certain kinds of spatial information (for example the work of Meerwaldt) and possibly also in limiting certain aspects of linguistic development (Jeeves and Temple, 1987; Temple and Ilsey, this volume).

There is, however, a second function of the forebrain commissures which has been noted as far back as the 1968 Johns Hopkins conference on Cerebral Asymmetries. In attempting to sum up the presentations at that conference, the neurophysiologist, Richard Jung, noted that in his judgment, most, if not all, of the findings reported could only come about as results of the inhibitory role of the forebrain commissures. This same point has been reiterated in a different context (Berlucchi and Antonini, 1990) who point out that the local inhibitory neurons are activated by the callosal input. As mentioned above, in a behavioral context, Dennis (1976, 1981), after studying two acallosal patients, argued that the course of normal development involves the progressive suppression of uncrossed sensory and motor projections. Thus, the acquisition of fine distal motor control involves the progressive inhibition of motor pathways other than crossed pyramidal ones. In the context of her discussion, the notion of inhibition is loosened somewhat and is taken to mean some form of suppression. Dennis (1981) followed up her earlier observations by further studies of callosal agenesis patients as a result of which she argued that they showed an inability to suppress the ipsilateral component of an auditory stimulus when there is spectral temporal overlap of the stimuli presented.

This same theme was taken up by Preilowski (1972) who noted that there was an important inhibitory role for the callosum, the interference with which resulted in the lowered quality and speed of performance that he demonstrated in split-brain patients when performing a bimanual coordination task. Zaidel (1976; 1986) has also suggested that the efficient performance of certain tasks may depend upon a unique hemispheric processing style and that this may at times be inhibited by the other hemisphere. Most recently Zaidel (1986) has discussed callosal dynamics and argued that 'the corpus callosum is a complex dynamic channel of both facilitatory and inhibitory codes' (p. 319 Zaidel, 1986). Others who have made use of this notion of an inhibitory or suppressive role of the corpus callosum in normal activity include Moscovitch (1977) who argued that the corpus callosum played a crucial role in establishing hemispheric asymmetry, since each hemisphere exerts an inhibitory influence on the other and in this way prevents the dual establishment of specialized functions in both hemispheres and thus avoids redundancy of representation. As we have seen in our earlier discussions, subsequent research has however failed to support this hypothesis as regards bilateral representation of functions but that does not mean that the possible inhibitory or suppressive role of the corpus callosum should be minimized.

Finally, Lassonde has repeatedly emphasized the need to take account of the facilitatory influence of the corpus callosum, not only in its interhemispheric aspect but also as regards intra hemispheric processing (Lassonde, 1986). Thus in a task requiring acallosals and matched I.Q. normals to make a same/different judgment of pairs of visual stimuli (letters, numbers, colors, or forms) Lassonde (1986) found that acallosals were able to effect inter hemispheric comparisons and that their accuracy level in making both inter and intra hemispheric judgments did not differ from matched I.Q. controls. However, a different result emerged when reaction times were compared. Acallosals required twice as long to respond as the matched controls. Lassonde believes that the corpus callosum participates in the bilateral activation of those regions specifically implicated in the neural activity required by her tasks. She argues that in the absence of the corpus callosum, neural activity in the regions concerned may be reduced, and hence the delayed responsiveness.

Kuks also emphasizes the manner in which the corpus callosum, forming as it does an interhemispheric connection, allows both an inhibitory and a facilitatory interaction between homotopic areas of the cerebral cortex.

THE WAY AHEAD

The contributions of this volume make it clear that research, on a broad front, into callosal agenesis is poised to take off at an accelerating pace in the next few years. With the widespread use of modern techniques of brain imaging, earlier detection of the condition combined with greater accuracy will be ensured and it is therefore highly probable that an increased number of cases will come to light. Moreover, it is likely that those that come to attention will not, as in the past, be restricted solely to patients with major mental retardation and gross associated abnormalities. It is likely that more above-average intelligence subjects will come to light and this will present a new challenge to devise suitable techniques to detect subtle abnormalities which can otherwise be readily disguised by intelligent people.

It seems certain that an increasingly consistent picture of the major features of the neuropsychological profile, both behavioral and cognitive, will be mapped out and placed on a more secure basis in the coming decade. With this more detailed specification of the behavioral phenotype, as well as the neuropsychological profile of callosal agenesis, will come, hopefully, an increasing sensitivity to the problems that such people are likely to encounter. And with this increased sensitivity and awareness will come a fresh opportunity for better counseling of the patients themselves, of their parents and of their teachers. The cumulative effect of small cognitive and sensory deficits can, if unrecognized and unattended to, result in general behavior problems at home and at school as well as in specific educational difficulties which should be susceptible, in some cases, of remedial education.

From the viewpoint of those interested in understanding the nature of hemispheric interactions in normal people, the increasing sophistication of brain imaging leading to much more precise description of the extent of callosal dysgenesis or hypoplasia, the presence or absence of the anterior commissure and the presence or absence of associated abnormalities, will make the careful neuropsychological study of these patients increasingly possible and important. As 'experiments of nature' they hold out enormous scientific potential which hopefully will be realized in future extensions of on-going research. More specifically, from the formal presentations at the meeting, as well as in the discussions following the papers, a number of promising lines for future research are already identifiable. These include: (i) the use of postnatal sonographic findings to identify a cohort of patients with varying degrees of dysgenesis for systematic longitudinal, cognitive and behavioral studies; (ii) correlative neuropsychological studies of patient groups with varying degrees of dysgenesis identified by MNR but paying attention, in interpreting the neuropsychological data, to the possible presence of associated cortical abnormalities; (iii) a specific study of acallosal patients both with and without the anterior commissure; such a study should clarify the ongoing debate of the relative roles of the anterior commissure and subcortical pathways in compensatory mechanisms; (iv) studies to discover whether the apparently normal topography of callosal fibers observed in dysgenesis in animals also applied to humans; (v) the need to extend the investigative techniques used with acallosals to include, for example, where appropriate, the use of evoked potential studies to elucidate the effects of the absence of presumed transcallosal excitation and inhibition present in normals, and cerebral blood flow studies to shed further light on the question of whether normal cerebral asymmetry of functions applies to acallosals; (vi) to clarify more precisely the nature and extent of deficits in language, memory and midline processing in acallosals; (vii) to encourage screening of 'learning disabled' children for possible minimal callosal pathology using both cognitive, behavioral and imaging techniques; (viii) to extend and evaluate the better management and rehabilitation of acallosal children and to identify times of maximum plasticity when intervention may be most effective.

REFERENCES

Berlucchi, G. and Antonini, A., 1990, The role of the corpus callosum in the representation of the visual field in cortical areas, *in:* "Brain Circuits and Functions of the Mind. Essays in honour of Roger W. Sperry," C. Trevarthen, ed., CUP, Cambridge.

Clark, C.R. and Geffen, G.M., 1989, Corpus callosum section and recent memory, *Brain* 112, 165-175.

Cowey, A.,1985, Disturbances of stereopsis by brain damage, *in:* "Brain Mechanisms and Spatial Vision, NATO Advanced Study Institute Series," D. Ingle, ed., Martimas Nijhoff, The Hague.

Cronin-Golomb, A.,1986, Subcortical transfer of cognitive information in subjects with complete forebrain commissurotomy, *Cortex* 22:499-519.

Dennis, M., 1976, Impaired sensory and motor differentiation with corpus callosum agenesis: a lack of callosal inhibition during ontogeny, *Neuropsychologia* 14:455-469.

Dennis, M., 1981, Language in a congenitally acallosal brain, *Brain Lang.* 12:33-53.

Hamilton, C.R. and Vermeire, B.A., 1986, Localization of visual functions with partially split-brain monkeys, *in:* "Two Hemispheres-one Brain: Functions of the Corpus Callosum," F. Lepore, M. Ptito, and H.H. Jasper, eds., Alan R. Liss, New York.

Innocenti, G.M., 1981, Growth and reshaping of axons in the establishment of visual callosal connections, *Science.* 212:824-827.

Jeeves, M.A., 1965, Psychological studies of three cases of congenital agenesis of the corpus callosum, *in:* "Functions of the Corpus Callosum", CIBA Foundation Study Groups, Vol 20, E.G. Ettlinger, ed., Churchill, London.

Jeeves, M.A., 1979, Some limits to interhemispheric integration in cases of callosal agenesis and partial commissurotomy, *in:* "Structure and Function of the Cerebral Commissures," I.S. Russell, M.W. van Hof, and G. Berlucchi, eds., Macmillan, London.

Jeeves, M.A.,1986, Callosal agenesis: neural and developmental adaptations, *in:* "Two Hemispheres-one Brain: Functions of the Corpus Callosum," F. Lepore, M. Ptito, and H.H. Jasper, eds., Alan R. Liss, New York.

Jeeves, M.A., 1991, Hemispheric Interactions, in: "Encyclopedia of Human Biology", Vol 4, R. Dulbecco, ed., Academic Press, New York.

Jeeves, M.A., 1991, Stereoperception in callosal agenesis and partial callosotomy, *Neuropsychologia* 29:19-34.

Jeeves, M.A. and Milner, A.D.,1987, Specificity and plasticity in interhemispheric integration: evidence from callosal agenesis, *in:* "Duality and Unity of the Brain - Unified Functioning and Specialization of the Hemispheres," D. Ottoson, ed, Macmillan, London.

Jeeves, M.A. and Rajalakshmi, R., 1964, Psychological studies of a case of congenital agenesis of the corpus callosum, *Neuropsychologia* 2:247-252.

Jeeves, M.A. and Silver, P.H., 1988a, The formation of finger grip during prehension in an acallosal patient, *Neuropsychologia* 26:153-159.

Jeeves, M.A. and Silver, P.H., 1988b, Interhemispheric transfer of spatial tactile information in callosal agenesis and partial commissurotomy, *Cortex* 24:601-604.

Jeeves, M.A., Silver, P.H., and Jacobson, I., 1988, Bimanual co-ordination in callosal agenesis and partial commissurotomy, *Neuropsychologia* 26:833-850.

Jeeves, M.A., Silver, P.H., and Milner, A.B., 1988, Role of the corpus callosum in the development of a bimanual motor skill, *Dev. Neuropsychol.* 44:305-323.

Jeeves, M.A. and Temple, C.M., 1987, A further study of language function in callosal agenesis, *Brain Lang.* 32:325-335.
Jeeves, M.A. and Wilson, A.F., 1969, Tactile transfer in neonatal callosal section in the cat, *Psychon. Sci.* 16:235-237.
Johnson, L.E., 1984, Vocal responses to left visual stimuli following forebrain commissurotomy, *Neuropsychologia* 22:153-166.
Laget, P., d'Allest, A.M., Fihey, R., and Lortholary, O., 1977, L'interet des potentials evoques somesthesiques homolateraux dans les agénésies du corps calleux, *Rev. E.E.G. Neurophysiol. Clin.* 7:498-502.
Laget, P., Raimbault, J. d'Allest, A.M., Flores-Guevara, R., Mariani, J., and Thieriot-Prevost, G., 1976, La maturation des potentiels evoques somesthésiques, PES, chez l'homme, *Electroencephalogr. Neurophysiol. Clin.* 40:499-515.
Lassonde, M., 1986, The facilitatory influence of the corpus callosum on intrahemispheric processing, *in:* "Two Hemispheres-one Brain: Functions of the Corpus Callosum," F. Lepore, M. Ptito, and H.H. Jasper, eds., Alan R. Liss, New York..
Martin, A., 1985, A qualitative limitation on visual transfer via the anterior commissure, *Brain* 108:43-63.
Meerwaldt, J.D., 1983, Disturbance of spatial perception in a patient with agenesis of the corpus callosum, *Neuropsychologia* 21:161-165.
Milner, A.D. and Jeeves, M.A., 1981, The functions of the corpus callosum in infancy and adulthood, *Beh. Brain Sci.* 4(1):30-31.
Moscovitch, M., 1977, The development of lateralization of language functions and its relation to cognitive and linguistic development: a review and some theoretical speculations, *in:* "Language Development and Neurological Theory," S.J. Segalowitz and F.A. Gruber, eds, Academic Press, New York.
Myers, J.J. and Sperry, R.W., 1985, Interhemispheric communication after section of the forebrain commissures, *Cortex* 21:249-260.
Njiokiktjien, C., 1991, A Historical Perspective, *in:* "The Child's Corpus Callosum," G. Ramaekers and C. Njiokiktjien, eds, Suyi, Amsterdam.
Perenin, M.T., 1978, Visual function within the hemianopic field following early cerebral hemidecortication in man - II. Pattern discrimination, *Neuropsychologia* 16:697-708.
Perenin, M.T. and Jeannerod, M., 1978, Visual function within the hemianopic field following early cerebral hemidecortication in man - I. Spatial localization, *Neuropsychologia* 16:1-13.
Preilowski, B.F.P., 1972, Possible contributions of the anterior forebrain commissures to bilateral motor coordination, *Neuropsychologia* 10:267-277.
Ptito, M. and Lepore, F., 1983, Interocular transfer in cats with early callosal section, *Nature* 301:513-515.
Ramaekers, G., 1991, Functions of the corpus callosum in adults, *in:* "The Child's Corpus Callosum," G. Ramaekers and C. Njiokiktjien, eds, Suyi, Amsterdam.
Risse, G.L., Ledoux, J., Springer, S.P., Wilson, D.H., and Gazzaniga, M.S., 1978, The anterior commissure in man: functional variation in a multisensory system, *Neuropsychologia* 16:23-31.
Sauerwein, H.C. and Lassonde, M., 1983, Intra- and interhemispheric processing of visual information in callosal agenesis, *Neuropsychologia* 21:167-171.
Sergent, J., 1986, Subcortical coordination of hemisphere activity in commissurotomized patients, *Brain* 109:357-369.
Sergent, J., 1990, Furtive incursions into bicameral minds: Integrative and coordinating role of subcortical structures, *Brain* 109:537-568.
Sperry, R.W., 1990, Forebrain commissures and conscious awareness, *in:* "Brain Circuits and Functions of the Mind. Essays in Honour of Roger W. Sperry," C. Trevarthen, ed., CUP, Cambridge.
Sperry, R.W., Zaidel, E., and Zaidel, D., 1979, Self recognition and social awareness in the disconnected minor hemisphere, *Neuropsychologia* 17:153-166.
Temple, C.M., Jeeves, M.A., and Vilarroya, O.O., 1989, Ten pen men: Explicit phonological processing in two children with callosal agenesis, *Brain Lang.* 37:548-564.
Temple, C.M., Jeeves, M.A., and Vilarroya, O.O., 1990, Reading in callosal agenesis, *Brain Lang.* 39:235-253.
Timney, B. and Lansdown, G., 1988, Binocular depth perception, visual acuity and visual fields in cats following neonatal section of the optic chiasm, *Exp. Brain Res.* 207.
Timney, B., Elberger, A.J., and Vandewater, M.L., 1985, Binocular depth perception in the cat following early corpus callosum section, *Exp. Brain Res.* 60:19-26.
Trevarthen, C. and Sperry, R.W., 1973, Perceptual unity of the ambient visual field in human commissurotomy patients, *Brain* 96:547-570.
Weiskrantz, L., 1986, "Blindsight: A Case Study and Implications," Oxford University Press, Oxford.
Zaidel, E., 1976, Language, dichotic listening and the disconnected hemispheres, *in:* "Conference on Human Brain Function," D.O. Walter, L. Rogers, and J. M. Finzi-Fried, eds., Brain Information Service/BRI Publications Office, UCLA.
Zaidel, E., 1986, Callosal dynamics and right hemisphere language, *in:* "Two Hemispheres-one Brain: Functions of the Corpus Callosum," F. Lepore, M. Ptito, and H.H. Jasper, eds., Alan Liss, New York.

CONTRIBUTORS

Jean Aicardi
Département de Pédiatrie
Hôpital Necker-Enfants Malades
149 rue De Sèvres
757143 Paris, Cedex 15, France

Eva Andermann
Neurogenetics Unit and
Departments of Neurology and
Neurosurgery, McGill University
The Montreal Neurological Institute
and Hospital, 3801 University St.
Montreal, Que. H3A 2B4, Canada

Frederick Andermann
Departments of Neurology and
Neurosurgery and Department of
Pediatrics, McGill University
The Montreal Neurological Institute
and Hospital, 3801 University St.
Montreal, Que. H3A 2B4, Canada

Laura Bartalena
Neonatal Intensive Care Unit
University of Pisa
Pisa, Italy

Denis Bergeron
Departments of Neurology and
Neurosurgery, McGill University
The Montreal Neurological Institute
and Hospital, 3801 University St.
Montreal, Que. H3A 2B4, Canada

Enrico Biagioni
Institute of Child Neurology and
Psychiatry, University of Pisa and
Stella Maris Foundation
Via dei Giacinti, 2
I-56018 Calambrone (Pisa)
Italy

Antonio Boldrini
Neonatal Intensive Care Unit
University of Pisa
Pisa, Italy

Stirling Carpenter
Departments of Neurology and
Neurosurgery, McGill University
The Montreal Neurological Institute
and Hospital, 3801 University St.
Montreal, Que. H3A 2B4, Canada

Jean-Jacques Chevrie
Département de Pédiatrie
Hôpital Necker-Enfants Malades
149 rue De Sèvres
757143 Paris, Cedex 15
France

Giovanni Cioni
Institute of Child Neurology and
Psychiatry, University of Pisa and
Stella Maris Foundation
Via dei Giacinti, 2
I-56018 Calambrone (Pisa)
Italy

Raquel del Carpio-O'Donovan
Department of Radiology
The Montreal Neurological Institute
and Hospital, 3801 University St.
Montreal, Que. H3A 2B4, Canada

Luc Forest
Groupe de Recherche en
Neuropsychologie Expérimentale
Département de Psychologie
Université de Montréal
C.P. 6128, Succ. A
Montréal, Qué., H3C 3J7
Canada

Glenys M. Forrester
Research Unit, Julia Farr Centre
Fullarton and Flinders University
of South Australia
Adelaide, Australia, 5063

Gina M. Geffen
Cognitive Psychophysiology Laboratory
Psychology Department and Clinical
Studies Unit, University of Queensland
and Queensland Health Department
Brisbane, Australia, 4072

Guy Geoffroy
Service de Neurologie, Département de
Pédiatrie, Hôpital Sainte-Justine pour les
Enfants, 3,175, Chemin Côte Ste.
Catherine, Montréal, Qué., H3T 1C5
Canada

Joanne Ilsley
Developmental Neuropsychology Unit
Department of Psychology
University of Essex, Wivenhoe Park
Colchester C04 3SQ, U.K.

Giorgio M. Innocenti
Institut d'Anatomie
9 rue du Bugnon
1005 Lausanne
Switzerland

Ivan Jacobson
Department of Surgical Neurology
Dundee University
Dundee, Scotland, DD1 9ND
U.K.

Malcolm A. Jeeves
Psychological Laboratory
University of St. Andrews
St.Andrews, Fife
Scotland, KY16 9JU, U.K.

Joseph S. Jeret
Institute for Basic Research in
Developmental Disabilities
1050 Forest Hill Road, Staten Island
New York, 10314-6399, U.S.A.

J. Randy Jinkins
Neuroradiology Section, Department of
Radiology, The University of Texas
Health Science Center
San Antonio, Texas 78284-7800, U.S.A.

Dean L. Jones
Cognitive Psychophysiology Laboratory
Psychology Department and Clinical
Studies Unit, University of Queensland
and Queensland Health Department
Brisbane, Australia, 4072

Jan B.M. Kuks
Department of Developmental Neurology
University Hospital, Oostersingel 59
9713 EZ Groningen
The Netherlands

Pierre Langevin
Departments of Neurology and
Neurosurgery, McGill University
The Montreal Neurological Institute
and Hospital, 3801 University St.
Montreal, Que. H3A 2B4, Canada

Maryse Lassonde
Groupe de Recherche en
Neuropsychologie Expérimentale
Département de Psychologie
Université de Montréal
C.P. 6128, Succ. A
Montréal, Qué., H3C 3J7, Canada

Franco Lepore
Groupe de Recherche en
Neuropsychologie Expérimentale
Département de Psychologie
Université de Montréal
C.P. 6128, Succ. A
Montréal, Qué., H3C 3J7, Canada

Roger Marcotte
Centre d'Etude du Sommeil
Clinique du Sommeil
Hòpital du Sacré Coeur
5400 Boul. Gouin O.
Montréal, Que., H4J 1C5, Canada

Jean Mathieu
Departments of Neurology and
Neurosurgery, McGill University
The Montreal Neurological Institute
and Hospital, 3801 University St.
Montreal, Que. H3A 2B4, Canada

Denis Melanson
Department of Radiology
The Montreal Neurological Institute
and Hospital, 3801 University St.
Montreal, Que. H3A 2B4, Canada

A. David Milner
Psychological Laboratory
University of St Andrews
St Andrews, Fife
Scotland, KY16 9JU, U.K.

Jacques Montplaisir
Centre d'Etude du Sommeil
Clinique du Sommeil
Hòpital du Sacré Coeur
5400 Boul. Gouin O.
Montréal, Que., H4J 1C5, Canada
and Département de Psychiatrie
Université de Montréal, Montreal
Canada

Richard Nagy
Center for Human Genetics
McGill University
1205 Docteur Penfield
Montreal, Que. H3A 1B1
Canada

Tore A. Nielsen
Centre d'Etude du Sommeil
Clinique du Sommeil
Hòpital du Sacré Coeur
5400 Boul. Gouin O.
Montréal, Que., H4J 1C5
Canada

Judith Nilsson
South Australia Head Injury Service
Julia Farr Centre
Adelaide, Australia, 5063

Pierre Nolin
Département de Psychologie
Université du Québec à Trois-Rivières
Trois-Rivières, Qué., G9A 5H7
Canada

Gregory O'Brien
Department of Psychiatry
Morpeth, Northumberland
NE61 3BP, U.K.

Jaime Olavarria
Department of Psychology, NI-25
University of Washington
Seattle, WA 98195, U.S.A.

Hiroki S. Ozaki
Department of Anatomy
Kagawa Medical School
Kagawa, Japan

Pierre Poirier
Groupe de Recherche en
Neuropsychologie Expérimentale
Département de Psychologie
Université de Montréal
C.P. 6128, Succ. A
Montréal, Qué., H3C 3J7, Canada

Ronald A. Rauch
Neuroradiology Section
Department of Radiology
The University of Texas
Health Science Center
San Antonio, Texas 78284-7800
U.S.A.

Armando Salazar
Department of Radiology
The Montreal Neurological Institute
and Hospital, 3801 University St.
Montreal, Que. H3A 2B4, Canada

Hannelore C. Sauerwein
Groupe de Recherche en
Neuropsychologie Expérimentale
Département de Psychologie
Université de Montréal
C.P. 6128, Succ. A
Montréal, Qué., H3C 3J7, Canada

Alessandra Schiavetto
Groupe de Recherche en
Neuropsychologie Expérimentale
Département de Psychologie
Université de Montréal
C.P. 6128, Succ. A
Montréal, Qué., H3C 3J7, Canada

Sergio L. Schmidt
Instituto de Biofísica Carlos
Chagas Filho, Universidade Federal
do Rio de Janeiro and Departamento
de Ciências Fisiológicas.Universidade
do Estado de Rio de Janeiro,Cidade
Universitaria-21949, Rio de Janeiro
Brazil

Marcos M. Serra-Oller
Remei: 10-12, Esc. B Atica 2B
08014 Barcelona, Spain

Priscilla H. Silver
Psychological Laboratory
University of St. Andrews
St.Andrews, Fife
Scotland, KY16 9JU, U.K.

Donald A. Simpson
NH and MRC Road Accident
Research Unit, University of Adelaide
and Department of Neurosurgery
Adelaide Children's Hospital
Adelaide, Australia, 5001

Nicola Strobelt
Department of Obstetrics Gynecology
Ospedale San Gerardo di Monza
Via Solferino, 16, Monza, Italia

Christine Temple
Developmental Neuropsychology Unit
Department of Psychology
University of Essex, Wivenhoe Park
Colchester C04 3SQ U.K.

Michel Vanasse
Service de Neurologie, Département de
Pédiatrie, Hôpital Sainte-Justine pour les
Enfants, 3175, Chemin Côte Sainte-
Catherine, Montréal, Qué., H3T 1C5
Canada

Richard C. Van Sluyters
School of Optometry
University of California
Berkeley, CA 94720, U.S.A.

Nathalie Veillette
Groupe de Recherche en
Neuropsychologie Expérimentale
Département de Psychologie
Université de Montréal
C.P. 6128, Succ. A
Montréal, Qué., H3C 3J7, Canada

Patrizia Vergani
Department of Obstetrics Gynecology
Ospedale San Gerardo di Monza
Via Solferino, 16, Monza, Italia

Johan E. Vos
Department of Developmental Neurology
University Hospital, Oostersingel 599713
EZ Groningen, The Netherlands

Douglas Wahlsten
Department of Psychology
University of Alberta
Edmonton, AB, T6G 2E9, Canada

Krystyna E. Wisniewski
Institute for Basic Research in
Developmental Disabilities
1050 Forest Hill Road, Staten Island
New York, 10314-6399, U.S.A.

Kathleen T. Yee
Department of Anatomy
University of Cambridge
Downing St.
Cambridge CB2 3DY, U.K.

INDEX

MIX
Papier aus verantwortungsvollen Quellen
Paper from responsible sources
FSC® C105338
FSC
www.fsc.org

If you have any concerns about our products,
you can contact us on
ProductSafety@springernature.com

In case Publisher is established outside the EU,
the EU authorized representative is:
Springer Nature Customer Service Center GmbH
Europaplatz 3, 69115 Heidelberg, Germany

Printed by Libri Plureos GmbH
in Hamburg, Germany